Fly Agaric:
A Compendium of History, Pharmacology, Mythology, & Exploration

Edited by

Kevin M. Feeney

Published by:

Fly Agaric Press
Ellensburg, Washington

Copyright © 2020 by Kevin Feeney et al.

All rights reserved.

Printed in the United States of America

ISBN: 978-0-578-71442-4

Library of Congress Control Number: 2020917396

Cover photo by Rinus Motmans (https://www.flickr.com/photos/rinusmotmans/)
Back cover art by Christopher Yurkanin
Cover design by Emma Mueller (https://sequanastudio.com/)
Book design by Kevin Feeney

DISCLAIMER: The information contained herein is intended for general informational and educational purposes only. The publisher and authors are not responsible for any damages or negative consequences that may arise from the use or misuse of the information provided in this volume.

In Remembrance of:

Stephan de Borhegyi

Gary Lincoff

Emanuel Salzman

Tjakko Stijve

R. Gordon & Valentina Wasson

For their work and pioneering efforts in the study of archaeomycology, ethnomycology, and mycology.

TABLE OF CONTENTS

Acknowledgements — ix

Introduction
Kevin Feeney — xi

PART I: MUSHROOM HUNTING & IDENTIFICATION

1. **Mushroom Hunting**
   Kevin Feeney — 3

2. **Amanita Basics**
   Kevin Feeney — 7

3. **Psychoactive *Amanitas* of North America**
   Kevin Feeney — 19

PART II: RELIGION, CULTURE, & FOLKLORE

4. **Soma's Third Filter: New Findings Supporting the Identification of *Amanita muscaria* as the Ancient Sacrament of the Vedas.**
   Kevin Feeney & Trent Austin — 51

5. **Travels with Santa and his Reindeer**
   Lawrence Millman — 63

6. **A Search for Soma in Russia's Kamchatka Peninsula**
   Jason Salzman, Emanuel Salzman, Joanne Salzman & Gary Lincoff — 71

7. **In Pursuit of Yaga Mukhomorovna: The Finno-Ugric Connection and Beyond**
   Frank M. Dugan — 93

8. Magical Potions: Entheogenic Themes in Scandinavian Mythology
   Steven Leto — 101

9. An Attempt to Explain the Battle-Fury of the Ancient Berserker Warriors through Natural History
   Samuel Ödman — 129

10. The Berserkers: Odin's Warriors & the Mead of Inspiration
    Mark A. Hoffman & Carl A. P. Ruck — 135

11. Speckled Snake, Brother of Birch: *Amanita muscaria* Motifs in Celtic Legends
    Erynn Rowan Laurie & Timothy White — 143

12. Fly Agaric Motifs in the Cú Chulaind Myth Cycle
    Thomas J. Riedlinger — 177

13. Bride of Brightness & Mother of all Wisdom: An Ethnomycological Reassessment of Brigid, Celtic Fertility Goddess and Patron Saint of Ireland
    Peter McCoy — 195

14. Mail-Order Mushrooms: An Interview with Mark Niemoller
    Kevin Feeney & Mark Niemoller — 221

15. Glückspilz: The Lucky Mushroom
    Kevin Feeney — 241

16. The Lucky Mushroom: A New Fairy Tale Story
    Marie Meissner, Karl Schicktanz, Sandra Grecki — 247

PART III: ARCHAEOLOGICAL EVIDENCE

17. Mushroom Effigies in Archaeology: A Methodological Approach
    Giorgio Samorini — 269

18. Beyond the Ballgame: Mushrooms, Trophy Heads, and the Great Maya Collapse
    Carl de Borhegyi — 297

PART IV: DIET & CUISINE

19. The Fly Amanita
Frederick Coville — 331

20. Amanitas in the Family: "Brownie Seats for dinner… again?"
Danny Curry — 335

21. Cooking with Fly Agaric
Kevin Feeney — 339

PART V: PHARMACOLOGY & PHYSIOLOGICAL EFFECTS

22. *Amanita muscaria* Chemistry: The Mystery Demystified?
Ewa Maciejczyk — 351

23. Re-examining the role of Muscarine in Fly Agaric inebriation
Kevin Feeney & Tjakko Stijve — 367

24. Agaricus Muscarius: the use of Fly Agaric in Homeopathy
Kevin Feeney & Bill Mann — 377

25. Agaricus Muscarius (1894)
Horace P. Holmes — 391

26. Fly Agaric as Medicine: From Traditional to Modern Use
Kevin Feeney — 397

27. How to make Medical Preparations
Kevin Feeney — 419

28. The Experience
Kevin Feeney — 425

29. The Formula?
Kevin Feeney — 445

References — 459

Index — 483

Acknowledgements

The book you hold in your hands would not have been possible without the contributions and support of dozens of people. Indeed, any attempt at a comprehensive volume on the enigmatic fly agaric would be practically impossible without the perspectives, experiences, research, and knowledge of multiple scholars from diverse backgrounds. This book is my attempt to bring these diverse voices together to provide an expansive, if not comprehensive, look at this peculiar fungus. In this regard, my greatest thanks goes to each of the individuals that contributed chapters to this volume, including Trent Austin, Lawrence Millman, Jason Salzman, Emanuel Salzman, Joanne Salzman, Gary Lincoff, Frank Dugan, Steven Leto, Mark Hoffman, Carl Ruck, Erynn Laurie, Timothy White, Tom Riedlinger, Peter McCoy, Giorgio Samorini, Carl de Borhegyi, Danny Curry, Ewa Maciejczyk, Tjakko Stijve, and Bill Mann. Thank you all for your thoughtful and provocative writings and for your responsiveness to my questions and editorial feedback. Thanks also to Mark Niemoller for his interview and contributions of images and content from his company catalog. Additional thanks go to Lawrence Millman for introducing me to Frank Dugan, and also to contributors Peter McCoy, Mark Hoffman, and colleague William Rubel for their feedback and guidance on matters of publishing. And of course, special thanks to my co-authors (also listed above): Trent Austin, Tjakko Stijve, and Bill Mann.

Further, no book on the fly agaric would be complete without a splash or two of color. I feel incredibly fortunate to be able to include contributions from a number of talented artists and photographers. For their photographic contributions I would like to thank Rinus Motmans, Martika Lyle, Ron Wolf, Matt Gardner, Jason Hollinger, Darvin DeShazer, Christine Young, Danny Curry, Eva Skific, Lisa Kimmerling, Renate Rohmann, Spike Mikulski, Mervi Alavuotunki, Alan Rockefeller, Jon Rapp, Walt Sturgeon, Paul Kroeger, Amanita Lloydii, Mark Niemoller, Gerald

Morin, Leah Kalasky, Jason Salzman, Emanuel Salzman, Tom Riedlinger, Giorgio Samorini, Carl de Borhegyi, Ilona de Borhegyi, and Inge Geurts. The artwork presented in these pages comes from several talented artists working in different mediums. For their permissions and contributions, I would like to thank Christopher Yurkanin, Timothy White, Peter McCoy, Giorgio Samorini, Joy Muller, and John W. Allen. I would also like to thank Robert Forte for the single photo of R. Gordon Wasson included in this volume.

Many people provided thoughts, feedback, comments, and support during the three-plus years spent developing this book, and many of these people are already included in the list above. The list could go on and on, but I would like to include a specific thanks to Alan Rockefeller for his comments and feedback on the field guide and to Emma Mueller for her wonderful cover design. If I have missed anyone, please forgive me. I am incredibly thankful to everyone who contributed to this book – it couldn't have happened without you – and I hope I have created a book that will make all of these wonderful people proud to have been a part of this project.

-KMF

Introduction

The Fly Agaric (*Amanita muscaria*), with its snow-speckled crimson cap, has inspired the human imagination for centuries, perhaps for millennia. Its otherworldly appearance has made it a frequent feature in the images and paintings of fairy tales, and its beauty has made it a subject of various fetishes, trinkets, and apparel. It is easily one of the world's most recognizable mushrooms, and yet many have only ever seen it in video games, cartoons, or children's books. Consequently, many are surprised to discover that it is in fact a *real* mushroom. This is one of the great paradoxes of the fly agaric; despite its ubiquity in nature and human culture it remains largely an enigma.

Quite apart from its psychoactive effects the fly agaric has a propensity for inspiring confusion, controversy, and passion among humans. While most agree that the mushroom is stunningly photogenic, this tends to be where agreement ends. Almost everyone familiar with this mushroom has some opinion about it, good or bad. Assertions regarding its virtue or danger are largely based on anecdotal experiences or hearsay but are nevertheless proclaimed as gospel. The fly agaric, some claim, is at the foundation of all human religions, while others counter that it is historically insignificant and, not even one of the more interesting species of fungi (certainly a point up for debate). For some it is a profound hallucinogen, capable of inducing mystical and spiritual death-rebirth experiences, while for others it is a "horrible poison", causing gut-wrenching vomiting leading to prayers for a quick (but improbable) death. Some know it to be a delightful culinary, following proper preparation, while others believe it should never be touched, lest one thoughtlessly brush their contaminated hands across their mouth and thus become poisoned.

The passion surrounding the fly agaric runs so high, in fact, that social media posts featuring photos of children holding handsome specimens of this mushroom are inevitably followed by comments of both praise and scorn; praise for artistry and engaging kids with the natural world and scorn for "child endangerment." Whatever the opinion someone has, there appears to be little neutral ground. Which all leaves one to ponder; how strange it is that a simple fungus should elicit such strong reactions, and could be perceived, experienced, and understood in such disparate and various ways.

The impetus underlying this *Compendium* is multi-layered, but at its core is an effort to explore this exceptional fungus from multiple angles, to address confusion and misconceptions, and to create an understanding of this mushroom that moves beyond the confining black-and-white categories of "poison" or "hallucinogen." Accordingly, this book is organized into five parts, each addressing a different and fascinating facet of the fly agaric. These include: (I) Mushroom Hunting & Identification; (II) Religion, Culture, & Folklore; (III) Archaeological Evidence; (IV) Diet & Cuisine; and (V) Pharmacology & Physiological Effects. The organization is intended to make the book more reader friendly, and to help you locate the topics and chapters that you, the reader, are most interested in. The book is not intended to be read front to back, but rather to allow you to follow your interests. Hopefully, the contents of this volume will help to add additional details and layers of complexity to the passionate debates and disagreements that are likely to continue regarding the aesthetic, culinary, and hallucinatory qualities of this singular mushroom.

Part I of the book introduces the reader to the basics of mushroom hunting and identification. For the novice, many important questions are addressed in chapter 1, such as: how to connect with local experts; what to bring on a mushroom hunt (foray); whether one needs a permit, as well as where to get one; how to pick mushrooms; and, in addition, some legal considerations are also addressed. Chapter 2 lays out some of the basics of mushroom identification, including how to recognize and differentiate distinguishing features of *Amanita* mushrooms. Chapter 3 is the heart of this section. Here, you will find a one-of-a-kind photo illustrated field guide with details for identifying over a dozen different psychoactive *Amanita* species, subspecies, and varieties.

There is a common belief that the fly agaric is easy to identify, which is true when one knows what one is looking for, but there is always the potential for confusion. On the other hand, there are those who warn novices against picking or consuming the fly agaric due to "possible" confusion with the "Death Cap" (*A. phalloides*) or "Destroying Angel" (*A. ocreata*), a claim that borders on the absurd. Nevertheless, look-alikes do exist, and this chapter provides the reader with practical guidance for distinguishing the fly agaric, and its relatives, from both benign and potentially harmful look-alikes.

Part II of this book delves into the murkier territory of religion, culture, history, and folklore. The use of the fly agaric among some Siberian tribes is well known and documented, but evidence for its use outside of Siberia is less clear and more circumstantial. Theories regarding fly agaric use among different cultural groups and within specific religious and cult practices have proliferated since the publication of R. Gordon Wasson's (1968) *Soma* over fifty years ago. In Wasson's pioneering work he proposed that the sacred inebriant Soma, central to the religious practices of early Indo-Aryan peoples living in the Indus Valley (present-day India and Pakistan), was none other than the fly agaric mushroom. While Wasson was a professional banker and an amateur mycologist and scholar, he presented a complex and multi-layered argument drawing on ethnographic, historical, semantic, pharmacological, and biological evidence. More importantly, however, was that Wasson actively pursued and engaged with scholars from different disciplines who could help him bring the various threads of his thesis together.

Here, in the spirit of Wasson, scholars, explorers and authors of various backgrounds have been brought together to examine both the substantial *and* circumstantial threads pointing to potential historical and cultural uses of the fly agaric and to determine whether these strands, like mycelial hyphae, may ultimately lead to a *fruiting* fly agaric! This undertaking begins in chapter 4, where Trent Austin and I revisit Wasson's theory identifying the fly agaric as the ancient Vedic Soma. Central to Wasson's theory are the three filters of Soma, which correspond to different steps in its preparation, as outlined in the Rig Veda. Wasson's interpretation of the first two filters, a filter of sunlight (sun-drying or desiccation) and a woolen filter (to remove solids from aqueous preparations), are generally uncontested, but his proposal for the third filter, the human body, which produces urine as the final form of Soma, has raised controversy. In this chapter, Austin and I propose an alternate third filter, a filter of milk, and detail how mixing aqueous extractions of *Amanita muscaria* with unpasteurized milk might lead to significant pharmacological changes potentiating the final beverage.

Over the last few decades several scholars have drawn some surprising parallels between the mythos of Santa Claus and ethnographic accounts of *Amanita muscaria* use among Siberian reindeer herders. In chapter 5, mycologist and adventurer Lawrence Millman details his travels to Siberia to investigate these connections. Eventually, Millman's investigations lead him away from Siberia to Europe, where he finds a more compelling foundation for the relationship between Santa Claus and *Amanita muscaria* among the Saami reindeer herders of Finnish Lapland.

In chapter 6, intrepid travelers Jason, Emanuel, and Joanne Salzman, along with Gary Lincoff, recount their 1994 and 1995 visits to the Kamchatka Peninsula in Eastern Siberia, where several Siberian tribes are reputed to use the fly agaric mushroom. R. Gordon Wasson attempted to travel to Siberia to test and confirm aspects of his Soma theory before he died but was repeatedly denied entry by the Soviet Government. After the fall of the Soviet Union in 1991, the authors of this chapter organized two trips to the region to learn about local uses of the fly agaric and determine whether additional evidence could be gathered to further substantiate Wasson's theory.

In chapter 7, Frank Dugan explores modern and historical associations between the famous Russian witch, Baba Yaga, and the fly agaric mushroom. After supporting the hypothesis of a Uralic (non-Russian) origin for Baba Yaga and her connection to *Amanita muscaria*, Dugan poses questions regarding the geographic and temporal extent of the witch-*Amanita* motif both in Eurasia and in the Northern Hemisphere more broadly.

In chapter 8, Steven Leto examines the "magical drinks" of Germanic and Nordic mythologies as well as descriptions of the actions and behaviors of gods, particularly Odin, that evoke shamanic activity, such as: divination, shape-shifting and astral projection. A chance encounter with a *Psilocybe* mushroom growing at the archaeological site *Tofta Högar* in Sweden, as well as parallels between the *Mead of Poetry* and the Vedic hallucinogen Soma, lead Leto to explore the parallels between these mythical sacraments and to examine Nordic mythology for evidence of potential historical uses of *Amanita muscaria* and *Psilocybe* mushrooms in Scandinavia.

Chapter 9 provides a rare English translation of Samuel Ödman's 1784 essay which explores the possibility that the behavioral and physiological features of the infamous berserker-rages of the early Vikings was drug induced. Relying on accounts of fly agaric use in Siberia, Ödman develops an argument that the fly agaric is the most likely substance to explain the berserker phenomenon, a theory that has remained popular to this day. In chapter 10, Mark Hoffman and Carl Ruck investigate and expand upon Ödman's hypothesis, exploring the historical, mythical, pharmacological, and etymological evidence for connections between *Amanita muscaria* and Odin's warriors, the *berserksgangr*.

Another investigation of magical brews and foods is presented by Erynn Rowan Laurie and Timothy White in chapter 11. References to magical brews and foods abound in Celtic legends dealing with journeys to *Tir Tairngire* (Land of Promise) or into the *sidhe* (faery mounds). The abundance of Celtic legends about crimson foods which induce mystical experiences, inspire extraordinary knowledge, and impart the gift of prophecy, is highly suggestive and Laurie and White contend that the motifs of magical foods can best be explained as metaphoric references to *Amanita muscaria*.

The Celtic focus continues in chapters 12 and 13. In chapter 12, Thomas Riedlinger explores the tales of Celtic hero Cú Chulaind for fly agaric motifs, and potential evidence of historical use among the early Celtic peoples. The ancient myths of Ireland include many fabulous tales regarding Cú Chulaind, a warrior of the Ulaid clan who lived in the province of Ulster. Some of these portray Cú Chulaind's legendary mood-swings: at one extreme, a battle-fury so intense that it terrified even his family, friends, and fellow warriors; at the other, a torpor with vivid, prophetic dreams in which he languished for a year. Both states, as described in the tales, resemble the effects of *Amanita muscaria* inebriation. In chapter 13, radical mycologist Peter McCoy examines the cultural and symbolic significance of the Celtic goddess Brigid and her association with *Amanita muscaria*. An important goddess in the Celtic pantheon, Brigid is a triple-aspect deity of fire, water, poetry, wisdom, renewal, omens, fertility, and healing. From these associations, McCoy draws parallels with similar traits found in goddess worship and *Amanita muscaria* consumption practices around the world.

Chapter 14 features an interview with Mark Niemoller, one of the first major purveyors of dried *Amanita muscaria* and *pantherina* mushrooms in the United States. Niemoller's pioneering business, first established in 1986, became one of the most successful ethnobotanical companies in the U.S. until legal expenses related to a DEA investigation led the company to close its doors in 2005. Niemoller discusses the provenance and history of JLF: *Poisonous Non-Consumables* as well as his interests and personal research into the properties of the fly agaric and its close relatives.

Chapters 15 and 16 both concern the fly agaric as a symbol of luck. In chapter 15 the history of the fly agaric as a "good luck" charm is explored. The etymology of *glückspilz* (luck mushroom) is examined as well as the association of the fly agaric with several other good luck charms, including chimney sweeps', four-leaf clovers, horseshoes and *glücksschweinchen*. Chapter 16 provides a colorful and fanciful departure from previous chapters. Here, an English adaptation is provided of an early 20th century German children's book that follows the adventures of Hänsel and Gretel after they get lost in the woods while searching for the lucky fly agaric mushroom. Beautiful color illustrations from the original book are reproduced here to accompany this rare adaptation.

The next section of the book, Part III, brings us to a more concrete exploration of the historical importance and use of the fly agaric through an examination of the archaeological evidence. Giorgio Samorini begins this section with chapter 17, where he delineates a "contextual outline" for identifying representations of *Amanita muscaria* and *pantherina* in petroglyphs, pictographs, and other types of archaeological evidence. Samorini draws on evidence and artifacts from Eurasian, African and American cultural domains in outlining the "killer details" for identifying archaeological representations of the fly agaric and its psychoactive relatives.

Carl de Borhegyi, son of Stephan F. de Borhegyi, the Mesoamerican archaeologist who collaborated with Gordon and Valentina Wasson (1957) on *Russia, Mushrooms and History*, continues our archaeological investigations in chapter 18. Here, de Borhegyi extends his father's research by pulling out and illustrating, with words and images, a few threads in the complex fabric of Maya art and mythology. These threads illustrate a relationship between the "mushroom stones" and hallucinogenic mushrooms, principally *Amanita muscaria*, with the Jaguar-Bird-Serpent god, Quetzalcóatl.

Part IV of the book offers several views and investigations on the underexplored subject of the fly agaric as a culinary mushroom. Originally written in 1898 by Frederick Coville, chapter 19 provides a remarkable account of *Amanita muscaria* use as food within African American communities of Washington, D.C. at the turn of the 20th century. Coville discusses its preparation and provides a historical perspective on the "poisonous" quality of this mushroom. In chapter 20, Danny Curry offers a surprising contrast to Coville by providing a personal account of his youth in 1960s rural Ohio, when his family frequently consumed fly agaric mushrooms as part of their regular diet. Chapter 21 concludes this section with a discussion of *Amanita muscaria*'s nutritional properties and culinary qualities, how to detoxify the mushroom for culinary use, as well as providing a handful of recipes for the curious foodie.

In Part V, a review of *Amanita muscaria*'s unique pharmacology is provided, as well as a look into its therapeutic and inebriating properties. In chapter 22, Ewa Maciejczyk traces the history of investigations into *Amanita muscaria*'s pharmacological properties and surveys current literature dealing with the isolation, structural elucidation, and biological activities of natural products from the *Amanita muscaria* fruiting body. This is followed by chapter 23, where Tjakko Stijve and I attempt to address the persistent confusion and controversy over muscarine, and the degree to which it is generally present in fly agaric fruiting bodies. Currently, it is widely (and mistakenly) believed that muscarine does not occur in *Amanita muscaria* in pharmacologically active levels. While muscarine is unable to cross the blood-brain barrier, and contributes no psychoactive effects to *Amanita muscaria* inebriation, the evidence gathered by the authors suggests that muscarine is present in sufficient quantities to have a physiological effect when moderate amounts of this mushroom are consumed.

In chapter 24, Bill Mann and I provide a brief history of the homeopathic use of *Amanita muscaria*, under the name of Agaricus muscarius, detail its various homeopathic applications, and provide a case study of a young man treated with this remedy. Chapter 25, originally written in 1894 by homeopathic doctor Horace

Holmes, provides an interesting and accessible look into the use of this mushroom within homeopathy just before the turn of the 20th century, including details of Holmes own successes using Agaricus muscarius to treat nervous disorders, sexual dysfunction, dyspepsia, chorea and related muscle-twitching, and afflictions of the skin.

In chapter 26, I explore a broader history of the therapeutic use of the fly agaric as well as potential modern applications, including for treatment of pain, inflammation, anxiety, cognitive decline and cancer. Following this, in chapter 27, I provide some broad guidelines for working with *Amanita muscaria* therapeutically, including information about the types of ailments that might be treated, how much to use, and how to prepare the mushroom for use as a medicinal.

In chapter 28 the focus veers away from the strictly therapeutic and delves into the otherworldly effects of the fly agaric. A contrast is drawn for the reader between the effects of isoxazole-containing mushrooms like *Amanita muscaria* and those produced by *Psilocybe* mushrooms. A handful of distinctive features of *A. muscaria* inebriation are identified and discussed and experience reports are also provided to help illustrate some of the more distinctive effects. The book closes with chapter 29, which covers some ethnographic information about how the fly agaric has been prepared in Siberia and other locales, and examines preparations outlined in the Rig Veda for the hallucinogen Soma, which is widely believed to have been *Amanita muscaria*. Several recipes, based on these ethnographic accounts, are provided.

In this volume I have attempted to compile and provide the reader with a comprehensive and exhaustive overview of the iconic red-and-white spotted fairy tale mushroom, the fly agaric, but addressing every intriguing aspect of this singular fungus in one book would be impossible. There are more stories to tell, more theories and hypotheses to develop and argue over, more lost histories, and more pharmacological secrets to be uncovered! Indeed, it has taken the work of over two dozen authors, researchers, explorers, academics, and entrepreneurs to assemble the book in your hands. Collectively, it is our hope that the following pages will inspire interest and passion for mushrooms and mycology, for nature and its preservation, for religion and spiritual experience, and for history and all her mysteries. Additionally, it is hoped that this volume will stimulate your curiosity while also stimulating friendly discussions and debates on the various ideas, thoughts, theories and hypotheses contained within these pages. With any luck, these pages will lay a foundation for future investigations, research, and explorations of this remarkable fungus.

Wonderfully, the fly agaric is cosmopolitan throughout the Northern Hemisphere, and also occurs throughout much of the Southern Hemisphere (though, there, it is considered invasive). As a result, this is a mushroom that belongs to *everyone*; one that is available to anyone willing to put in the time to look and discover its beauty either in nature or by the side of the road. Anyone can find it, study it, taste it, photograph it, paint it, or meditate with it. So, grab your book and your boots and head outdoors. The world awaits. And perhaps, just perhaps, you will encounter the mushroom that is the subject of the following pages… the Fly Agaric!

Part I:
Mushroom Hunting & Identification

Chapter 1

Mushroom Hunting

Walking through the woods on a cool and damp autumn morning in search of hidden fungal treasures, one might find themselves feeling like a child on an Easter egg hunt. Looking under bushes, behind logs and under clumps of dirt one finds, instead of eggs, mushrooms of different shapes and sizes with colors ranging from deep purples to bright reds and golden yellows. Instead of chocolates and candies one fills their basket with the delectable bounty of the forest floor. When tromping through the woods looking for mushrooms with my own family, I feel the weight of day-to-day responsibilities gently lift from my shoulders, the fresh air clears my head, and I feel like a kid again. Perhaps, like a child on Easter morning.

Mushroom hunting is a wonderful hobby and pastime, one that can be enjoyed with friends and family of all ages and one that, hopefully, results in some delicious shared meals. Mushrooming is not a hobby that requires a lot of equipment, nor one that requires extensive training. All that it requires is curiosity, a dose of caution, and some guidance. For those just starting out it is recommended to find a good field guide that covers the region where you live and to connect with your local mycological society. A quick Google search should tell you whether any mushroom

Fig. 1: Basket of *Amanita muscaria* and *Lactarius sp.*, picked in Iisalmi, Finland (Photo by Martika Lyle).

groups are present in your community and where to find them.[1] Typically, mycological societies have regular meetings, particularly in the fall, and they also organize mushroom forays. The group mushroom foray is a wonderful place for the newbie to begin and learn about their new hobby and to meet others with shared interests. Forays are typically led by experienced and knowledgeable members of the local mycology club who can help with identification and who also provide tips and guidance for novice and intermediate mushroomers.

So, what should you bring on your first foray? Here are a few tips for items you might bring with you:

- Basket
- Paper or wax paper bags
- Small knife
- Brush
- Pocket field guide
- Compass
- Whistle
- Water
- Snacks
- Warm or water-resistant clothes
- Boots
- Extra clothes and shoes to keep in the car
- Camera
- Mushroom picking or wilderness permits, if necessary
- A companion

The primary essentials are your knife, basket, brush, and several baggies for storing your mushrooms. You will want a basket that will be big enough to hold what you think you can collect but not so big that it's burdensome (Figure 1). When picking mushrooms, you also want to keep your mushrooms separate from one another to prevent mixing and potential confusion later. If you suspect that two mushrooms are not the same species, place them in separate bags. You might even use a marker to label your bags, particularly if someone is assisting you with your identifications. The bags should be paper or wax rather than plastic, which will suffocate your mushrooms and cause them to sweat. The knife is important for digging up mushrooms from their base; this is particularly important with *Amanita* species since the base of the mushroom may hold the key to a correct identification. There continues to be some controversy within mushrooming circles regarding whether mushrooms should be dug up or cut off at the base, a controversy that is rooted in the belief that picking mushrooms from the base may damage the mycelium and prevent future growth. This belief has been shown to be unwarranted (Egli et al., 2005; Norvell, 1995) and the picker will lose important information for an accurate identification if they leave part of the fruiting body behind. The knife, along with the brush, can also be used for cleaning specimens before putting them in your bag. For those collecting edibles, it can be a pain to come home and find your "dinner" covered in dirt. Keeping your mushrooms clean as you go will save you time later in the kitchen when mouths are watering and tummies are grumbling.

For safety purposes it is recommended that mushroomers bring a companion, compass, whistle, water, and clothes appropriate for the weather and environment where they will be picking. When one has their eye on the ground, searching for their elusive prey, it becomes easy to lose track of where one is and how far one has traveled. One may even lose track of time depending on how busy one finds oneself. Having a companion is always a good idea and a whistle can help you quickly locate your companion if you have wandered in opposite directions. Because one may also lose track of their direction a compass is also a good idea but be sure to note the direction of your car or the trailhead before you set out. Food and water are a must, particularly if you plan on staying out for any significant length of time, and dressing appropriately will keep you comfortable. Even if it is not raining one can get wet very quickly walking through damp brush, and one does not want to be the "wet blanket" that insists on ending the hunt early because he or she was unprepared. Even with appropriate clothing one may want to change into lighter and dryer clothing after the hunt, so coming prepared with an extra change of clothes is recommended.

Another important consideration will be to determine whether you need a mushroom hunting permit or a forest pass. Requirements and restrictions may vary from state to state and may vary from park to park. Be sure to check with your State Parks Department or with U.S. Forest Services before venturing out on the hunt. A simple Google search for "Mushroom Permit" along with the name of the forest or park you plan on visiting should be sufficient to discover the information you are looking for. Permits, if required, should be available for purchase online or at your local Forest Service or Parks Department office. Typically, there are personal use exceptions – though there may be a one to three-gallon limit on personal harvests – and a physical permit may or may not be required. Any commercial harvesting one plans to do will likely require a permit.

One final consideration concerns the legality of mushrooms themselves. *Psilocybin*-containing mushrooms are considered a Federally Controlled Substance and are also illegal in all fifty states; however, one is not likely to run into trouble if they have mistakenly picked one or have gathered a small collection along with a number of other innocuous mushrooms. Individuals with bags full of *Psilocybe* mushrooms found picking on private land or where law enforcement know *Psilocybes* occur may run into legal trouble. Typically, state law will be applied rather than Federal, so make sure you are familiar with local laws and penalties if you choose to knowingly pick *Psilocybe* mushrooms. Remember, this is at your own risk.

While there are a number of psychoactive *Amanitas*, which are the focus of this book, these are not regulated like *Psilocybes*. The exception is the state of Louisiana which prohibits the possession and distribution of *Amanita muscaria* mushrooms. Fortunately, for Louisiana natives, the local psychoactive variety is *Amanita persicina* not *Amanita muscaria*. There are a number of reasons why psychoactive *Amanitas* are not and have not been regulated, but primarily it seems to be that: (1) they are generally considered poisonous; (2) psychoactive effects are elusive and difficult to achieve without knowledge of dosing and preparation techniques; and (3) the active compounds (ibotenic acid, muscimol) are both important research chemicals. It seems unlikely that the legal status of these mushrooms will change any time soon, but when it comes to psychoactive substances the political winds can change quickly.

In any case, you now have the basic knowledge necessary to get yourself started with your new hobby of mushrooming. Remember to pack the essentials when you hunt and make use of your local resources, including your local mycological society. Once you have found and identified your first mushroom, whether it be edible, psychoactive, or medicinal, you will discover that you have also found a hobby that will stick with you for life. So, happy trails and good luck on the hunt!

Notes

1. A list of NAMA (North American Mycological Society) affiliated Mushroom Clubs organized by state and region can be found on NAMA's website, here: https://namyco.org/clubs.php

Chapter 2

Amanita Basics

The genus *Amanita* is known for mushrooms with a handsome and sturdy stature and for several exceptional mushrooms, including some of the most visually stunning mushrooms (*A. cokeri, A. muscaria*), some of the most delectable (*A. caesarea, A. calyptroderma*), and several of the most deadly mushrooms known (*A. phalloides* [Figure 1], *A. virosa*). The presence of deadly mushrooms within the genus has resulted in a great deal of caution within mycological circles, particularly when instructing and guiding novice mushroomers. The paranoia caused by potentially fatal poisonings has resulted in individuals being discouraged from picking or learning about edible *Amanitas*, and in those interested in the psychoactive *Amanitas* being warned about the potential for liver failure and death should they pick and

consume the wrong mushroom. While caution is warranted, the atmosphere that has developed around the *Amanita* genus is unfortunate, particularly when a little skill and knowledge can quickly mitigate potential dangers. Whenever picking a mushroom for the first time, however, it is always good to have someone more experienced with you who can help point out the primary distinguishing features.

Fig. 1: *Amanita phalloides*, the "Death Cap" mushroom. Though distinctive this mushroom continues to be a source of poisonings and fatalities in California and elsewhere. Wunderlich County Park. Woodside, San Mateo County, California (Photo by Ron Wolf).

Notable Physical Features

The *Amanita* genus has some very unique features that are common among a number of *Amanita* species. Features such as warts, an annulus, volva and a basal bulb are common among the psychoactive *Amanita*, though hardly unique to these species. The typical *Amanita* starts life as an "egg." It emerges from the ground whitish and bulb-like and is frequently mistaken for edible puffballs at this stage of its life cycle. Eventually, these mushrooms "hatch." As the stem grows it pushes the cap up breaking the universal veil, which is equivalent to the eggshell in this analogy. But unlike an eggshell the universal veil is made up of soft tissue which frequently breaks apart as the mushroom cap expands leaving patches of material on the cap or small fragments known as "warts," though some species simply emerge with bald caps. The breaking of the universal veil may also result in fragments of the veil being left atop the basal bulb; these remains are referred to as the volva. These veil remnants typically take specific forms that can help identify the mushroom to a specific section, subsection, or species of *Amanita* (Figure 2).

Another important feature is the partial veil which covers the mushroom's gills until it reaches maturity. As the cap expands the partial veil tears, frequently leaving an annulus, or ring, on the stalk of the mushroom. In some cases, veil remnants may also adhere to the margin of the mature cap. The annulus can be an important identifying feature. While the annulus is most frequently white, depending on the species it may have some yellow, pink or other coloration. The annulus

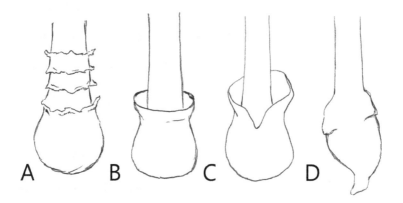

Fig. 2: Common volva and base formations in the genus *Amanita*. **A.** Ascending concentric rings of shaggy volval material – typical of *A. muscaria*. **B.** Basal bulb with volval collar – typical of *A. pantherina*. **C.** Sack-like volva – found in *A. phalloides*, among others. **D.** Spindle-shaped bulb – found in *A. cokeri* and *A. smithiana*.

also typically takes a skirt-like shape, but in some instances may be funnel shaped as with *A. velatipes* and *A. multisquamosa*. The location of the annulus on the stem can also be an important identifying feature as some mushrooms retain a superior annulus, found on the upper-stem, a median annulus, in the middle of the stem, or an inferior annulus, found towards the bottom of the stem. Of course, some species will not have an annulus at all, or the annulus may be fragile and fall away in the rain or on its own. As you become more familiar with *Amanitas* these important distinctions will begin to stand out more prominently.

Amanita Taxonomy

The *Amanita* genus can be broken down into subgenera and further into sections and subsections which can help the mushroomer understand the relationships between different species as well as how closely or distantly two species within the *Amanita* genus are related (Table 1). Currently, two subgenera are recognized within the *Amanita* genus; these are subgenus *Amanita* and subgenus *Lepidella*.[1] To date all of the psychoactive *Amanita* have been identified within subgenus *Amanita*, section *Amanita*, which is quite distant from subgenus *Lepidella*, section *Phalloideae*, which contains several of the deadliest *Amanita* species including the notorious "Death Cap" (*A. phalloides*). These taxonomic distinctions are important because there is a persistent belief among amateur mushroomers that psychoactive *Amanitas* either contain or "might" contain liver toxins. This belief likely stems from the fact that the deadly *Amanitas* contain hepatotoxic amatoxins and that the genus has come to be associated with these deadly compounds. However, amatoxins are not known to occur in subgenus *Amanita*, section *Amanita*, nor are any other hepatotoxins known from this section, making this particular fear an unwarranted one. The primary concern here, of course, is proper identification.

Several species of *Amanita* from subgenus *Lepidella* have previously been identified as psychoactive, including *A. cokeri* (Hall et al., 1987), *A. smithiana* (Hall & Hall, 1994), *A. solitaria* (Sumstine, 1905) and *A. strobiliformis* (Takemoto, 1964), however, any identification of a psychoactive *Amanita* outside of subgenus *Amanita*, section *Amanita* should be considered suspect. Both *A. smithiana* and *A. solitaria*, for example, are nephrotoxic and known to cause liver damage and renal failure (Barman et al., 2018; West et al., 2009). Subsequent studies on *A. strobiliformis* have failed to identify either ibotenic acid or muscimol (Benedict,

Brief Summary of Amanita Subgenera and Related Sections

SUBGENUS AMANITA	SUBGENUS LEPIDELLA
Section Amanita **Subsection Amanita** A. *chrysoblema* A. *frostiana* A. *muscaria* var. *formosa* A. *muscaria* var. *muscaria* A. *muscaria* var. *guessowii* A. *muscaria* subsp. *flavivolvata* A. *parcivolvata* A. *persicina* A. *regalis* **Subsection Amanitella** A. *crenulata* **Subsection Gemmatae** A. *alpinicola* A. *aprica* A. *gemmata* A. *pantherinoides* A. *russuloides* **Subsection Pantherinae** A. *albocreata* A. *hallingiana* A. *multisquamosa* A. *pantherina* A. *praecox* A. *velatipes* **Section Caesareae** A. *caesarea* A. *calyptroderma* **Section Vaginatae**	**Section Lepidella** A. *cokeri* A. *smithiana* **Section Amidella** **Section Phalloideae** A. *phalloides* A. *virosa* **Section Validae** A. "*amerirubescens*" A. *augusta* A. *brunnescens* A. *citrina* A. *flavoconia* A. *novinupta*

Table 1: The taxonomic table above provides a general outline of how the genus *Amanita* is divided. The list of species is incomplete and includes only those that are specifically discussed in this volume.

1966; Chilton & Ott, 1976; Michelot & Melendez-Howell, 2003), the psychoactive compounds known from the fly agaric, and suggest that it is edible (Kirchmair et al., 2012). The picture surrounding *A. cokeri* is more ambiguous. It continues to be identified as psychoactive in some field guides (Miller & Miller, 2006) though most address the status of this mushroom ambiguously. Gary Lincoff (1981: p. 532), for example, cautions against eating it but does not detail why the mushroom should be avoided. The general take-away from this, however, is that one should be wary of any suggestion that an *Amanita* outside of subgenus *Amanita*, section *Amanita* is psychoactive. It is, of course, possible, but without a positive assay on properly identified specimens such identifications should be considered with the utmost skepticism.

Potency of Psychoactive Amanitas

While most think of *Psilocybes* when discussions of magic mushrooms arise, there is a whole separate category of "magic" mushrooms within the *Amanita* genus. However, it is important to note that the effects of these two classes of magic mushrooms are quite different (*see* Ch. 28). While the psychedelic effects of *Psilocybe* mushrooms are caused by psilocybin and psilocin, serotonergic compounds, the psychoactive effects produced by some *Amanita* species are caused by the isoxazole compounds ibotenic acid and muscimol, both of which are GABAergic (*see* Ch. 22). The effects of these compounds, their potency (Table 2), and variations in their concentrations require one to approach these mushrooms differently than one might approach their *Psilocybe* cousins.

Unlike *Psilocybes*, whose potency tends to vary within limited and identifiable ranges, estimating the potency of specimens of psychoactive *Amanita* can feel

Psychoactive doses of Isoxazole Compounds	
Ibotenic Acid	50 – 100 mg
Muscimol	10 – 15 mg

Table 2: Psychoactivity of Isoxazole Compounds (Ott, 1993: 328).

like a haphazard endeavor. The potency varies greatly within individual species and varieties, which has led to much speculation about potential factors that might influence potency, including season of growth, life-cycle stage, weather patterns, tree hosts, soil type and other considerations. Few studies have been conducted to determine what factors, if any, are the primary determinants of potency, leaving those with an interest in these mushrooms with a fair amount of guesswork.

One factor that sets the psychoactive *Amanitas* apart from *Psilocybes* is that *Amanitas* are mycorrhizal. This means that the mushroom grows in a symbiotic relationship with host trees and leaves open the possibility that the pharmacological makeup of the mushroom is determined, in part, by the identity of the tree host. The honey mushroom (*Armillaria mellea* group), though parasitic rather than symbiotic, might serve as an example. The honey mushroom, which parasitizes trees, is generally considered a good edible, however, specimens parasitizing hardwoods have been known to cause illness and gastrointestinal distress. It is possible that differences in tree hosts for psychoactive *Amanita* might similarly influence the pharmacological makeup of fruiting bodies and the effects experienced by those who ingest them. I am unaware of any studies that have investigated whether host trees impact the pharmacology or edibility of different *Amanita* species, though there is a popular belief that *Amanita muscaria* that grow in association with birch are the most potent. This belief appears to arise from the fact that the birch is the most common tree host in Siberia, where *Amanita muscaria* has been used as a medicine and inebriant for centuries among several indigenous Siberian peoples. There does not appear to be any evidence, anecdotal or otherwise, however, that would support this belief.

One area that *has* been studied is how potency varies throughout the lifecycle of *Amanita muscaria* mushrooms as well as how its psychoactive compounds are distributed throughout the fruiting body. It has been found that the cap is the most potent part of the mushroom and that the most potent part of the cap is the yellowish-orange tissue found immediately underneath the skin, or cuticle, of the cap (Gore & Jordan, 1982; Michelot & Melendez-Howell, 2003: 134, *citing* Catalfomo & Eugster, 1970; Tsunoda et al., 1993b). While muscimol appears to be fairly uniform throughout the cap, with the exception of the cuticle which is significantly weaker than the rest, the distribution of ibotenic acid in the cap is more variable. The content of ibotenic acid in the yellow flesh beneath the cap may be as much as three times more concentrated than in the white parts of the cap and ibotenic acid concentrations in the whole cap may be up to five times higher, or more, than what is found in the stem. Similarly, muscimol is more concentrated in the cap with levels up to four times higher than found in the stem (Gore & Jordan, 1982: 327). The basal bulb of the mushroom also has higher concentrations of the mushroom's psychoactive compounds than the stem, though significantly less than is found in the cap (Tsunoda et al., 1993b).

The concentrations of psychoactive compounds also vary depending on when in the lifecycle the mushroom is picked. Interestingly, it is the smaller mushrooms, where the partial veil is still intact or just breaking, that are the most potent. Concentrations of both muscimol and ibotenic acid appear to peak during the 2nd and 3rd phases of growth, as illustrated in Figure 3. Large specimens with fully opened caps, while quite impressive to find, are in fact in their least potent stage of the mushroom lifecycle. Notably, this finding appears to support the preference among some Siberian peoples for smaller fly agaric mushrooms. There also appears to be a slight difference in when concentrations peak depending on whether the mushroom is found growing alone or in a group. Mushrooms growing in a group tend to hit peak concentrations earlier, during phase two, while mushrooms growing alone reach peak concentrations during phase three (Tsunoda et al., 1993b).

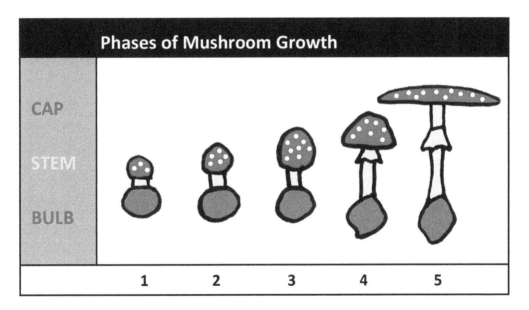

Fig. 3: Phases of Mushroom Growth. Mushrooms in the second and third phases of growth demonstrate the highest concentrations of ibotenic acid and muscimol (Tsunoda et al., 1993b).

While the study on correlations between potency and lifecycle were conducted on specimens of *Amanita muscaria* var. *muscaria*, it is likely that these same patterns will be observed with other muscarioids (Subgenus: *Amanita*; Section: *Amanita*; Subsection: *Amanita*; Series: *Amanita*; Stirps: *Muscaria*). It is also possible that these patterns will be found with other psychoactive *Amanita*, outside of the muscarioids, but additional studies are necessary to determine whether these patterns are typical of psychoactive *Amanitas* more broadly.

In terms of general potency, there are several studies that have been done to determine concentrations of ibotenic acid and muscimol in *Amanita muscaria* var. *muscaria* and in *Amanita pantherina* (Table 3), but studies on other isoxazole containing *Amanitas* are either obscure or non-existent. Based on anecdotal reports it appears that most of the muscarioid *Amanitas* show a similar range of psychoactivity, with threshold doses starting around five grams (Table 4). Pantheroid (Subgenus: *Amanita*; Section: *Amanita*; Subsection: *Pantherinae*) *Amanitas* also appear to be similarly active based on anecdotal reports, and these are consistently more potent than their muscarioid counterparts.

Based on the chart below, one can see a great deal of variability with the concentrations of these compounds in both *A. muscaria* and *A. pantherina*. Based on a psychoactive dose of 10 mg for muscimol (Table 2) we can estimate that six

Mushroom species	Source	Ibotenic Acid	Muscimol
Amanita muscaria	Poliwoda et al., 2014	Min: 0.292 mg/g Max: 6.565 mg/g Ave: 2.409 mg/g	Min: 0.073 mg/g Max: 3.561 mg/g Ave: 1.051 mg/g
	Stijve, 1981; 1982	No data.	Min: < 0.01 mg/g Max: 2.2 mg/g Ave: 0.76 mg/g
Amanita pantherina	Poliwoda et al., 2014	1 Sample: 3.367 mg/g	1 Sample: 1.228 mg/g
	Stijve, 1981; 1982	No data.	Min: 0.25 mg/g Max: 3.1 mg/g Ave: 1.75 mg/g

Table 3: Comparisons of potency between *A. muscaria* and *A. pantherina*.

grams of dried *A. pantherina* with average muscimol concentrations would produce a psychoactive effect compared to 10 g for *A. muscaria*. Of course, this estimate does not account for the co-occurrence of ibotenic acid, which will also affect the overall potency of the mushroom. Anecdotal reports also suggest that *A. pantherina*, and relatives, are more potent than their muscarioid counterparts though the range in concentrations in both species results in an overlap, where *A. muscaria* may exhibit equal or greater potency in some instances.

Table 4, below, has been constructed based on both chemical analyses and anecdotal reports in order to give a very general view on the levels of potency found within the psychoactive *Amanitas*. In the field guide that follows, information is provided regarding whether a particular mushroom is mildly, moderately, or highly active, though typically a range is given. Most of the psychoactive *Amanita* have not been chemically analyzed to any significant degree, and very little experimentation has been conducted outside of the various muscarioid and pantheroid species, leaving several species where there is little to no data about potency. Most of the data on these species come from poisoning reports, and it is difficult to glean much data from these other than to identify the effects as characteristic of ibotenic acid and muscimol. Still, the psychoactive *Amanita* appear to generally fall within the potency ranges outlined in Table 4.

A Note on Nomenclature

The names of mushroom species and their classifications as species, subspecies, or varieties are frequently revised as the field of mycology advances and more

Psychoactive Level	Dose		
	Low	Moderate	High
Mild	10 – 15 g	15 – 20 g	20 – 25 g
Moderate	5 – 10 g	10 – 15 g	15 – 20 g
High	1 – 5 g	5 – 10 g	10 – 15 g

Table 4: Potency ranges of psychoactive *Amanita*. Weights provided refer to dried mushrooms.

precise information becomes available. The field of mycology in North America has historically applied European classifications and species names to mushrooms found in North America. While these species frequently appear identical on a macroscopic level, studies have generally found that these North American varieties are genetically distinct. For example, *Amanita muscaria* is believed to originate from the Siberian-Beringian region but as the species spread three separate geographic clades developed, resulting in a series of varieties, subspecies and in some cases, distinct species (Geml et al., 2006). As you will see in the field guide that follows, various forms of the "fly agaric" occur in North America but the "true" fly agaric, *Amanita muscaria* var. *muscaria*, only occurs in Alaska, near its Siberian-Beringian origins.

In the field guide you will notice that alternative names are given for many of the mushrooms described. One of the reasons for this is that many popular field guides use the older recognized names of many of these species. These alternate names are provided to help prevent confusion and allow for the reader to consult some of these older field guides, many of which are otherwise very good sources of information. You will also notice that some species names are provided in quotes, such as *A.* "*pantherina*" and *A.* "*amerirubescens.*" Here, quotes are used to indicate that a specific or formal designation has yet to be officially adopted and indicates that the names in quotes are the names that are commonly used in North American field guides and in mycological circles.

Fig. 4: *Amanita brunnescens*. While the cap color can vary from a pale greenish-yellow to dark brown the brown cap with white warts in the featured specimen is reminiscent of *A.* "*pantherina.*" However, *A. brunnescens* is only found in Eastern North America while *A.* "*pantherina*" is found west of the Rockies. Nevertheless, the star-shaped, or clefted bulb of *A. brunnescens* is a conspicuous distinction (Photo by Matt Gardner).

Look-Alikes

In the following chapter descriptions of over a dozen species, subspecies and varieties of psychoactive *Amanitas* are provided, including important details about look-alike species and how to distinguish them from the mushrooms presented in this book. Look-alikes are typically mentioned only if they share a geographic range with the described species. There are other mushrooms that share some physical features with the species described here but since they do not share a geographic range it would be impossible to confuse them (For example, *A. brunnescens* [Figure 4] shares *some* features with *A.* "*pantherina*," but the species are geographically distinct). Location and habitat are important factors when identifying mushrooms and information is provided about the general range and distribution of each of the psychoactive *Amanitas* described herein.

It is always possible that there exist additional look-alike mushrooms that are not covered here, though the most common and likely look-alikes have been identified. The field guide addresses these look-alikes briefly, identifying them by name and describing the features that set them apart from their psychoactive counterparts but does not provide individual entries for these species. Additional information about these mushrooms can be found in a number of different mushroom field guides, and also online. The website <u>Amanitaceae.org</u>, in particular, is a great resource on the *Amanita* genus and is regularly kept up to date. In any case, it's always good practice to consult several sources when identifying mushrooms for the first time.

Having now covered some important basics you should be able to proceed to the Field Guide with the necessary information and context for interpreting the guide and beginning on your own psychoactive *Amanita* forays!

Notes

1. Mushrooms within the *Amanita* genus are separated into subgenera *Amanita* and *Lepidella* based on how their spores react when exposed to Melzer's reagent or to a solution of iodine. If the spores stain or darken blue or black they are amyloid and the mushroom belongs to the subgenus *Lepidella*. Spores that do not react this way are inamyloid, or nonamyloid, and the mushroom belongs to the subgenus *Amanita*. Thus, if you are looking for psychoactive *Amanita* and encounter a mushroom that may be a look-alike, such as *A. augusta* or *A. flavoconia*, then checking for an amyloid reaction would be one way to eliminate look-alikes. Unfortunately, Melzer's reagent is difficult to come-by since it contains chloral hydrate, a controlled substance. A solution of iodine can be used but this occasionally results in false "negatives," indicating that an *Amanita* mushroom is in the *Amanita* subgenus rather than *Lepidella*. For tips on how to procure Melzer's reagent *see* Leonard (2006).

Chapter 3

Psychoactive Amanitas of North America

The Muscaria Group

Subgenus: *Amanita*; Section: *Amanita*; Subsection: *Amanita*; Series: *Amanita*; Stirps: *Muscaria*

Active Muscarioids

Fly Agaric
Amanita muscaria var. *muscaria* (Fig. 1)
AKA: *Agaricus muscarius*
Fruiting Body: Small to large, with caps ranging from 3 to 8 inches in diameter and stems measuring up to 8 inches tall and 1 inch wide. **Cap:** The shape of the cap ranges from convex to plane, with the margin occasionally turning upward in age. The cap is typically bright red to reddish-orange, sometimes orangish-yellow, though the colors may fade in the sun. Warts are white to creamy white to slightly yellow. Flesh is white except for a thin layer of yellowish-orange tissue just below the cuticle. **Gills:** The gills are white, free from the stem, and crowded. **Stem:** The stem is white with a bulbous base and a skirt-like annulus. A series of ascending concentric rings at the top of the basal bulb is considered a defining feature. **Spores:** Spores are white, elliptical in shape, and inamyloid. **Edibility:** Mildly to moderately psychoactive. Edible with proper preparation (*see* Ch. 21). **Habitat:** Found solitary to scattered in mixed forests but preferring birch and pine hosts. **Region:** Occurrence in North America is limited to the state of Alaska. This variety is common and well known in Europe and Asia and has been introduced in Australia and parts of South America, where it is considered invasive. **Season:** Occurs

primarily in the Fall. **Looks like:** *A. muscaria* subsp. *flavivolvata* (Fig. 3), *A. regalis* (Fig. 11). **Comments:** *A. muscaria* var. *muscaria* overlaps with **A. muscaria subsp. *flavivolvata*** in Alaska. The two are nearly indistinguishable since many of their features overlap, however, a "typical" specimen of *muscaria* var. *muscaria* will appear red with white warts whereas a "typical" *flavivolvata* will be orangish-red with yellow warts. ***A regalis*** also occurs in Alaska and was once considered a variety of *A. muscaria*. These are easily distinguished by cap color, with *regalis* sporting a rich nut-brown cap in contrast to the red of *muscaria* var. *muscaria*.

White Fly Agaric
Amanita chrysoblema (Fig. 2)
AKA: formerly classified as *Amanita muscaria* var. *alba*
Fruiting Body: Small to large, with caps ranging from 2 to 8 inches in diameter and stems measuring up to 6 inches tall and 1 inch wide. **Cap:** The shape of the cap ranges from convex to plane. The cap is typically white to grayish-white or pallid in color with a faintly striate margin. The flesh is white and despite the color of the cap a thin layer of yellowish tissue can be found directly below the cuticle. Warts are typically buff to tan in color. **Gills:** The gills are white to pale cream, free from the stem, and crowded. **Stem:** The stem is white and equal to tapering upwards and has a bulbous base. A cream to pale yellow skirt-like ring, or annulus, can typically be found on the middle to upper stem. Ascending rings of volval material at the top of the basal bulb are also typically present. **Spores:** Spores are white, elliptical to elongate in shape, and inamyloid. **Edibility:** Mildly to moderately psychoactive. Edible with proper preparation (*see* Ch. 21). **Habitat:** Solitary to gregarious in mixed woods and at forest edges. **Region:** Coast to coast in the northern regions of the U.S. (though documented as far south as Northern California) and north into Canada. Jenkins (1986) reports populations from ID, IN, MI, NY, PA, WA. Rare. **Season:** Summer and fall. **Looks like:** *A. alpinicola*, *A. muscaria* subsp. *flavivolvata* (Fig. 3), *A. muscaria* var. *guessowii* (Fig. 4). **Comments:** *A. chrysoblema* is a close member of the Fly Agaric family and may turn out to be a simple variant of *A. muscaria* subsp. *flavivolvata* rather than a distinct species of its own. *A. chrysoblema* can be distinguished from **subsp. *flavivolvata*** and **var. *guessowii*** predominantly by its pallid whitish-gray cap color. Another potential look-alike is

North American Field Guide • 21

FIG. 1: *Amanita muscaria* var. *muscaria*. Autumn fruiting under Italian stand of beech trees (Photo © Fabrizio Robba: ID 131765027 | Dreamstime.com).

FIG. 2: *Amanita chrysoblema*. Found and photographed mid-summer in George Washington National Forest (Photo by Jason Hollinger).

FIG. 2B: *Amanita chrysoblema*. Monte Bello Open Space Preserve. Santa Clara Co., California (Photo by Ron Wolf).

the recently described ***A. alpinicola*** (Cripps et al., 2017). *A. alpinicola* is a close relative to the *A. muscaria*-group and may also be psychoactive. *A. alpinicola* is a small to medium sized mushroom with a dirty-white to pale yellow cap. *A. alpinicola* fruits in the late spring and summer in subalpine habitats and appears to be specific to five-needle pines or "White" pines.

American Fly Agaric

Amanita muscaria subsp. *flavivolvata* (Fig. 3)

Fruiting Body: Small to large, with caps ranging from 2 to 12 inches in diameter and stems measuring up to 8 inches tall and 1 inch wide. **Cap:** The shape of the cap ranges from hemispheric to convex to plane, with the margin occasionally turning upward in age. The cap is typically red, orangish-red, or orangish-yellow, though the colors may fade in the sun. Occasionally, paler colors are found. Warts are white to yellow. Flesh is white except for a thin layer of yellowish-orange tissue just below the cuticle. **Gills:** The gills are white and typically free from the stem but may be slightly attached. The spacing of gills is close to crowded. **Stem:** The stem is white, with a bulbous base and tapers towards the top. A significant defining characteristic is a series of ascending concentric rings at the top of the bulbous base. Additionally, there is typically a white, skirt-like ring, or annulus, found on the middle to upper stem, which may feature some yellow coloration. **Spores:** Spores are white, elliptical in shape, and inamyloid. **Edibility:** Mildly to moderately psychoactive. Edible with proper preparation (*see* Ch. 21). **Habitat:** Occurs solitary to gregarious in mixed coniferous and deciduous forests, though it is not as picky about tree hosts as some guidebooks would lead one to believe. This particular variety has been found in association with birch, cottonwood, madrone, oak, pine, spruce, and among various conifers. This variety can also be found in yards, parking lots, and other landscaped areas. **Region:** Subspecies *flavivolvata* has been found as far north as Alaska and as far south as Costa Rica. Within these boundaries it ranges from the Pacific Coast to the Rockies. **Season:** Most common in fall, though the season can start as early as late summer and continue through early spring depending on weather and climate. **Looks like:** *A. aprica* (Fig. 15)*, A. muscaria* var. *muscaria* (Fig. 1)*, A. muscaria* var. *formosa*. **Comments:** Unlike its Eastern relatives the number of look-alikes is limited, and the primary look-alikes

FIG. 3: *Amanita muscaria* subsp. *flavivolvata*. These scarlet specimens would have once been considered *A. muscaria* var. *muscaria* but are now recognized as genetically distinct. Youth Community Park, Santa Rosa, California (Photo by Darvin DeShazer).

FIG. 3B: *Amanita muscaria* subsp. *flavivolvata*. This peachy-orange capped specimen gives a sense of the degree to which cap color can vary within subspecies *flavivolvata*. Found under black cottonwood in Central Washington state (Photo by Kevin Feeney).

are also psychoactive. Subspecies *flavivolvata* shares a range with **A. muscaria var. *muscaria*** in Alaska and is typically distinguished from the true fly agaric by its orangish cap-color and its yellow warts. There also appear to be pockets of the European Yellow Fly Agaric, **A. muscaria var. *formosa***, along the coast and coastal interior of the Pacific Northwest and British Columbia. Subspecies *flavivolvata* can be distinguished by its deeper and darker cap color that tends towards a reddish-orange rather than the yellow color typical of var. *formosa*. **A. aprica** is another look-alike but is typically a spring mushroom. *A. aprica* has more of a sunny-apricot color and its base is more club shaped. Another distinguishing feature is that the warts on subspecies *flavivolvata* should be easily removed with a wet finger whereas the warts and volval remnants on *A. aprica* appear to be embedded in the cuticle of the cap, making them difficult to remove.

Eastern Yellow Fly Agaric
Amanita muscaria var. *guessowii* (Fig. 4)
AKA: formerly classified as *Amanita muscaria* var. *formosa*
Fruiting Body: Small to large, with caps ranging from 2 to 7 inches in diameter and stems measuring up to 6 inches tall and 1 ¼ inch wide. **Cap:** The shape of the cap ranges from convex to plane. The cap is typically yellow to orangish-yellow. Warts are yellowish-buff to pale tan. The margin of the cap tends to become striate with maturity. Flesh is white except for a thin layer of yellowish tissue just below the cuticle. **Gills:** The gills are white to pale cream and typically free from the stem but may be slightly attached. The spacing of gills is crowded. **Stem:** The stem is white to yellowish cream, with a bulbous base and tapers upwards before slightly flaring at the top. A series of ascending concentric rings is typically found at the top of the bulbous base. Additionally, a buff colored skirt-like ring is typically present on the upper stem, which may feature some pink coloration. **Spores:** Spores are white, elliptical in shape, and inamyloid. **Edibility:** Mildly to moderately psychoactive. Edible with proper preparation (*see* Ch. 21). **Habitat:** Found solitary to gregarious in boreal, coniferous and deciduous forests. **Region:** Found from Michigan to North Carolina and north into Quebec. **Season:** Late summer through fall. **Looks like:** *A. crenulata* (Fig. 13), *A. flavoconia* (Fig. 8), *A. frostiana* (Fig. 5), *A. gemmata* (Fig. 16), *A. persicina* (Fig. 7), *A. praecox* (Fig. 6), *A. wellsii* (Fig.

FIG. 4: *Amanita muscaria* var. *guessowii.* Growing on the edge of a pond among moss and ferns in a mixed forest. Here, the cap color is lemon-like (Photo by Christine Young).

FIG. 4B: *Amanita muscaria* var. *guessowii* with a golden yellow cap. Notice the concentric rings rising from the basal bulb (Photo by Danny Curry).

10). **Comments:** Of all the Fly Agaric varieties *guessowii* shares its range with the highest number of look-alikes. For *guessowii*, its size and ascending concentric rings forming at the top of the basal bulb set it apart from many of the look-alikes, which are mostly similar due to cap color. Another distinguishing feature is the yellowish tissue below the cuticle of the cap. ***A. frostiana*** shares a yellowish cap color with *guessowii* but can be distinguished by its yellowish warts, annulus, and collar. ***A. flavoconia*** is another look-alike though it is more often confused with *A. frostiana* than *A. muscaria* var. *guessowii*. *A. flavoconia*, known as the "Yellow Dust Amanita," is often found with light yellow dust on the gills and top part of the stem. The cap is yellow, the warts and annulus are typically yellow, as is the upper part of the stem. *A. flavoconia* is also typically smaller than *guessowii*, with the cap only reaching 4 inches, and lacks the concentric rings at the top of the basal bulb that is typical of muscarioid species like *A. muscaria* var. *guessowii*.

Another similar looking species is ***A. praecox***, the "Early Spring Amanita." This *Amanita* is a spring and summer mushroom, so its season only partially overlaps with that of *guessowii*. One distinguishing factor is that *A. praecox* does not have warts; the cap is either bare or may feature several random cottony patches of volval material. *A. praecox* is also a smaller mushroom, with the cap only reaching up to 3 inches in diameter and the coloration of the cap is typically a dull or pale yellow. ***A. wellsii***, the "Salmon Amanita," is another potential look alike, though smaller than *guessowii* with a salmon-colored cap reaching only 4 inches. *A. wellsii* has a yellowish stem with a yellow annulus and wooly volval remnants can sometimes be found adhering to the margin of its cap. The typical muscarioid concentric rings are also missing from the apex of the basal bulb in this species.

The remaining look-alikes are all psychoactive *Amanitas* but it is still necessary to accurately identify your mushrooms and to understand what distinguishes one from the other. Of the remaining look-alikes, *guessowii* and ***perscina*** are the most similar; *A. persicina* was once considered a variety of *A. muscaria*. The distinctions here will primarily be based on geographic regions, with *persicina* primarily occurring in the South, and on cap color, with *persicina* tending towards the more orangish and reddish tones. ***A. crenulata*** also shares some similarities but is typically smaller than *guessowii*. The color of the cap is pale yellowish with hints of grey and not brightly colored. The gills of *crenulata* are attached to the stem and the ascending concentric rings common among muscarioids is absent in this species.

FIG. 5: *Amanita frostiana*. Turkey Point Provincial Park, Ontario, Canada (Photo by Eva Skific).

FIG. 6: *Amanita pracecox*. Notice how the "warts" are random and appear only precariously attached. Found near Port Dover, Ontario, Canada (Photo by Eva Skific).

Mushrooms from the *Amanita gemmata* **group** could also be tricky based on the yellowish coloring of the cap. Mushrooms in this group are small to medium in size and feature a distinctive collar or free rim at the top of the basal bulb, distinct from *guessowii's* ascending rings.

Southern Peach Fly Agaric
Amanita persicina (Fig. 7)
AKA: formerly classified as *Amanita muscaria* var. *persicina*
Fruiting Body: Small to large, with caps ranging from 2 to 12 inches in diameter and stems measuring up to 8 inches tall and 1 ¼ inch wide. **Cap:** The shape of the cap varies from hemispheric to convex to plane. The cap typically has a peachy color, which is richest in the center of the cap and may transition to a yellowish color as it approaches the margin, which is faintly to moderately striate. Warts range from pale yellow to light tan. Flesh is white except for a thin layer of yellowish to pinkish tissue just below the cuticle. **Gills:** The gills are white with a pinkish tint and are typically free from the stem but may be slightly attached. The spacing of gills is close to crowded. **Stem:** The stem is white, with a bulbous base and is equal to tapering towards the top. A significant defining characteristic is a series of ascending concentric rings at the top of the bulbous base. There is typically a white, skirt-like ring found on the middle to upper stem, which may feature some yellow coloration. Stem may bruise yellowish with handling. **Spores:** Spores are white, elliptical to elongate in shape, and inamyloid. **Edibility:** Mildly to moderately psychoactive. Edible with proper preparation (*see* Ch. 21). **Habitat:** Appears to prefer oak and pine, particularly Virginia Pine (*Pinus virginiana*) and grows solitary to gregarious in mixed forests, at the forest edge and among planted trees. **Region**: Primarily found in Southern states (AL, FL, GA, MS, NC, TN, VA) but has been found as far north as Long Island. **Season:** Fall but may occur during a warm winter. **Looks like:** *A. flavoconia* (Fig. 8), *A. muscaria var. guessowii* (Fig. 4), *A. parcivolvata* (Fig. 9), *A. wellsii* (Fig. 10). **Comments:** *A. persicina* and ***A. muscaria* var. *guessowii*** are closely related, both psychoactive, and share a range from North Carolina to Long Island. Some key distinctions include cap color which is typically yellow in *guessowii* and peachy-orange in *persicina*. The pink tint of the gills and yellowish-pinkish flesh under the cuticle of *persicina* also distinguishes it from *guessowii*.

FIG. 7: *Amanita persicina*. Growing in a grassy field in Floyd County, Northwest Georgia (Photo by Lisa Kimmerling).

FIG. 8: *Amanita flavoconia* group. Found in June in NE Tennessee (Photo by Renate Rohmann).

A flavoconia is another mushroom that could be mistaken for *A. persicina*. Known as the "Yellow Dust Amanita," *A. flavoconia* is often found with light yellow dust on the gills and top part of the stem. The cap is yellow, the warts and annulus are typically yellow, as is the upper part of the stem. *A. flavoconia* is also typically smaller than *persicina*, with the cap only reaching 4 inches, and lacks the concentric rings at the top of the basal bulb that is typical of muscarioid species like *A. persicina*. Another look-alike that, at first glance, may seem like a dead ringer for a fly agaric is **A. parcivolvata**. This mushroom has a striking red cap with white warts, but closer inspection reveals important distinctions. The cap of *parcivolvata* has tall, distinct striations along the margin. It has no skirt or annulus, has yellow gills, and lacks the concentric rings atop the basal bulb. Fortunately, *parcivolvata* is known to be edible. The final look-alike of concern is **A. wellsii**, also known as the "Salmon Amanita." *A. wellsii* overlaps *A. persicina* in its upper range, from North Carolina and further north. The cap colors are very similar with *A. wellsii* sporting a pinkish-orange cap to *A. persicina's* peach colored cap. *A. wellsii* is smaller than *persicina*, with a cap reaching only 4 inches, and has a yellowish stem. The annulus is also yellow and wooly volval remnants can sometimes be found adhering to the margin of the cap. The typical muscarioid concentric rings are also missing from the apex of the basal bulb.

Royal Fly Agaric
Amanita regalis (Fig. 11)
AKA: formerly classified as *Amanita muscaria* var. *regalis*
Fruiting Body: Medium to large, with caps ranging from 3 to 8 inches in diameter and stems measuring up to 8 inches tall and 1 ¼ inch wide. **Cap:** The shape of the cap ranges from convex to plane while its color varies from a golden to nut-brown. The margin of the cap is striate. Warts range from yellow to creamy yellow but may appear slightly grayish in older specimens. Flesh is white except for a thin layer of reddish to yellowish-brown tissue just below the cuticle. **Gills:** The gills are cream to pale yellow and typically free from the stem. The spacing of gills is crowded. **Stem:** The stem is white to pale yellow and may brown with handling. It has a bulbous base and tapers upward, though slightly flaring at the top. Above the bulb there are several yellowish scaly rings and there is typically a white to pale yellow

North American Field Guide • 31

 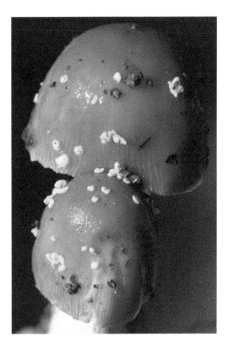

FIG. 9: *Amanita parcivolvata*. Found at the edge of a dense mixed wood (hardwood/coniferous) growing among herbs and wildflowers. Gordon County, NW Georgia (Photos by Lisa Kimmerling).

FIG. 10: *Amanita wellsii*. Notice the woolly volval remnants at the margin of the cap. Found mid-September near Smithfield, Rhode Island (Photo by Spike Mikulski).

skirt-like ring found on the upper stem. **Spores:** Spores are white, subglobose to elliptical in shape, and inamyloid. **Edibility:** Mildly to moderately psychoactive. Edible with proper preparation (*see* Ch. 21). **Habitat:** In Alaska, *A. regalis* can be found solitary to subgregarious along the timberline of coniferous forests. **Region:** Alaska. **Season:** Summer and fall. **Looks like:** *A. muscaria* var. *muscaria* (Fig. 1), *A. novinupta* (Fig. 12; "Blusher"), *A. "pantherina"* (Fig. 19). **Comments:** *A. regalis* was once considered a variety of ***A. muscaria*** but is now recognized as a separate species. They are both psychoactive and share many features, but the contrasting cap color should prevent confusion. The yellowish hints in the gills, stem, and volval remnants as well as the scaly character of the concentric rings above the basal bulb further distinguish *A. regalis* from *A. muscaria*. It is more likely that one would confuse *A. regalis* with one of the North American varieties of "Blusher" mushrooms. *Blushers* are known for their tendency to bruise pink or red. The cap of *A. regalis* should not stain and any bruising on the stem will be brownish in color, presenting a clear contrast to any of the *blusher* mushrooms. The closest look-alike is ***A. "pantherina,"*** which shares a range in Alaska. Both mushrooms sport brown caps but *"pantherina"* does not have the yellow coloring that is sometimes found on the stem, skirt, and warts of *regalis*. *A. "pantherina"* can also be distinguished by the prominent collar at the top of the basal bulb which contrasts with the yellowish scaly rings found at the apex of the basal bulb in *A. regalis*.

FIG. 11: *Amanita regalis.* Notice the distinct striations on the cap margins (Photo © Elisa Putti: ID 120063023 | Dreamstime.com).

Fig. 11B: *Amanita regalis.* Found during summer in Finland. Notice the golden nut-brown coloration of the caps as well as the cream-yellow color of the warts and stems (Photo by Mervi Alavuotunki).

FIG 12. *Amanita novinupta.* Notice the red staining on the cut specimen and reddish bruising on the others. Found April in Redwood Regional Park, Oakland, California (Photo by Alan Rockefeller).

The Crenulata Group

Subgenus: *Amanita*; Section: *Amanita*; Subsection: *Amanitella*; Series: *Crenulatae*

Champagne Amanita
Amanita crenulata (Fig. 13)
Fruiting Body: Small to medium, with caps ranging from 1 to 3 ½ inches in diameter and stems measuring up to 4 inches tall and up to ½ inch wide. **Cap:** The shape of the cap ranges from convex to plane while its color varies from whitish-gray to pale tan to yellowish-tan. The margin of the cap is striate. The warts have been described as "champagne" colored. **Gills:** The gills are white and typically free from the stem but may be narrowly attached. The spacing of gills is close to crowded. **Stem:** The stem is white and equal to tapering upwards and has a bulbous base. The stem may be hollow or stuffed. The annulus, when present, will be found on the upper stem, however, it is frequently absent in older specimens. At the top of the basal bulb there is typically a wooly ring of volval material. Up to an inch of the stem immediately above the wooly ring may be slightly scaly. **Spores:** Spores are white, subglobose to elliptical in shape, and inamyloid. **Edibility:** Psychoactive. Potency unknown. **Habitat:** Found solitary to subgregarious in mixed coniferous and deciduous woods. **Region:** Occurring primarily in the Northeast U.S. and Southeast Canada. *A. crenulata* has been reported specifically from: CT, MA, NH, NJ, NY, PA, and VT. **Season:** Fall. **Looks like:** *A. gemmata* group (Fig. 16), *A. muscaria* var. *guessowii* (Fig. 4), *A. russuloides* (Fig. 14). **Comments:** The ***Amanita gemmata* group** is one of the primary look-alikes for *A. crenulata*, however, despite *crenulata* carrying hints of yellow in the cap the cap color is typically a pale white to gray color, different from the indisputable yellow coloration found in the *gemmata* group. *A. crenulata* is also missing the boot-like collar typical of mushrooms in the Gemmatae and Pantherinae subsections of *Amanita*. Another look-alike, ***A. russuloides***, also features a boot-like collar which similarly distinguishes it from *A. crenulata*, which typically has a light ring of wooly volval material at the top of the basal bulb. ***A. muscaria* var. *guessowii*** is another possible look alike but can be distinguished by its bright yellow cap color and the series of ascending rings at the apex of the basal bulb.

FIG. 13: *Amanita crenulata*. Found growing on the side of a dirt road in the Northeast United States, under Oak (Photo by Christine Young).

FIG. 14: *Amanita russuloides*. Found mid-September near Puxico, Missouri (Photo by Jon Rapp).

The Gemmata Group (gemmatoid)

Subgenus: *Amanita*; Section: *Amanita*; Subsection: *Gemmatae*

Active Gemmatae

Sunshine Amanita
Amanita aprica (Fig. 15)
Fruiting Body: Small to large, with caps ranging from 2 to 6 inches in diameter and stems measuring up to 4 inches tall and up to 1 ¼ inch wide. **Cap:** The shape of the cap ranges from spherical to convex to plane. The margin may or may not be faintly striate. The cap color ranges from yellow to egg-yolk yellow to apricot. With *A. aprica* the warts are less distinct and often occur as part of a layer of volval material on the cap and occur most distinctly as warty protuberances. This volval layer appears to be connected to the cuticle and cannot be easily wiped off or removed. The flesh of the cap is white with a thin layer of yellow tissue just below the cuticle of the cap. **Gills:** The gills are white to creamy and typically free from the stem but may slightly secede. The spacing of gills is close to subdistant. **Stem:** The stem is white to creamy tan and may bruise a darker tannish color when handled. The stem is stuffed but may become partially hollow with age. The annulus is white to cream and occurs on the mid to upper stem. The basal bulb is often indistinct and may appear more club-like. A free collar like margin, sometimes out-turning, can be found at the top of the bulb. **Spores:** Spores are white, elliptical to elongate in shape, and inamyloid. **Edibility:** Mildly active (Lindgren, 2014 *citing* Colby, et al., 2013), though probably ranging between mildly and moderately active. **Habitat:** Prefers sunny locations with disturbed soil. Occurs in sunny patches along hiking trails and in campgrounds, or in the woods where there are breaks in the canopy. **Region:** Found from California to British Columbia, particularly in the cascade mountain range of the Pacific Northwest and in the Sierra Nevada. **Season:** Spring to early summer (June). **Looks like:** *A. gemmata* group (Fig. 16), *A. muscaria* subsp. *flavivolvata* (Fig. 3). **Comments:** *A. aprica* is a species that has only recently been described (Tulloss & Lindgren, 2005) and is thought to have previously been identified as either *A. gemmata* or **A. muscaria subsp. *flavivolvata***. While *A. aprica* sometimes has rings at the apex of the basal bulb reminiscent of muscarioid

FIG. 15: *Amanita aprica*. Eastern Shasta-Trinity National Forest, Siskiyou County, California (Photo by Alan Rockefeller).

FIG. 16: *Amanita "gemmata"* group. Notice how one cap looks warted while the other looks patchy (Photo © Toshihisa Shimoda: ID 125720048 | Dreamstime.com).

mushrooms the volval material present on the cap, either as warts or a volval layer, is connected to the cuticle and cannot be easily removed as with the muscarioids. *A. aprica* is more closely related to the ***Amanita gemmata* group**, though is typically more robust. *A. aprica* may or may not have a booted collar and its base may be more club-like than bulbous.

Gemmed Amanita
Amanita gemmata group (Fig. 16)
Fruiting Body: Small to medium, with caps ranging from 1 to 4 inches in diameter and stems measuring up to 5 inches tall and up to ¾ inch wide. **Cap:** The shape of the cap ranges from spherical to convex to plane. The margin may or may not be striate. The cap color ranges from a butter or golden yellow to a light or grayish-yellow. Warts are cream or dingy-white colored and may merge to form a patch of volval material. Flesh is white. **Gills:** The gills are white to cream and typically free from the stem but may be slightly attached. The spacing of gills is crowded. **Stem:** The stem is white to cream, with a bulbous base and tapers towards the top. The stem is stuffed or hollow with a fragile annulus occurring on the middle to upper part of the stem. The top of the basal bulb features a distinct free collar. **Spores:** Spores are white, elliptical in shape, and inamyloid. **Edibility:** Inactive to highly psychoactive. **Habitat:** Occurs solitary to scattered in mixed coniferous and deciduous forests, sometimes occurring in urban parks. **Region:** Throughout North America. **Season:** Late spring to fall and mild winters. **Looks like:** *A. albocreata* (Fig. 17), *A. citrina, A. muscaria* var. *guessowii* (Fig. 4), *A. pantherinoides* (Fig. 18), *A. russuloides* (Fig. 14). **Comments:** This species group is complicated for a number of reasons, first of which is that it is described as "*gemmata*," a European taxon and has not received a species designation for its North American counterpart. Another part of the problem is that several historical collections appear to have been misidentified which has led to confusion regarding the true characteristics of this "species." Others believe that "*gemmata*" may hybridize with the North American *Amanita "pantherina,"* leading to a spectrum of forms. Currently, this "mushroom" is considered to be a complex of closely related mushrooms, or a "group."

    Amanita albocreata, *A. russuloides*, and the *gemmata* group are all "booted" *Amanitas*, meaning they all have a distinct collar, like the rim of a boot, atop

Fig. 16B: *Amanita "gemmata"* group. Found under Atlas Cedar (*Cedrus atlantica* var. *glauca*) in Oakland, CA (Photo by Alan Rockefeller).

FIG. 17: *Amanita albocreata*. Port Dover, Ontario, Canada (Photo by Eva Skific).

Figure 17B: *Amanita albocreata*. Found during summer in the Wolf Creek Narrows in Slippery Rock, Pennsylvania (Photo by Walt Sturgeon).

the basal bulb. ***A. albocreata*** occurs in the Northeastern U.S. and Southeastern Canada and should not be considered a look-alike outside of this region. The cap of *albocreata* is typically white with a yellow center and it lacks an annulus entirely. Mushrooms in the *gemmata* group may lose their annulus, but the presence of an annulus would rule out *albocreata*. It is possible that *albocreata* is similarly psychoactive, but more data is currently needed. ***Amanita russuloides*** tends to have a straw-colored cap with a distinctively striate margin and occurs east of the Great Plains. ***Amanita muscaria* var. *guessowii*** can be distinguished from the *gemmata* group based on its larger size and by the presence of ascending concentric rings at the apex of the basal bulb. Another potential look-alike is ***A. pantherinoides***, which is also a small to medium sized mushroom. *A. pantherinoides* has a honey-brown colored cap and has a persistent annulus in contrast to the fragile annulus found in the *gemmata* group, which frequently disappears with age.

Western False Panther
Amanita pantherinoides (Fig. 18)
AKA: formerly classified as *Amanita pantherina* var. *pantherinoides*
Fruiting Body: Small to medium, with caps ranging from 1 to 4 inches in diameter and stems measuring up to 4 ½ inches tall and up to ½ inch wide. **Cap:** The shape of the cap ranges from convex to plane while its color varies from a light honey to tan with a brownish center. The margin of the cap is smooth but becomes faintly striate with age. The warts are white to cream colored. The flesh is white and should not stain with handling. **Gills:** The gills are white and typically free from the stem. The spacing of gills is crowded. **Stem:** The stem is whitish, with a bulbous base and tapers towards the top. The stem may be hollow or stuffed. The annulus occurs on the upper part of the stem, is large, white, and persistent. The top of the base features a thin wooly collar made up of volval material. **Spores:** Spores are white, elliptical to elongate in shape, and inamyloid. **Edibility:** Psychoactive. Potency unknown. **Habitat:** Typically found in coastal forests, among conifers or deciduous trees. **Region:** West coast, from California to British Columbia. **Season:** Can be found spring, summer, and fall. **Looks like:** *A. gemmata* group (Fig. 16), *A. "pantherina"* (Fig. 19). **Comments:** The honey-brown color of its cap along with a persistent annulus are two features that help distinguish *A. pantherinoides* from

North American Field Guide • 41

FIG. 18: *Amanita pantherinoides*. Found growing under *Carpinus betulus* in Vancouver, British Columbia, Canada (Photo by Paul Kroeger).

FIG. 19: A*manita "pantherina"*. Found during spring in Washington State (Photos by Amanita Lloydii).

the ***gemmata* group.** Another potentially distinguishing feature is the thin wooly collar that contrasts with the more distinctive free collar found in the *gemmata* group. **Amanita "pantherina"** is another potential look-alike though "pantherina" frequently grows to a larger size and typically has a dark brown cap.

The Pantherina Group (pantheroid)

Subgenus: *Amanita*; Section: *Amanita*; Subsection: *Pantherinae*

Active Pantherinae

American Panther Cap
Amanita "pantherina" (Fig. 19)
AKA: *Amanita "ameripanthera"*
Fruiting Body: Small to large, with caps ranging from 1 to 8 inches in diameter and stems measuring up to 7 inches tall and 1 inch wide. **Cap:** The shape of the cap ranges from rounded to convex to plane. The cap is typically dark to light brown but can be tannish-brown. Warts are white to cream or buff colored. Flesh is firm and white except for a thin layer of yellowish-tan tissue just below the cuticle. The flesh does not bruise. **Gills:** The gills are white to off-white and typically free from the stem but may be slightly attached. The spacing of gills is close to crowded. **Stem:** The stem is white, with a bulbous base and tapers towards the top. The stem may be stuffed or hollow and tends to brown with handling. The annulus is white, and it appears on the middle to upper stem. At the apex or top of the basal bulb there is a distinct free collar circling the bottom of the stem. **Spores:** Spores are white, elliptical in shape, and inamyloid. **Edibility:** Moderately to highly psychoactive. **Habitat:** Occurs solitary to gregarious in mixed coniferous and deciduous forests. This species appears to have a particular affinity for Doug Fir. **Region:** Occurs from California to Alaska and west through the Rocky Mountains. **Season:** Fruits primarily in spring and fall but may also appear during mild winters. **Looks like:** *A. augusta* (Fig. 20), *A. gemmata* group (Fig. 16), *A. hallingiana* (Fig. 21), *A. novinupta* (Fig. 12), *A. pantherinoides* (Fig. 18), *A. regalis* (Fig. 11). **Comments:** Amanita "pantherina" has a number of look-alikes though many of these are easily

FIG 19B: *Amanita "pantherina"*. Found during fall in the Puget Sound, Washington State. Notice the distinct collar atop the basal bulb (Photo by Kevin Feeney).

FIG. 20: *Amanita augusta*. The yellow coloration in these specimens is stunning but is frequently more subtle. Mt. Tamalpais State Park. Marin County, California (Photo by Ron Wolf).

distinguished based on region, cap color, and bruising. ***A. augusta*** is notable for its yellow warts and annulus which contrast with *A. "pantherina's"* white warts and annulus. The yellow coloration in *A. augusta*, however, can vary from pale to bright yellow. Similarities can also be found with the ***Amanita gemmata* group** though *A. "pantherina"* tends to be larger and the deep brown to tan-brown color is distinct from the yellows typical of the *gemmata* group. ***A. hallingiana*** is a notable species that shares many features with *A. "pantherina"*, including psychoactivity, but their regions do not overlap. *A hallingiana* is known from Mexico and Costa Rica and is thought to occur in other parts of Latin America. ***A. novinupta*** is a variety of "blusher" that occurs in Western North America. This species is known for staining red, a characteristic that clearly distinguishes it from *A. "pantherina"* which may brown slightly with handling but doesn't stain to a significant degree and does not stain red. ***Amanita pantherinoides*** is typically smaller than *A. "pantherina"* and its cap is a light honey-brown. Small and light colored *"pantherinas"* could be confused with this species. Finally, ***Amanita regalis*** shares habitat with *A. "pantherina"* at the very top of its range in Alaska. Both mushrooms sport brown caps but *"pantherina"* does not have the yellow coloring that is sometimes found on the stem, skirt, and warts of *regalis*. *Amanita "pantherina"* can also be distinguished by the prominent collar at the top of the basal bulb which contrasts with the yellowish scaly rings found at the apex of the basal bulb in *A. regalis*.

Small Funnel-Veil Amanita
Amanita multisquamosa (Fig. 22)
AKA: formerly known as *A. cothurnata* and *A. pantherina* var. *multisquamosa*
Fruiting Body: Small to medium, with caps ranging from 1 to 4 ½ inches in diameter and stems measuring up to 5 inches tall and ½ inch wide. **Cap:** The shape of the cap varies from hemispheric to convex to plane. The disk or center of the cap is tannish to brownish but otherwise the cap is pallid or whitish in color. The margin of the cap is striate, and the warts are white. The flesh is also white. **Gills:** The gills are white and are freely to remotely attached. The spacing of gills is crowded. **Stem:** The stem is white and typically hollow. The annulus, located on the middle to upper part of the stem, is typically pulled up giving it a funnel-like shape rather than a skirt-like shape. At the top of the basal bulb there is a distinct collar, which

North American Field Guide • 45

FIG. 21: *Amanita hallingiana*. Balneario Maritaro, Las Azufres, Michoacan, Mexico (Photo by Alan Rockefeller).

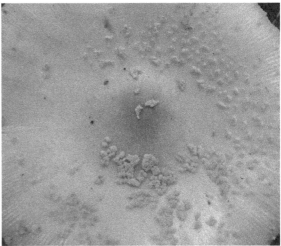

FIG. 22: *Amanita multisquamosa*. Found in Bartholomew County, Indiana, under Oak (Photos by Mark Niemoller).

is typical in *pantheroid* species. **Spores:** Spores are white, subglobose to elliptical in shape, and inamyloid. **Edibility:** Psychoactive. Most likely moderately to highly psychoactive. **Habitat:** Found solitary to subgregarious in coniferous and deciduous woods. **Region:** From Florida north to New York and west to Michigan. According to Jenkins (1986) this species has been found in the following states: AL, GA, IN, LA, MA, MD, MI, MS, NC, NJ, NY, PA, SC, TN, TX, VA, VT, and WV. **Season:** Mid-summer to fall. **Looks like:** *A. albocreata* (Fig. 17)*, A. gemmata* group (Fig. 16)*, A. russuloides* (Fig. 14), *A. velatipes* (Fig. 23). **Comments:** One of the distinctive features of *A. multisquamosa* is its funnel-shaped annulus, a feature it shares with **A. velatipes**. *Amanita velatipes* runs larger than *A. multisquamosa* and its cap is typically brown to yellow to cream contrasting with the pallid cap color of *multisquamosa*. The funnel-shaped veil of *multisquamosa* also clearly distinguishes it from **A. albocreata**, which lacks an annulus. **Amanita russuloides** is another similar mushroom, which tends to have a straw-colored cap with a distinctively striate margin. The warts of *A. russuloides* are sparse in comparsion to *multisquamosa* and, of course, it lacks a funnel-shaped annulus. Another potential look alike is the ***gemmata* group**. The funnel-shaped annulus again provides a clear distinction but the pallid cap color of *multisquamosa* is also dissimilar to the yellows typical of the *gemmata* group.

Great Funnel-Veil Amanita
Amanita velatipes (Fig. 23)
AKA: Formerly known as *A. pantherina* var. *velatipes*
Fruiting Body: Medium to large, with caps ranging from 3 to 7 inches in diameter and stems measuring up to 8 inches tall and 1 inch wide. **Cap:** The shape of the cap ranges from convex to plane and is typically brownish in the center of the cap but becoming yellow to cream color towards the margin of the cap, which is striate. Warts are typically white. **Gills:** The gills are white, free from the stem, and crowded. **Stem:** The stem is white, has a bulbous base and tapers upward. The stem may be hollow or stuffed. The annulus is white and funnel shaped, occurring at the middle of the stem or below. At the top of the basal bulb there is a distinct rim sometimes described as "booted" since it appears like the top of a boot in relation to the stem. **Spores:** Spores are white, elliptical to elongate in shape, and inamyloid.

North American Field Guide • 47

FIG. 23: *Amanita velatipes*. Found in Southern Ontario, Canada (Photo by Gerald Morin).

FIG. 23B: *Amanita velatipes.* Found in NE Ohio during the fall (Photo by Leah Kalasky).

Edibility: Psychoactive. Most likely moderately to highly psychoactive.
Habitat: Found solitary to subgregarious in coniferous and deciduous woods. **Region:** Occurring primarily in the Northeast U.S. and Southeast Canada. *A. velatipes* has been reported specifically from: CT, ME, MD, MI, NC, NJ, NY, PA, TN.
Season: Summer and fall. **Looks like:** *A. "amerirubescens"* (Fig. 24), *A. gemmata* group (Fig. 16), *A. multisquamosa* (Fig. 22). **Comments:** Like **Amanita multisquamosa**, the upturned funnel-shaped annulus is a primary distinguishing feature of this mushroom. This mushroom can be distinguished from *multisquamosa* by its larger size and by its cap color, which ranges from brown in the center to yellow and/or cream at the margins. Another look alike is the "Eastern American Blusher," **Amanita "amerirubescens"**. As with other "blushers," *A. "amerirubescens"* stains red when bruised while *velatipes* does not. The yellowish color of its cap also makes *A. velatipes* similar in appearance to the **gemmata group**. *Amanita velatipes* is typically larger than the *gemmata* group, has an upturned annulus, and the center or disc of the cap is frequently brownish.

FIG. 24: *Amanita "amerirubescens"*. New Milford, Connecticut (Photo by Christine Young).

Part II:
Religion, Culture, & Folklore

Chapter 4

Soma's Third Filter:
New Findings Supporting the Identification of Amanita muscaria as the Ancient Sacrament of the Vedas

Kevin Feeney & Trent Austin, MD

In 1968 R. Gordon Wasson first proposed his groundbreaking theory identifying Soma, the entheogenic sacrament of the Vedas, as the *Amanita muscaria* mushroom. Central to Wasson's theory are the three filters of Soma, which correspond to different steps in the sacrament's preparation, as outlined in the Rig Veda. Wasson's interpretation of the first two filters, a filter of sunlight (sun-drying or desiccation) and a woolen filter (to remove solids from aqueous preparations), are generally uncontested, but his proposal for the third filter has raised controversy and ire. Based on the practice of urine-recycling among *Amanita muscaria* using groups in Siberia, and a single reference to urinating Soma in the Rig Veda, Wasson proposed that the third filter was the human body, and that the urine produced by bemushroomed individuals was considered the purest form of Soma. Here, we examine the strengths and weaknesses of Wasson's argument, and in turn propose an alternate third filter, a filter of milk.

While the Rig Veda has little to say about urine, the practice of mixing the Soma-drink with milk is frequently mentioned. Importantly, the blending of dairy with dried *Amanita muscaria* to create an inebriating beverage is also supported by ethnographic data (Mochtar & Geerken, 1979). To test our hypothesis that milk is the third and final filter in the preparation of Soma, infusions of *Amanita muscaria* were treated with an enzyme (glutamate decarboxylase [GAD]) produced by various *Lactobacillus* bacteria, typically present in unpasteurized milk, to determine whether the addition of milk served any pharmacological purpose. Our findings suggest that the addition of milk could have had a significant potentiating effect on the final Soma beverage while simultaneously reducing some of the more unpleasant effects. While we identify a different third filter from Wasson, our findings support and further his overarching theory identifying Soma as the *Amanita muscaria* mushroom.

Before diving into our methods and results, it is first important to have a basic understanding of the pharmacological constituents of *Amanita muscaria* and of the three Soma filters, as proposed by Wasson, and why Wasson's identification of the third filter is ultimately problematic. We will begin with a brief review of *Amanita muscaria*'s primary psychoactive constituents: ibotenic acid and muscimol (for a more in-depth review of *Amanita muscaria*'s pharmacology *see* Ch. 22).

A Brief Pharmacological Review

While uncertainties remain, two compounds have been identified as primarily responsible for the psychoactive effects of *Amanita muscaria*: ibotenic acid and muscimol. Muscimol, the primary psychoactive agent in the fly agaric is GABAergic, meaning that it resembles and imitates the inhibitory neurotransmitter GABA (*gamma*-Aminobutyric acid) and acts upon the brain's GABA receptors, while ibotenic acid, the other psychoactive compound in fly agaric, resembles the neurotransmitter glutamate and binds with the brain's glutamate receptors. Ibotenic acid and muscimol are closely related compounds and only differ by a carboxyl group. This means that if ibotenic acid loses a carboxyl group (decarboxylates) it turns into the more highly psychoactive muscimol. Because muscimol is more potent and because ibotenic acid is more highly associated with gastrointestinal distress and other toxic effects, it is likely that human cultures familiar with and favoring the psychoactive properties of this mushroom would have developed methods for promoting decarboxylation. This is certainly true in Siberia, where the fly agaric is always dehydrated and where it is sometimes combined with acidic fruits (fermented bilberries) and consumed as a beverage. Both of these methods encourage decarboxylation and increase the potency of the mushroom.

While muscimol is structurally similar to GABA and ibotenic acid structurally similar to glutamate, there is another important parallel between the fly agaric's compounds and their neurotransmitter counterparts. As explained earlier the only thing that separates ibotenic acid from muscimol is a carboxyl group. Similarly, a carboxyl group is all that separates glutamate from GABA, which is what one

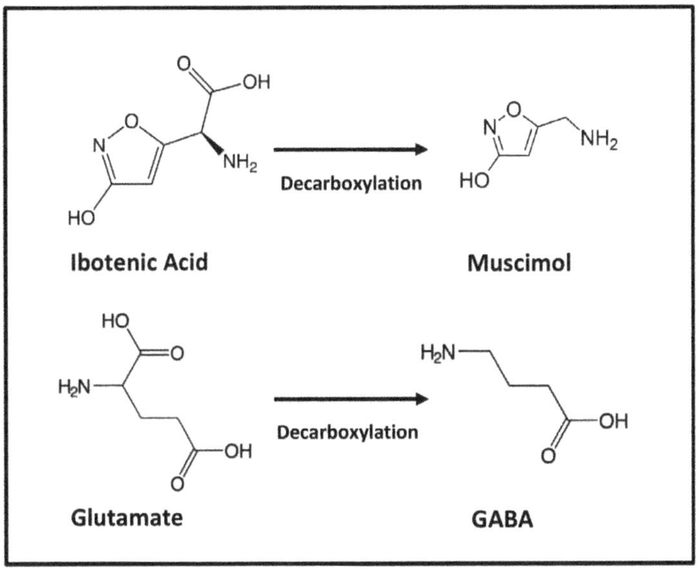

Fig. 1: Structures of ibotenic acid and glutamate and their decarboxylation products, muscimol and GABA.

gets following the decarboxylation of glutamate (Figure 1). These structural parallels will become important later in our discussion.

Wasson on Preparation

Wasson's identification of the three filters used in the preparation of the Soma beverage rests upon the mention of three filters in two separate passages from the Rig Veda:

> 9.73.8: The Guardian of *Rtá* [Soma] cannot be deceived, he of the good inspiring force; he carries three filtres inside his heart. (Wasson, 1968: 54)

> 9.97.55: Thou runnest through the three filtres stretched out; thou flowest the length, clarified. Thou art Fortune, thou art the Giver of the Gift, liberal for the liberal, O Soma-juice. (p. 55)

Only one of the three filters, the second, is easily identified, but for the sake of orderliness we will begin with the first filter. Wasson identifies the first filter as a filter of sunlight, or a *Celestial* filter. Wasson provides several passages to support his identification, including the following:

> 9.76.4: He [Soma] who has been cleansed by the sun's ray. (Wasson, 1968: 38)

> 9.86.32: The Soma envelopes himself all around with rays of the sun . . . (p. 54)

Wasson contends that this filter represents the drying or desiccation of the Soma plant, and notes that "the Rig Veda speaks on several occasions of water being added to the (presumably dry) Soma, so that it would swell up again" (p. 13), such as in the following passages:

> 8.9.19: When the swollen stalks were milked like cows with [full] udders . . . (p. 43)

> 10.125.2: I carry the swelling Soma . . . I bestow wealth on the pious sacrificer who presses the Soma and offers the oblation. (Doniger, 2005: 63)

While many plants are dried for later use this feature of the Soma plant provided an important connection for Wasson. *Amanita muscaria* is not known to occur in the Indus Valley though it has been identified in nearby mountains and would need to be dried for transport and preservation for use in later ceremonies. Importantly, the Soma plant is frequently identified in the Rig Veda as one that is found in the mountains, as indicated by the following passages:

> 9.46.1: …these Somas grown on the mountain top. (Wasson, 1968: 22)

> 9.62.15: …Born on the mountain top… the Soma juice is placed for Indra. (p. 22)

Most important for the identification of Wasson's first filter are the pharmacological changes brought about by drying the mushroom, though Wasson wouldn't discover this until later. When *Amanita muscaria* is dried a portion of the ibotenic acid decarboxylates into muscimol. While the amount of ibotenic acid which decarboxylates by drying is variable, somewhere around a 30 % conversion is usually achieved (Tsujikawa et al., 2006; Tsujikawa et al., 2007). A study on sun-drying *Amanita muscaria* demonstrated that, on average, mushrooms sun-dried for three days saw muscimol concentrations increase by a factor of twelve while ibotenic acid concentrations decreased by more than half (Tsunoda et al., 1993c: 155). Although only a portion of the ibotenic acid is converted to muscimol via drying, this represents a significant pharmacological change to the potency of these mushrooms and may help explain the significance of the *Celestial* filter, as well as bolstering Wasson's identification of *Amanita muscaria* as Soma.

Soma's Third Filter • 55

Fig. 2: R. Gordon Wasson sampling the pressed juice of *Amanita muscaria* in Japan, circa 1965 (Photo from the Wasson Archives, courtesy of Robert Forte).

The second filter identified by Wasson is a filter of woolen cloth. The process of straining the Soma juice through wool is clearly and explicitly described in the Rig Veda, but Wasson largely ignores this step. He appears to take for granted that a wool strainer would eliminate "pulp and fibrous elements" from the final beverage and does not consider whether this step may also be significant in proving his theory that Soma is *Amanita muscaria*. Significantly, however, there is evidence that this step, when applied to *Amanita muscaria*, results in a beverage that produces a "cleaner" effect than the dried mushrooms themselves. A review of over five hundred anecdotal reports of ingestion of either *Amanita muscaria* or *Amanita pantherina* demonstrated that preparing the mushrooms as tea significantly reduced the frequency of nausea and vomiting in comparison to fresh mushrooms, and that tea was more effective at reducing these symptoms than dried preparations of the mushrooms alone (Feeney, 2010). This suggests that there are clear pharmacological reasons for preparing *Amanita muscaria* as a beverage, which should also be seen as adding strength to Wasson's overall theory (Figure 2).

This brings us to the third filter, which is the crux of our current investigation.

The Third Filter

Wasson identifies the third filter as the human body, which presumably acts to filter out some of the Soma plants more toxic effects and produce the purest form of Soma, urine. In Wasson's investigations of Siberian cultures that use *Amanita muscaria* he discovered the practice of urine recycling, in which one bemushroomed individual could share his urine, still containing the active principle of the mushroom, with another who would then also become inebriated. It seems that Wasson believed that if he could persuasively argue that the Vedic Soma ceremony involved the ritual consumption of urine that his theory would be indisputable.

The case Wasson makes for his identification of the third filter as the human body is complex, confusing, and ultimately unconvincing, but we will attempt a brief summary. The Rig Veda appears to describe Soma (the sacramental beverage) as having two forms. Based on his readings of Macdonell (1897) and other Vedic scholars, Wasson explains the consensus identification of these forms as follows: "the *First Form* is the simple juice of the Soma plant, and the *Second Form* is the juice after it has been mixed with water and with milk or curds" (Wasson, 1968: 26). While acknowledging that "the matter is considered settled," Wasson makes the interesting argument that "unmixed" Soma is not actually unmixed (pp. 26-27). Instead, Wasson contends that the Soma juice is always mixed with milk and that "[i]n the many instances where the poet [authors of the Rig Veda] does not mention the milk or curds, the omission seems accidental" (p. 27). While it is possible that the Vedic poets simply understood milk and/or curds to be an inherent part of the Soma drink and perhaps omitted its mention as unnecessary, something more concrete is necessary to corroborate such an interpretation. Here is where Wasson provides us with the possibility that Soma-infused urine, having passed through the third filter (the human body), is the second form of the Soma beverage.

In support of his theory, Wasson offers a single verse from the Rig Veda:

> 9.74.4: Soma, storm cloud imbued with life, is milked of ghee, milk. Navel of the Way, Immortal Principle, he sprang into life in the far distance. Acting in concert, those charged with the Office, richly gifted, do full honor to Soma. The swollen men piss the flowing [Soma]. (p. 29)

The passage is explicit but singular. It also does not mention anything about drinking urine. Out of more than a thousand hymns in the Rig Veda no other passage explicitly corroborates Wasson's interpretation that the writer of the hymn intended to convey that urine was one of the forms of Soma, and according to one critic of Wasson's theory "the verb to urinate is used in connection with the word soma only twice in the Rig Veda" (Ingalls, 1971: 189). By comparison, the word milk is used over three hundred times in the whole of the Rig Veda. There are, however, several passages that suggest that Soma is filtered through the body of the God Indra, and Wasson suggests that the Vedic priests impersonated the God in order to produce the second and final form of Soma during the Soma ritual. Some of the suggestive passages follow:

> 9.80.3: In the belly of Indra the inebriating Soma clarifies itself. (p. 56)
>
> 9.70.10: Purify thyself in Indra's stomach, O juice! As a river with a vessel, enable us to pass to the other side... (p. 55)

The primary problem here is that there is simply no evidence in the "Rigveda that priests ever impersonate the gods in any capacity" (Ingalls, 1971: 189).

While we find Wasson's evidence unpersuasive, we do believe that an identified pharmacological basis for each of the filters is important in establishing Soma's true identity. While the textual evidence Wasson relies upon to build his theory regarding the third filter is lacking, the pharmacological argument is interesting. We know that significant amounts of ibotenic acid, and to a lesser degree muscimol, pass through the human body unmetabolized and are present in the urine of the ingester and that this can be "recycled" to prolong the mushroom's effects (Deja et al., 2014; Stříbrný et al., 2012; Wasson, 1968). It is also possible that filtration through the human body may eliminate some of the less desirable effects, such as those produced by muscarine. However, when Wasson proposed this element of his theory he believed that the ability to recycle the inebriating compounds of the fly agaric was unique to this mushroom, which was probably why he pushed this idea so hard (ie: *if his interpretation of the third filter was correct, Soma could be no other substance!*). Fifty years later, however, we know that consumption of *Psilocybe* mushrooms and mescaline-containing cacti also result in significant amounts of their inebriating compounds, psilocin and mescaline, passing unmetabolized in the urine (Grieshaber et al., 2001; Kalberer et al., 1962; Ott, 1998). This means that consumers of either of these substances could also recycle their urine to prolong the effects of the ingested substance, making the fly agaric less unique in this regard and, as a result, urine recycling becomes a less crucial component of Wasson's theory.

Milk, A New Candidate

Milk, sour milk, curds, and ghee are all mentioned as Soma *admixtures* throughout the Rig Veda. We also learn from MacDonell (1897: 106) that the term "*mrj*, 'to cleanse', is not only applied to the purification of Soma with the strainer [woolen filter], but also to the addition of water and milk." While MacDonell categorizes milk as an admixture the fact that the addition of milk is associated with "cleansing" suggests that it is an important part in the preparation of the Soma beverage. A number of passages from the Rig Veda also attest to the importance of milk in the preparation of Soma, including the following:

> 9.8.5: When through the filter thou art poured, we clothe thee with a robe of milk to be a gladdening draught for gods. (Griffith, 1891: 370)

> 7.21.1: Pressed is the juice divine with milk commingled: thereto hath Indra ever been accustomed. (Griffith, 1891: 30)

We contend, given the numerous references to milk being added to the Soma beverage as well as passages that suggest that milk plays a "cleansing" or purifying role in the preparation of Soma, that milk, and not the human body as argued by Wasson, is the third and final filter used in the preparation of Soma. While Wasson pushed ethnographic accounts of urine-recycling in support of his theory, ethnographic evidence is also available to support our contention; specifically, it has been reported, from the Shetul Valley of Afghanistan, that dried and powdered *Amanita muscaria* was once "boiled with whey from goat's milk cheese" to produce a beverage called *bokar* (Ott, 1993: 340-341, *citing* Mochtar & Geerken, 1979), ultimately, a drink that may have been a lot like Soma.

We also contend, as with the first two filters, that the third filter should produce some discernible pharmacological change when applied to the "true" Soma plant. Since pasteurization of milk wasn't developed until 1864 the milk used in Vedic ceremonies would have likely contained some variety of *Lactobacillus* bacteria which produces the enzyme glutamate decarboxylase (GAD). The frequent mention of curdled milk also supports this contention as curdled milk is milk that has been naturally fermented and typically contains high amounts of *Lactobacillus* bacteria and GAD. Several verses that mention blending of Soma juice with milk curd follow:

> 9.63.15: Over the cleansing sieve have flowed the Somas, blent with curdled milk, Effused for Indra Thunder-armed.
>
> 9.103.2: Blended with milk and curds he flows on through the long wool of the sheep. The Gold-hued, purified, makes him three seats for rest.
>
> 9.109.15: All Deities are wont to drink of him, pressed by the men and blent with milk and curds. (Griffith, 1896)

GAD catalyzes decarboxylation of glutamate to GABA, a process that parallels the decarboxylation of ibotenic acid to muscimol, compounds which are respectively similar in structure to glutamate and GABA (Figure 1). If GAD catalyzes decarboxylation of ibotenic acid in the same way it does with glutamate, the presence of GAD in unpasteurized milk that's blended with an infusion or decoction of *Amanita muscaria* should result in a significant increase in muscimol content in the final beverage. Such an increase would make this a vital step in the preparation of Soma as a sacramental and entheogenic beverage.

Methods

In order to support our contention that milk constitutes the third filter of Soma it is necessary to demonstrate that, like the first two filters, the addition of milk produces pharmacologically significant changes to the final beverage. To test whether the addition of milk might produce noteworthy pharmacological changes we focused on the potential role of glutamate decarboxylase (GAD) on the ibotenic acid and muscimol content of infusions of *Amanita muscaria* var. *guessowii*.

Samples of *Amanita muscaria* var. *guessowii* were dried in a dehydrator for two days at 125 degrees Fahrenheit. The caps were selected, ground to a powder, and mixed. Two hundred and forty grams (240g) of powder was infused in 60 ounces of distilled water at 45 degrees Fahrenheit for 24 hours, then filtered to remove the solid particles. Thereafter, a portion of the filtrate was set aside and frozen for later analysis. This portion was retained as an untreated, or control sample.

A second sample was created using 2 ml of the undiluted filtrate, to which 14 mg of purified glutamate decarboxylase was added. Additionally, 0.3 mg of pyridoxal phosphate (P-5-P) was added as a co-factor, and the sample was maintained at 37 degrees Celsius for 4 hours and then refrigerated. These samples were then analyzed utilizing high performance liquid chromatography (HPLC), following derivatization using dansylation reaction (Tsujikawa et. al, 2007), to determine their respective ibotenic acid and muscimol levels.

Results

Analysis of the control sample yielded 22.4 ppm muscimol compared to 77.9 ppm ibotenic acid, a ratio of 0.29 muscimol to 1.0 ibotenic acid. The sample treated with GAD demonstrated excellent conversion of ibotenic acid to muscimol, yielding 94.6 ppm muscimol compared to 1.0 ppm ibotenic acid. While the yield in muscimol nearly quintupled, levels of ibotenic acid decreased nearly 80-fold (Figure 3).

Discussion

While the fact that ibotenic acid, and to a lesser extent muscimol, are present in human urine following ingestion of the fly agaric is interesting, there is nothing to suggest that the resulting product is more potent or less "toxic" than when

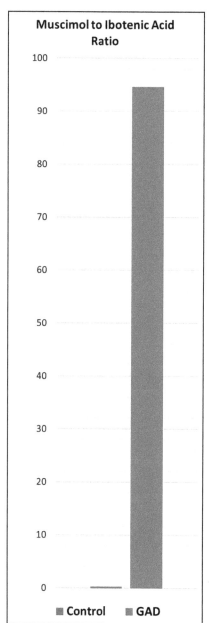

Fig. 3: Muscimol to Ibotenic Acid Ratio. The chart at left illustrates "parts" muscimol in relation to 1-part ibotenic acid within the analyzed "control" and "GAD" samples.

initially ingested. This is in stark contrast to the remarkable conversion of ibotenic acid to the more potent muscimol in the presence of an enzyme produced by *Lactobacillus* bacteria, as demonstrated by this study. *Lactobacillus* is common in unpasteurized milk, and plays a key role in the production of sour and curdled milk,[1] both of which are mentioned in the Rig Veda as "admixtures," or which we would argue are really the third Soma filter. The almost certain presence of *Lactobacillus* bacteria in the milk used by Vedic priests means that the juice pressed from the Soma plant would be exposed to GAD when "cleansed" by the addition of milk. If Soma is *Amanita muscaria*, as we contend, this would result in the conversion of ibotenic acid to muscimol, thereby potentiating the final beverage and reducing toxic effects associated with ibotenic acid, thereby giving the third and final filter a clear pharmacological purpose.

Limitations

While the methods used in this study demonstrate the dramatic effects that GAD can produce on solutions containing ibotenic acid our methods do not reflect the process that would have been employed by Vedic priests. Our results provide concrete data and allow us to speculate on the pharmacological changes that are possible when *Lactobacillus* containing milk is combined with solutions containing ibotenic acid, however, a research study that used *Lactobacillus* containing milk rather than pure GAD would provide results that would more closely approximate the beverage that could have been produced by Vedic priests over 3,000 years ago.

Conclusion

While Wasson's focus on the preparation of Soma and in finding parallels between cultural preparations of *Amanita muscaria* and the methods described for the preparation of Soma in the Rig Veda were highly intuitive, his focus was ultimately misplaced. Instead of examining the obvious solution, presented clearly and repeatedly in the text of the Rig Veda, Wasson opted for a radical interpretation of the text in order to draw connections between Siberian urine-recycling and the Soma ceremony. Wasson's argument may have helped bring attention to and popularize his theory, a theory which people continue to debate today – as here – but it is also possible that this facet of his argument negatively affected his theory's reception among some Vedic scholars. Wasson's basic premise that Soma was a psychoactive substance with hallucinatory qualities is widely accepted, though many now reject his specific contention that *Amanita muscaria* was Soma (Ott, 1998; Rätsch, 2005). Here, we have revisited Wasson's *Amanita muscaria* theory and through identification of an alternate third filter, a filter of milk, we have provided evidence that supports Wasson's identification. While debate will likely continue, the incredible effect of glutamate decarboxylase on ibotenic acid and muscimol levels within infusions of *Amanita muscaria* cannot be ignored and adds another important layer of data and support for Wasson's groundbreaking theory.

Notes

1. *Lactobacillus* consume lactose and convert it to lactic acid, which causes milk to sour. As levels of lactic acid increase milk proteins begin to coagulate and turn into curd.

Chapter 5

Travels with Santa and his Reindeer
Lawrence Millman

In the summer of 2003, I was traveling through the Chukotka Autonomous Okrug in northeastern Siberia, and I happened to eat Santa Claus. Or maybe I should say that I ate the mushroom traditionally eaten by shamans before they become Santa Claus. Or maybe I should just say that the composite figure of Santa Claus includes an anonymous shaman high on a certain mushroom.

Before you accuse me of profaning a popular Christmas icon already profaned by Coca Cola and the poet Clement Moore (author of "The Night Before Christmas"), consider the no less iconic mushroom in question, *Amanita muscaria,* or the Fly Agaric. In its most dramatic color phase, it has a robust red cap with concentric rings of white, wart-like spots on it. Even in its immature button stage, it's still very robust. If it were a human being, you might refer to it as obese.

You might recognize in my description a mushroom depicted in numerous greeting cards (Figure 1), paintings, posters, and stoner websites. Contrary to popular belief, it's not the mushroom depicted in *Alice in Wonderland*, though. If you look at the early illustrations in Lewis Carroll's book, you'll see that the mushroom

Fig. 1: Vintage Estonian postcard (ca. 1920s) wishing the recipient a "Happy New Year."

looks like some sort of lawn mushroom (Figure 2). Likewise, Alice's trip, taken under the tutelage of the hookah-smoking caterpillar, was more typical of a *Psilocybe* trip than an *A. muscaria* trip. Alice got bigger and smaller, respectively; if she had eaten *A. muscaria*, she probably would have felt like she was flying, which is how I felt.

Gazing at me, Dmitri, my Chukchi companion, said: "With a *wapak* [the Chukchi word for *A. muscaria*], you don't need a ticket or a boarding pass."

Initially, I felt nauseated as well, but nausea comes with the biochemical territory. For while *A. muscaria* contains muscimol, the alkaloid that gave me the illusion of flying, it also contains ibotenic acid, a very potent compound that can give your stomach a turbulent ride. If you dry the mushroom, most of the ibotenic acid decarboxylates (i.e., degrades) into muscimol... most, but not all. Here I should mention that *A. muscaria* has not been implicated in a single death, except in the works of detective writers like Dorothy Sayers (see her novel *The Documents in the Case*, for example).

Dmitri seemed slightly circumspect about eating the mushroom, or at least eating it in public. Part of his reticence probably came from the fact that he was a schoolteacher, and it wouldn't have been appropriate for his students to see him eating this mushroom. But part of it might have been a survival from Stalinist times, when any indigenous person who ate *mukhomor* (the Russian word for *A. muscaria*) was considered an enemy of the state. Repeat offenders reputedly were herded into airplanes, and when the plane was airborne, the cargo door would be opened.

"You say you can fly," a

Fig. 2: Illustration by John Tenniel, published in the original *Alice in Wonderland* (Carroll, 1865).

Stalinist henchman would announce. "Okay – then fly!"

Whereupon he would push the victim out of the plane.

Meanwhile, the Soviets were flooding Chukotka with cheap vodka, the better to wean the locals from their time-honored traditions. The ritual use of *A. muscaria* might have been a *very* time-honored tradition: carved into the rock-face near the mouth of the Pegtymel River are a number of ancient petroglyphs that seem to be of shamans whose heads are crowned by *A. muscarias* (*see* Ch. 17, Fig. 6).

According to Dmitri, Chukchi shamans would ingest three dried mushrooms to sustain them during long hours of drumming and singing. As muscimol goes through the human kidneys more or less unaltered, the shaman's acolytes would gather around him and drink his urine. In bad mushroom years, Dmitri said, there was quite a lot of recycling of shamanic urine.

When we encountered an *A. muscaria*, Dmitri gave me precise instructions on how to pick it. Be especially careful with the cap, he told me. If I damaged it, I might end up with some sort of head injury. If I removed the warts from the cap, I would end up losing all of my hair. And if I injured the stem, something unpleasant would happen to one of my legs.

"How unpleasant?" I asked.

"You might need to have the leg amputated," he replied.

Is there any mushroom that's been so anthropomorphized as *A. muscaria*? I wondered while cautiously picking the specimen for later consumption.

In a 1986 article in *The New Scientist,* the English historian Ronald Hutton dismissed the notion that Santa Claus, or Father Christmas, might have been a mushroom-taking Siberian shaman. Such shamans never wore red-and-white clothes, he argues. Nor did they climb in and out of smoke-holes in a trancelike state. Nor were they ever inclined toward gift-giving – instead, their constituents gave them gifts.

Of course, Santa Claus wasn't a Siberian shaman. To insinuate himself into European folk memory, he would have needed to travel an unconscionably long distance. What's more, contact with many groups in Siberia didn't occur until well after Santa had done his insinuating. It would have been much more appropriate for him to enter European lore from a European place, like, for instance, Lapland.

In fact, I'd like to propose that Santa (or at least part of him) might have been a Sami (Lapp) shaman. Indeed, many Europeans used to regard all Samis as shamans: they believed that the women could curdle milk or give a person smallpox just by thinking about it, and their men could chant away bad weather or disease with their *joiks* (ritual songs).

Only small shreds of traditional Sami culture still survive, but let's go back several hundred years, before Christian missionaries assaulted the Sami way of life. Let's also say that you're a Sami reindeer herder, and that you've taken to your sleeping skins with a mysterious ailment which makes your whole body ache. Through the taiga telegraph, you've put in an urgent call to the local *noaidi* (shaman). Soon the fellow pulls up in front of your lodge in his reindeer-drawn sled, then enters the premises via the smoke-hole. He couldn't have come through the door owing to the pile-up of snow.

"What's wrong with me?" you ask your visitor.

Prior to his visit, the noaidi had eaten a few dried *karpassienis* (the Sami word for *A. muscaria*), and he now peers at you with an expression that indicates he's entered a *gievvot,* or altered state. After a minute or two, he offers his diagnosis of your condition: you've been cursed by one of your neighbors, a guy who seems to think your reindeer are his reindeer.

Damn that little scuzzball, you say to yourself; he'll do anything to get my reindeer.

"But I can get rid of the curse," the noaidi tells you.

Stooped over, he starts beating a reindeer chamois *kobda* (drum) ringed with bear teeth and decorated with images of the sun and moon. While he's doing this, he's chanting a joik that consists of certain repeated phrases: *Awaken, O my nature...O awaken...go down...down into the Underworld...* Every once in a while, he leans over and spits on you. Since his saliva contains the healing particles of one or more karpassienis, you don't object to what would otherwise be insulting behavior.

Curiously enough, the noaidi himself seems to have turned into a karpassieni. For a shaman who has consumed this mushroom typically turns into a facsimile of it, or at least has taken on its distinctive red-and-white color scheme. Or so the Sami used to say.

Miraculously, you feel better, considerably better, after the noaidi's performance. You rise up from your sleeping skins with the idea of making your curse-wielding neighbor become a resident of his own sleeping skins. At the same time, you feel very grateful to your visitor.

"What a gift you've got!" you inform the noaidi.

"Thank the karpassieni, not me," he replies. "It transported me to the realm of Jabmiekka, Mistress of the Dead, and she permitted me to retrieve your missing soul, which in turn permitted me to heal you."

After you reward him with some reindeer meat and, especially reindeer fat (his ample belly suggests that he's been rewarded in this manner many times before), the noaidi clambers back up your lodge's smoke-hole. Soon you hear him whisking off with his reindeer to visit another client.

Here I might mention that Siberians never traveled in reindeer-drawn sleds, a fact that Mr. Hutton points out as well. But the Sami did travel in reindeer-drawn sleds, and some of them continue to do so today (Figure 3). A single reindeer can pull a sled and its driver over a ten-mile course in forty minutes, while a five-dog team will take an hour or more over the same course.

I'd also like to mention that while reindeer will eat almost any mushroom,

Fig. 3: Sami man in traditional dress, with sleigh and reindeer
(Photo © Potatushkina: ID 73970577 | Dreamstime.com).

especially during the fall rut, when they're actively seeking protein, they're especially fond of *A. muscaria*. If they eat it, they often end up stumbling around in what would appear to be a stoned manner. For Sami herders, it's not easy trying to move stoned reindeer from one grazing spot to another. Rather than wait until their charges get over the effects of mushroom ingestion, some herders have been known to collect *A. muscaria* fruiting bodies and create trails out of them, the better to make their reindeer go where they want them to go.

No one has interviewed any reindeer about their taste for *A. muscaria*, but if they could talk, I suspect they might say that they quite enjoy the sensation of flying. They might add that a reindeer with a red nose is afflicted with a parasite, usually a bunch of *boaro* (the Sami word for warble-fly larvae), and while the condition can be very painful, it doesn't usually result in one's nose glowing like a light bulb.

Unlike their reindeer, the contemporary Sami seem to have no interest in *A. muscaria*. In the fall of 2005, I was traveling north of Rovaniemi, the administrative capital of Finnish Lapland, and I saw some large fruitings of the mushroom. My Sami companions told me that it was highly poisonous, and they would never eat it. One elderly man added something to this in Davissamegiella, the Sami language. A younger person translated for me:

"He says our noiadi used to eat karpassienis, and then they would go to other worlds. But that was a long, long time ago."

It's the night before Christmas, and not a creature is stirring in your house. Not a creature, that is, except for you. You're putting some last-minute decorations on a spruce tree that will never reach maturity.

All of a sudden, there's a loud noise in your chimney, and an overweight man with a disheveled white beard lands in the fireplace with a resounding crash. His unusual method of entry suggests that he might be some sort of shaman. Or perhaps a burglar posing as a shaman. After all, a person innocent of any wrongdoing will usually enter a house through the front door rather than the chimney.

"Are you going to rob me?" you ask the fat man.

This idea strikes him as so funny that he bursts into a loud, somewhat vulgar laugh. Then he shows you a bag nearly bursting at the seams.

Truth to tell, your visitor is a hybrid concoction of many diverse parts, some commercialized, some fictionalized, and some reaching into the distant past. While certain aspects of his personality might be inclined to do some shamanizing, other aspects of it would be no more capable of shamanizing than, for example, putting together a weight-watcher's cookbook. The non-shaman aspects win. So he doesn't spit on you or beat on a reindeer skin drum while chanting in an unknown language. Nor does he seem to be in any sort of altered state, although you can't help but think that he's had a few beers.

The fat man unties his bag, and you see that he's brought all kinds of gifts for you and your family: video games for the kids, a new microwave for the kitchen, Chanel perfumes for your wife, some extremely up-to-date iDevices, and even (this may sound a bit far-fetched) a few books for you.

You stare at your beneficent visitor, and maybe, just maybe, you detect in his robust body, not to mention his bright red outfit with white buttons and trimming, a certain robust red-and-white mushroom.

Chapter 6

A Search for Soma in Russia's Kamchatka Peninsula

Jason Salzman, Emanuel Salzman, Joanne Salzman & Gary Lincoff

"If she doesn't drink it, she will lie in bed all day long," says Tatiana, explaining why her 80-year-old mother ingests a daily dose of fly agaric mushrooms. "She takes it to keep her going."

Standing near an ironing board covered with drying red-and-white mushrooms (Figure 1), Tatiana holds up a bottle of the gray liquid that her mother drinks every day (Figure 2). She recites the recipe: Mix five or six whole fly agarics (*Amanita muscaria* complex) with one liter of boiling water, cover and refrigerate at least one day before using. If fresh blueberries from the surrounding tundra are available, add them too. Take a shot a day.

This may sound like a potion from a New Age healer but it's not: Tatiana and her mother are native Koryak people, living in a remote village on the Kamchatka Peninsula, in the Russian Far East.

In Search of Soma

In early August 1994, our group of mushroom hunters went to Kamchatka seeking proof of Gordon Wasson's theory that the fly agaric is the long-lost identity of Soma, the ecstasy-inducing plant that inspired the hymns of the Rg Veda, the sacred Hindu religious text. The Rg Veda, which was first recorded in Sanskrit about 3500 years ago, is

Fig. 1: *Mukhomor* drying on an ironing board.

the oldest still-extant sacred book of the ancient Aryans, who began migrating from the steppes of Eurasia into the Indus Valley around 2000 B.C. In his 1968 classic, *Soma: Divine Mushroom of Immortality*, Wasson reviewed 17th and 18th century reports by western explorers that the tribal people of Siberia and the Russian Far East used the fly agaric as a shamanic and recreational inebriant. These reports, plus Wasson's detailed linguistic studies, led him to conclude that the fly agaric was used as a "divine inebriant" long before the Aryans migrated "from the North" to the Valley of the Indus, present-day Pakistan and India.

Wasson's critics, however, point out that the psychoactive effects of ingesting the western fly agaric – so named because extracts of the mushroom attract and stupefy or kill flies – do not match the ecstasy experience described in the Rg Veda. Indeed, in the U.S. the fly agaric is widely considered poisonous or, at best, as a second-rate recreational drug. Eating two to three caps is reported to cause loss of balance, muscle twitching, somnolence, sweating, nausea, occasionally vomiting, and hallucinations, including visual distortions and light patterns.

A Koryak's Divine Inspiration, an American Poison

Why would the fly agaric be a Koryak's divine inspiration and an American's poison? Maybe the Kamchatka variety of *A. muscaria* contains different compounds. Maybe its unique nutrient-exchanging mycorrhizal relationship with the endemic birch (*Betula er-*

Fig. 2: A Koryak woman displays a bottle of the *mukhomor*-blueberry tonic used by her mother.

manii) explains the difference. Maybe the disparity in the reaction to the mushroom is caused by cultural or even genetic influences.

Despite his best efforts, Wasson was never able to visit Russia to collect fly agaric, talk to native people, and respond to his critics. Under the Soviets, Siberia and the Russian Far East – which includes Kamchatka – were closed to foreigners to protect secret military installations and the gulag prison system. Wasson had repeatedly requested a visa to visit Kamchatka in order to study fly agaric, called "*mukhomor*" in Russian (mukho = fly, mor = death), but his applications were consistently denied by the Soviet government. As we now know, the Soviets were trying to eradicate the use of both *mukhomor* and shamanism in the region, and it is possible that they did not appreciate Wasson's interest in the shamanic and religious uses of *mukhomor*.

After the Soviet Union adopted its policy of *glasnot* (freedom of expression), Maret Saar (1991a), an Estonian ethnographer, published a paper on the use of *mukhomor* in Russian Asia. In addition to reviewing historic accounts mentioned by Wasson, Saar incorporated new information from her own field studies in 1971 among the Khanty of western Siberia as well as new data collected from Soviet ethnographers. Saar's (1991a) report confirmed Wasson's position that *mukhomor* had been used shamanically in both western and eastern Siberia, but it also took a more conservative Russian position that *mukhomor* "was used only by some magicians and not in all rites" (p. 159). Interestingly, Saar's research among the Khanty indicated that their shamans or wisemen sometimes used *mukhomor* in healing and divination – but that, when a wiseman was unavailable, Khanty individuals might also use *mukhomor* "to call forth a spell of second sight" and as an aid in communicating with the supernatural world (p. 162). One of Saar's contacts, Yuri Simchenko of the Moscow Institute of Ethnography, had indicated that *mukhomor* was still being used by the Koryak people in northern Kamchatka and the Chukchi in Chukotka, but Saar's report includes few specific details on its current usage in Kamchatka.

Our group of "fungophiles" was on a quest to see if we could find any evidence among the Koryaks that might confirm the basic theories of Wasson regarding the ecstatic and shamanic use of *mukhomor* and, with the end of the Cold War, it suddenly became possible to visit these areas and travel where Wasson could not go. *Fungophile*, the Denver-based mushroom education organization, sponsored trips to Kamchatka in 1994 and 1995 to collect fly agarics and interview native Koryak tribespeople about their use of the mushroom. If only Gordon Wasson could have joined us.

Wishing for seat belts, we float across Kamchatka

Early in August 1994, fifteen mushroom hunters huddled in a hay field, waiting for a Soviet-made M18 helicopter, allegedly dispatched from a military base on the outskirts of nearby Petropavlovsk, Kamchatka's largest city.

The configuration of cow pies on the grass does not inspire any confidence that the helicopter, already over 20 hours late, has ever landed here before – or ever will land here. Standing pathetically next to empty mushroom collecting baskets and luggage, most of us think we are the latest victims of a Russian scam: the helicopter will never show, and our hope of verifying Gordon Wasson's theories will be shattered. And we'll be bilked out of a lot of money. This pricey helicopter flight was pre-paid in randomly numbered U.S. hundred-dollar bills, carried into Russia literally under our shirts.

But just as the sun is sinking, making a helicopter flight across the Kamchatka Peninsula's spine of mountains and volcanoes a physical impossibility, the "heli" emerges from the horizon and, to our deep surprise, touches down in our hay field. With the blades spinning and the engines roaring, we take our seats on long wooden benches wedged between our luggage and the gas tank. It smells vaguely like a barn, which makes sense because these transport helicopters are the all-purpose vehicles of Kamchatka. Wishing for seat belts, we grab the bench, the *heli* lurches, and we float north.

Mukhomor!

Our first mushroom foray near Palana, a town of about 1,000 mostly Russian people on the west coast of Kamchatka, turned up about 50 species, including one fly agaric, or *mukhomor*. It looked very much like the western variety with a red cap and white patches (Figure 3). After our foray, we proudly showed our *mukhomor* specimen to the Russian townspeople. They reacted with some amusement, but mostly with disapproval accompanied by serious warning about its toxicity. "Poganka! Poganka!" Russians repeated (Sidestepping taxonomic wars, Russians use one common – and universally known – name for all poisonous mushrooms, "poganka.").

Yes, the Russians know that the indigenous people ate this mushroom. No, they could not explain why it didn't poison them – though they accepted this difference as part of the gulf that separated "them Koryaks" from "us Russians." One of the angriest responses to our suggestion that *mukhomor* could have a pleasant mind-altering effect came from a drunk Russian alcohol distributor (It's been observed that alcoholics express strong opposition to the use of all drugs except their own drug of choice).

The Russians' fierce hostility toward *mukhomor* surprised us because Russia is a great mycophilic society. Mushroom paintings adorn outdoor markets and mushrooms even appear as playground structures (Figure 4). People give you a look of admiration – not horror as in our country – if you're walking down the street with a large edible wild mushroom. In an incident that would never occur in the U.S., a member of our group was accosted by a Russian hotel receptionist who insisted on removing and discarding "poganka" specimens from his collecting basket. Conversely, a batch of white *Russula* mushrooms that we discarded under a bush near a hotel was retrieved joyously by a passer-by who lectured us on how to salt them for eating.

Joy and Envy at the Sight of *Mukhomor*

Later, in a school cafeteria in Lesnaya, north of Palana, five Koryak people reacted with the same gasp of joy at the sight of our *mukhomor* specimen as the Russian passerby did to the sight of the mushroom detritus we chucked under a

Fig. 3: Samples of *Amanita muscaria* found during our travels in Kamchatka.

Fig. 4: Author Jason Salzman crouches under a *mukhomor* umbrella at a playground in Korf, southwest of Khailino.

bush. "Where did you find this *mukhomor*?" asked one of the Koryaks, a question universally heard from mushroom hunters who want to pick their own collections of a wild mushroom gathered by someone else.

But one Koryak woman warns us that our mushroom was picked incorrectly. All *mukhomor* should be collected with the cup at the base and the accompanying dirt (Ours had been sliced toward the bottom of the stem by its collector, Rita Rosenberg, who was apparently gathering this mushroom as she would other edibles).

"If you eat this mushroom, you will be jumping on one leg," warns an 81-year-old Koryak fisherman. His comment reflects a Koryak belief, which we encountered in northern towns on our 1995 journey, that eating *mukhomor* with a damaged stem – or with the stem cut off completely – will paralyze a leg of the eater. As Western-trained mycologists, we were skeptical that cutting the mushroom could change its chemical constituents, but we were delighted to finally find some people who didn't treat *mukhomor* as *poganka*.

How to Pick and Prepare *Mukhomor* for Eating

On our trips in Kamchatka, we conducted formal interviews with a total of over 20 Koryak natives in eight towns (on the west coast: Palana, Lesnaya, and Manily; in the northern, central part of the Peninsula: Talovka, Khailino, and a reindeer herders' camp; and on the Bering Sea: Ossora and Tymlat; *see* Map 1). These interviews clearly revealed that the Koryak people have a history of eating the fly agaric and that some are still eating the mushroom today.

Map 1: Map of the Authors' travels in Kamchatka.

But across the Peninsula we heard vastly different opinions and traditions about how to pick, prepare, and eat *mukhomor* – and what effects can be expected. For example, contradicting the warning we received in Lesnaya, some Koryaks told us not to worry about eating a stemless *mukhomor* and becoming temporarily paralyzed. Any *mukhomor*, cut or collected in any manner, will do just fine.

Yet others insisted that *mukhomor* must be picked with a stick, eaten in pairs only, and never cut at all. We were also advised to eat only odd numbers of *mukhomor* and to dance before the mushroom is dug from the earth. One Koryak woman warned us not to eat *mukhomor* from mainland Russia, lending credence to the idea that the Kamchatka variety of the fly agaric is different from other varieties. Others, however, disagreed.

In all our interviews, we were told that the mushroom should be dried before being eaten. This was one of Gordon Wasson's conclusions. Koryak people offered these specific *mukhomor* preparations and doses:

- Mix five liters of water with 10 to 15 dried *mukhomor*. Add blueberries, some sugar, and boil for a half hour. Drink one-half to one cup.
- Eat two-to-four dried whole mushrooms (never more than four) with or without water.
- Apply *mukhomor* externally to treat arthritis and heal

wounds.
- Mix a half cup alcohol with three dried *mukhomor*. Leave in a bottle for two weeks. Apply externally for pain relief.
- Mix *mukhomor* with tobacco and birch polypore, toasted on coals to a white ash. This snuff can be used as a substitute for tobacco or to relieve toothaches.

The Koryaks also told us that finding *mukhomor* can be difficult, which is one reason that young people favor vodka as an inebriant. We had already noticed that signs of drunkenness were inescapable throughout the town of Lesnaya.

When we asked the Koryaks if they had any knowledge about *mukhomor* being used for shamanic purposes, they quickly told us that there were no shamans living in Lesnaya. We sensed that the Koryaks were reluctant to talk about the subject, so didn't press them further. It was clear that if any shamanizing was still going on in the area, such activities wouldn't readily be revealed to outsiders. We were just glad that the Koryaks weren't hesitant to talk about other uses of *mukhomor*.

Mukhomor's Dark Side

We heard many more positive stories about eating *mukhomor* than negative ones, but a few Koryaks said the mushroom caused poisonings. "It was total hell," said a 30-something Koryak of her experience on the mushroom. Another warned that eating too much *mukhomor* causes "convulsions" and "foaming at the mouth." We heard two vague reports of death caused by *mukhomor* – an outcome that would not be considered valid in western toxicology without thorough documentation.

Confirming another conclusion of Wasson's, we heard no reports of a be-mushroomed Koryak hurting himself or another person. However, one woman in Ossora told us that a Koryak man killed four reindeer after eating *mukhomor*.

The most common effects, reported by Koryaks who either ate the mushroom themselves or knew people who did, ranged from its inspiring general euphoria, dance, dreams, communication with ancestors, drumming and song to causing sleep, increased energy, "drunkenness," and hallucinations. A few Koryaks explained that all or some of these effects can be "transferred" from one person to another, sometimes by direct eye contact and, confirming an odd-sounding conclusion of Wasson's, we heard numerous stories of *mukhomor* commanding people using it.

Many Koryaks reported at least some increase in physical strength, if not a tremendous increase, after eating *mukhomor*. "When you eat *mukhomor*, you do not feel your backpack," one woman told us, describing a day when she ate two mushrooms and walked with a heavy pack 20-to-30 kilometers in four hours, "as if my legs were separate from me." She then fell asleep and awakened with bruises from her backpack. Another Koryak told us that reindeer herders, after eating *mukhomor*, have been known to "hop around the herd," chasing reindeer on foot with little food across long stretches of tundra.

The Mushroom will give you a Song

In most interviews, Koryaks emphasized the importance of drumming and song in the *mukhomor* experience and therefore the pleasure of its recreational use at parties or holidays, like the October end-of-the-harvest celebration in Lesnaya. The mushrooms we were told repeatedly, not only prompts long sessions of drumming and song but inspires people to compose their "own song."

In Palana, an accomplished Koryak folk guitarist and ethnographer, Ura Alotov, told us *mukhomor* inspires him to compose music and create improvisations of old songs. Similarly, in Khailino, one Koryak in her 40s said she does not eat *mukhomor* because she doesn't have her own song yet, but her elderly mother tells her that the mushroom will "give you a song." The relationship between musical creativity and taking the mushroom adds credence to Wasson's belief that the fly agaric inspired the hymns of the Rg Veda.

A River from Spit

Koryaks described a variety of hallucinations caused by *mukhomor*, ranging from "seeing sounds" and colors to full-fledged visual creations. A woman in Tymlat, wearing a purple and black sparkling sweater with brown pants and beaded skin boots, said that after eating cooked *mukhomor* with berries, she first slept then awakened with her mouth full of saliva. The *mukhomor* told her that she was the heroine of a fairy tale and that she should spit and make a big river, which she did. The mushroom then told her to spit out the rest of her saliva to drain other rivers.

Psychoactive experiences like this were so important to the Koryaks that they found ways to avoid wasting any of the mushroom, according to historic reports cited by Wasson. Since the drugs causing the effects are excreted in the urine, the experience could be obtained or prolonged by drinking the urine of a bemushroomed person, either one's own or another's. Particularly if the fly agaric was in short supply, poor Koryaks might drink the urine of the intoxicated rich to attain the *mukhomor* state of consciousness. The high price of *mukhomor* in the ancient Siberian cultures may have made it easier for some to relate to the need to drink urine: one mushroom, according to Wasson's citations, was equal in value to a reindeer.

At first, it was not easy to ask a Koryak, "Do you drink urine." But after timidly raising the urine-drinking issue a few times, it became apparent that the Koryak people do not seem embarrassed by the question, and a handful described, matter-of-factly, relevant stories or personal experiences:

- One man in Palana told us he saw his uncle drink the urine of another man who had eaten *mukhomor*. The urine did not produce a stronger effect, as reported in Wasson's book (1968), but it did generate the experience.
- A woman in Ossora explained that her father would drink his own urine two or three times during the day after he ate the mushroom. This, she said, helped with the hangover and prolonged the experience.
- Another woman reported that drinking one's own urine while on the mushroom can stop unwanted convulsions and hallucinations.

Do Reindeer Attack Urinating Koryaks?

Wasson claimed that, in addition to drinking urine to obtain the *mukhomor* experience, Koryaks would eat the drugged meat of reindeer who were intoxicated by the mushroom at the time of killing. Reindeer consider the fly agaric such a delicacy, according to Wasson's historical analysis, that not only would they eat the mushroom at any opportunity, but they would attack a bemushroomed Koryak, peacefully urinating on the tundra, for a taste of *mukhomor* urine.

Our interviews produced conflicting stories about reindeer and *mukhomor*. Knowledgeable Koryak reindeer herders agreed that their animals love mushrooms, often – to the intense frustration of the herders – bolting from the herd in the fall to feast on them. But some herders said the reindeer do not eat the fly agaric. Others said they do, describing how the animals act "drunk" or "sleepy" or even have convulsions after eating the mushroom. Regardless, none of the half dozen herders we interviewed reported seeing a reindeer so hungry for *mukhomor* that it would attack a urinating bemushroomed person.

After returning home to Denver, Colorado, we unexpectedly found an opportunity to confirm conclusively the reindeer's strong appetite for *mukhomor*. In a pre-Christmas Santa Claus/reindeer show on the downtown Denver Mall, we offered a Kamchatka *mukhomor* to a reindeer, a descendent of a herd of Siberian reindeer imported to Alaska in 1890 by

Fig. 5: Emanuel Salzman feeds *mukhmor* to an imported Siberian reindeer near Denver, Colorado.

the U.S. Government. The reindeer greedily ate the mushroom with obvious relish, preferring the mushroom to the bucket of grain in the corner of the corral (Figure 5). Santa Claus observed the reindeer for 24 hours after the ingestion and reported that consuming a single mushroom apparently had no effect.

Mukhomor and Shamanism

As we traveled by helicopter from town to town in Kamchatka, it became more and more clear that, among the Koryaks, vodka is replacing *mukhomor* as a recreational drug and causing widespread alcoholism. Vodka, an elderly Koryak man lamented, has made his Koryak kin "forget their songs." But it was equally clear from our interviews that before vodka was introduced to the culture, the mushroom was widely used in rituals as well as recreationally. This confirms information from ethnographer Maret Saar (1991a: 159), according to whom *mukhomor* was probably still being used in northwest Siberia, where the practice had been widespread among the Khanty, Mansi, Forest Nenets, Selkup, Nganasan, and Ket. While in northeast Siberia, *mukhomor* use has been documented among the Chukchi, Koryak, Itelmen, and Yukagir.

Most of these cultures attributed a conscious life and in-dwelling spirit to every material form of reality – animals, plants, stones, even earthquakes and thunderstorms. They believed in supernatural beings or spirits that control a world in which people can shape their lives by influencing these beings. Shamans, the priest-like tribal members, are the experts in communicating and influencing these supernatural beings through a trance-like "state of ecstasy." This state of shamanic consciousness is reached through the use of *mukhomor* or by non-drug techniques, such as drumming, chanting, or dancing.

In western Russian Asia, only the shamans use *mukhomor*, and in eastern Russian Asia the tribal people perform their own shamanic services, frequently with the aid of *mukhomor* (Saar, 1991a). Under the Soviets, the use of *mukhomor* was declared a criminal offense, and many shamans "disappeared." Others were driven underground.

Between the Soviet oppression and the introduction of vodka, the shamanic use of *mukhomor* in Kamchatka appears to have disappeared. No one we interviewed reported any current use of the mushroom by shamans. However, some Koryaks in northern Kamchatka said they use the mushroom to predict the future or answer questions – to locate lost objects or people, to determine where to take the reindeer herd, to forecast the weather or an upcoming death, or to choose a name for a child. One Koryak in the town of Manily said she takes the mushroom to help her heal people of various ills.

Mukhomor Shaman?

In Palana we encountered 72-year-old Tatiana Urkachan, a seventh-generation shaman from the Even tribe. Tatiana was a native-born Even, but she had lived much of her life with the Koryak. Although we had a letter of introduction from Yelena Batyanova, a Russian

Fig. 6: Tatiana dressed in a traditional Even shaman's outfit, drumming, in the hills outside Palana.

enthnographer who has worked with Tatiana, Tatiana was suspicious of our motives and reluctant to meet with us. It was only after our resourceful guides volunteered to clean her small apartment, carry her firewood, and run some errands for her that Tatiana finally agreed to meet with our group.

She met with us, leading our group early one morning up a muddy trail outside Palana to the place where she had spread her mother's ashes. There, with lush green hills stretching beneath her, Tatiana lectured for nearly two hours, stopping from time to time to pick and eat blueberries (Figures 6 and 7). At one point, she announced that she would play the role of a shaman. She began dancing rapidly back and forth, then, slipping a hunter's jump suit on over her shaman's outfit, she began to enact the role of a hunter. Holding a bow, she stalked back and forth across the hill, stopping occasionally to bend backwards and pull the bow strings as if shooting an arrow into the sky. In one playful dance, she asked Ulrich Danckers of our group to play the role of Prince Ememkuk, a mythical Koryak character. She played the role of *mukhomor*. The prince found her on the hillside, placed his hand on her head and made her human again.

Fig. 7: Tatiana enacts the role of Raven.

Based on Tatiana's performance, we asked her if she had ever used *mukhomor* in her shamanic work. Tatiana insisted that she does not eat *mukhomor*, because she has plenty of her own energy and power. Then, almost as if apologizing for being secretive, Tatiana explained that she had managed to escape the Stalinist purges of shamans only by going into hiding and practicing the utmost discretion. Suddenly, Tatiana announced that she had to leave us, to go mushroom hunting. We barely had time to thank her before she was gone.

Later, she met us wearing a red dress with white spots and carrying a bucket of *mukhomor* (Figure 8). She was dressed like a *mukhomor*! She gave the mushrooms to Gary and announced that he had convinced her of our group's sincerity during our initial meeting with her by identifying a number of poisonous species growing near her mother's grave.

Standing behind a table in a hotel like a wild mushroom identifier at a typical U.S. foray, she lectured fiercely about the mushroom. She obviously knew much more about *mukhomor* than she had let on at our previous meeting. She stated that we should gather only those *mukhomor* that grew alone, not in clusters, and that we should pick the mushrooms with our fingers. She indicated that smaller mushrooms are stronger. She explained how to dry the mushrooms in the shade, cap side up. She told us to ingest dried mushrooms in odd numbers (three, five, or in parts of three or five mushrooms), and to drink water with the dried mushrooms. She warned us that eating too many *mukhomor* could cause a lethargic state that could last hours, days, or even a year.

She indicated several ways that *mukhomor* could be used medicinally. She explained that an extract of *mukhomor* could be applied topically to heal wounds and to treat arthritis pain, and that chewing three small pieces of fresh *mukhomor* can heal sore throats (a few members of our group later tried this last remedy on their sore throats, with little noticeable effect).

Tatiana was so enthusiastic and knowledgeable about *mukhomor* that we found it hard to believe that she had never used the mushroom. She insisted that she does not eat *mukhomor*, but her flushed face, hyperactivity, irritability, rapid pace of talk, and ideas suggest that she was bemushroomed. Of course, we didn't push her about the matter. We certainly understood how years of Soviet persecution would make her reluctant to reveal any personal *mukhomor* use.

A Reindeer Herder's Camp

Toward the end of our 1995 visit to Kamchatka, we floated in our helicopter above the bright reds, oranges, and yellows of the fall tundra, touching down in mid-morning in a place relegated in the American mind to Santa Claus: a reindeer herder's camp.

In the center of the camp on nets strung across birch sticks, dozens of *mukhomor* were laid out carefully to dry (Figures 9 and 10). Large and small, they looked at first like strips of meat, but upon closer examination it was obvious they were a harvest of *mukhomor*. To our group of fungophiles, it was clear that we had arrived at the height of the *mukhomor* season.

Fig. 8: Tatiana gives an impromptu lecture about *mukhomor* at our Palana hotel (Photo by Emanuel Salzman).

Drinking vodka and eating salted salmon and dried reindeer meat inside a 50-foot-diameter yurt, the camp's dominant structure, five Koryaks told us that two elderly residents, who appeared to be in their 70s and 80s, eat the mushrooms regularly. The younger Koryaks in the remote yurt, even those in their 40s, have no interest in it. This confirmed our observations across Kamchatka that, today, it is predominately the oldest Koryaks who are still eating the *mukhomor*, apparently for its stimulant/therapeutic effects.

Like everyone else we interviewed, the Koryaks at the camp claimed they didn't know of any shamans still living in the area or of any ritualistic uses of the *mukhomor*.

What about Wasson's Theory now?

In a last helicopter stop before returning to Petropavlovsk, we descended for the evening near a spectacular bubbling hot springs in the shadow of a volcano (Figure 11). Cabins around this and other hot springs we visited were built by the Soviet army as playgrounds for elite Communist Party officials and top military brass. And now an American group of mushroom hunters…

We reflected on evidence we now had relating to Wasson's Soma theory. Does the Kamchatka variety of *A. muscaria* differ from the European and American subspecies? The various subspecies of *A. muscaria* found in Russian Asia, Europe,

Fig: 9: *Mukhomor* laid out on a net to dry.

and North America complicate solving the problem of the varying reactions to the mushroom.

Most of the *A. muscaria* in North Asia appears to be the "classic" European type (red with white patches), *A. muscaria* subspecies *muscaria*. Several subspecies are found in North America: ssp *muscaria*, ssp. *flavivolvata* (red caps with cream to yellow patches), ssp. *alba* (white caps with white patches), and ssp. *guessowii* (yellow orange caps with cream patches).

Although ibotenic acid (related to MSG) and muscimol are recognized as the psychoactive drugs in *A. muscaria*, and muscarine, named after being isolated from *A. muscaria* in Europe, as responsible for the sweating caused by the mushroom, no systematic study has been made comparing the concentration of the drugs in the various *A. muscaria* subspecies.

To identify possible anatomical differences, we provided specimens of the Kamchatka variety of *A. muscaria* to Rod Tulloss, an Amanita authority, who reported that the mushroom was identical microscopically to the North American variety.[2] The results of chemical analysis of the same mushroom collections submitted to Prof. Schott Chilton are not yet available.[3]

Fig. 10: A Koryak woman at Talovka sets out some *mukhomor* to dry on a lattice of branches.

Had we found current, ritualistic/shamanic use of *A. muscaria* in Kamchatka, we would have been able to conclude with more certainty whether Koryaks ingesting the fly agaric can attain the ecstatic state associated with Soma in the Rg Veda. Sadly, we did not encounter such use of the mushroom.

After much debate about the potency of *mukhomor*, some members of our group decided that the only way to see if the Kamchatka *mukhomor* could induce ecstatic visions was for us to try it ourselves. When our Russian guides learned of our plans, they demanded that we sign release forms protecting them from liability for any costs incurred due to sickness or death from the *mukhomor*. In contrast to the grave fears of our Russian guides, we had high expectations and gladly signed their waivers. We were inspired not only by the testimonies from the Koryaks but also by Mark Niemoeller's account of his experiments with the Kamchatka *mukhomor* that we had collected the previous year [Ed. Note: *see* Ch. 14]. He had reported a "very high" quality experience with "zero" unpleasant side effects.

Finally, one rainy day toward the end of our second expedition, five of us willing subjects and four "caretakers" gathered in a hotel room outside of Petropavlovsk to put our *mukhomor* to another test. After a brief, impromptu ceremony – involving a tobacco offering and some preliminary drumming – the five volunteers ate from two to eight grams of dried *mukhomor* each. According to Niemoeller, five grams of North American *A. muscaria* is considered to be a low dose, ten grams is a moderate dose, and fifteen grams is a high dose.

The results of our *mukhomor* experiment varied largely in proportion to the dosage. One two-gram eater said he noticed no effects other than feeling a surge in energy the next day. The other two-gram eater said that she felt more relaxed and that she experienced some hot and cold spells. Three hours into the experience, she drank two ounces of her own urine, and she then stayed awake most of the night, without feeling tired until the following afternoon. Neither experienced any visionary effects.

The eaters of five, six, and eight grams of *mukhomor* initially felt some queasiness but soon experienced general feelings of good will and euphoria, along with some lightheadedness and muscular imbalance. One person experienced some light sweating on his face and some increased salivation. Interestingly, all three of them slept off-and-on for about half of the seven or eight hours that the mushrooms affected them. Again, no one experienced any significant visionary effects.

After our return to the U.S., several members of our group chose to continue our experiential research using the Kamchatka *mukhomor*. On one sunny morning in western Colorado, one woman ate a four-gram dose of Kamchatka *mukhomor* and another ingested seven grams – two grams more than each had eaten in Russia. The woman who ate seven grams reported that, despite heavy sweating and some vomiting during the first hour, "everything was fine and wonderful." A little later, she described having a shamanistic experience of leaving her body and finding herself "on a shaft of bright light sailing through space."

Later that afternoon, both women drank their own urine in order to test the effect of recycling ibotenic acid in that way. The two soon began to talk about their lives and families. One commented that she "really understood how and why we had chosen our parents and all the lessons we had learned or come to learn on this journey." Their *mukhomor* experience lasted close to twelve hours; the second dose of ibotenic acid apparently extended the experience by about five hours.

Both participants slept peacefully through the night, and the next morning they both commented they were "thrilled to be alive." One of them, who is in her late seventies, observed that some chronic pain in her knees and back was gone, and

Fig. 11: A helicopter view inside one of Kamchatka's many large volcanoes.

that she had more energy and a better appetite than usual. She stated that she felt as if she were thirty or forty years old. This feeling of well-being lasted for about one week.

In order to compare the effects of Kamchatka *mukhomor* with North American varieties of *A. muscaria*, two members of our group conducted another experiment in the U.S. These two men each consumed seven grams of dried *A. muscaria* gathered in Telluride, Colorado. They noted that the Colorado mushrooms had a nutty taste similar to that of the Kamchatka *mukhomor*, but they both experienced significantly more intense negative side effects with the Colorado mushrooms than they had with the Kamchatka *mukhomor*. Both felt slightly more nausea than they had with the Kamchatka *mukhomor*. One experienced muscle tics and visual distortions (described as "visual tics or spasms"). Both men experienced sweating and chills, prompting them to wear ski parkas and to take hot showers – even though it was a warm, summer day. The men concluded that the Colorado mushrooms produced noticeably stronger mind-altering states, accompanied by a distinct feeling of euphoria, floating, and goodwill. One briefly experienced micropsia, an altered state in which objects appear much smaller than normal. As in the Kamchatka experiment, the effects subsided after seven hours.

Based on these limited comparative experiments, we would tentatively suggest that the less unpleasant side effects associated with the Kamchatka *mukhomor* may be due to a lower concentration of muscarine. However, based on the more visionary results achieved with the Colorado mushrooms, it may be possible that the Kamchatka *mukhomor* contains lower concentrations of all three psychoactive compounds found in the *Amanita* species. It is also possible that multiple variables may be involved: the strength of the specimens may vary with the season, the age of the mushroom, and the mycorrhizal host plants, as well as the eater's mindset. More extensive study needs to be done in order to explain the varying reactions to this mushroom.

A Powerful Mushroom in a Dying Culture

The conflicting, scrambled stories we heard in Kamchatka – not only about whether reindeer eat the fly agaric, but also about its stimulating or revelationary effects on the Koryaks – take on strange mythic character when reviewed together as in this chapter. For example, one Koryak woman told us, "I heard there is a Mother

Mukhomor somewhere on Earth, circling around. If you approach this Mother *Mukhomor*, you will be put in a trance."

What to make of this? In support of Wasson, the evidence is clear that *mukhomor* is associated among the Koryaks with a "non-western" way of thinking about human power and the relationship between people and the visible and invisible worlds. This way of thinking – which includes the belief that humans can transcend time, predict the future, communicate with ancestors, heal ourselves, and much more – is clearly linked to the shamanic/ecstatic state inspired by Soma in the Rg Veda.

But is the link strong enough to conclude that *mukhomor* could have been Soma? The answer is yes. Even with their shaman's dead and vodka and poverty fracturing their culture, the Koryaks retain a curious respect for *mukhomor*, seemingly unlike any other symbol or substance.

The scrambled, mythic, larger-than-life stories that endure in the shambles of Koryak culture testify to the powerful cultural force the mushroom must have exerted generations ago – powerful enough, in the Koryak cultural context, to generate any of the transcendent or ecstatic experiences produced by other psychoactive plants and surely strong enough to inspire the hymns of the sacred Hindu texts.

So *mukhomor* could easily be the legendary Soma. But, because Russian society has all-but destroyed the traditional cultures that nurtured the proper "set and setting" for the use of the mushroom, it is likely that the final answer of the mystery of Soma has been lost forever.

Notes

1. This chapter is a composite of two articles detailing the authors' travels that were originally published in 1996. The original publications include "A Search for Soma in Russia's Kamchatka Peninsula," published in *Mushroom, the Journal of Wild Mushrooming* and authored by Jason Salzman, Emanuel Salzman, Joanne Salzman and Gary Lincoff; and "In Search of Mukhomor, the Mushroom of Immortality," published in *Shaman's Drum: A Journal of Experiential Shamanism* and authored by Emanuel Salzman, Jason Salzman, Joanne Salzman and Gary Lincoff.
2. Ed. Note: Subsequent investigations by Rod Tulloss and others have demonstrated the existence of "several distinct phylogenetic species within *A. muscaria*," and it is now understood that North American and Siberian varieties are, in fact, distinct (Geml et al., 2008: 694).
3. Ed. Note: The results of Chilton's analysis, or whether the analysis was ever completed, are unknown.

Chapter 7

In Pursuit of Yaga Mukhomorovna:
The Finno-Ugric Connection and Beyond

Frank M. Dugan (retired)
USDA-ARS Plant Introduction
Washington State University
Pullman WA 99164 USA

Mukhomor and Baba Yaga

In Russian fairy tales, there is a wizard of abundant bad character. His name is Mukhomor, meaning "Poison Mushroom (Fly-agaric)" and he is sometimes equated with the equally villainous sorcerer, Koshchei the Deathless (Haney, 2013: pp. xiii, 183, 186). Koshchei is a close relative (sometimes her son, the exact relationship varies) of the famed Russian fairy tale witch, Baba Yaga (Shapiro, 1983). The name Yaga Mukhomorovna (i.e., Baba Yaga's patronymic) appears in some contemporary Russian cultural productions (Sveta Yamin-Pasternak, personal communication). These connections of Baba Yaga with fly agaric (*Amanita muscaria*) and other mushrooms have been traced in detail elsewhere, along with speculation regarding the origins of Baba Yaga (Dugan, 2017). Provided here is additional related information on plausible relationships between Baba Yaga and female folkloric personas of the Russian North, further indications of connections between these figures and *A. muscaria*, and the place of these personas, fly agaric, and the related *A. pantherina*, within Northern Hemispheric folk traditions.

The leading book-length academic references on Baba Yaga are *Baba Yaga: The Wild Witch of the East in Russian Fairy Tales* (Forrester, Goscilo, and Skoro, 2013, with several complete tales) and *Baba Yaga: The Ambiguous Mother and Witch of the Russian Folktale* (Johns, 2004). The witch is at the center of numerous fairy tales (e.g., 'Baba Yaga and the Little Girl with the Kind Heart' in Ransome,

1916), and is either a dreadful cannibal (interested primarily in children as dinner, as in 'Little Girl'), or when treated with the proper combination of assertiveness and respect, a very witchy fairy godmother, as in 'The Frog Princess' (Afanas'ev, 1945), where she helps Prince Ivan reclaim his bride. She is of hideous appearance, always found deep in the forest, living in a hut that rotates on outsized chicken legs, and may fly about in a mortar while brandishing a pestle and broom. Her incarnations in contemporary fiction are not addressed here, but readers can note the thousands of records for Baba Yaga in Google Books.

The above philological evidence on 'Mukhomor,' in the context of late nineteenth and early twentieth century Russian art (Dugan, 2017), strongly indicates a connection between Baba Yaga and *A. muscaria*. It is also well documented that *A. muscaria* played a role in Siberian shamanic practices, including those of Finno-Ugric peoples in the Russian far north (e.g., Mägi & Toulouze, 2002; Wiget & Balalaeva, 2001), and along with other fungi, in Finno-Ugric folk medicine (Saar, 1991b). Early travelers in Siberia noted that the "moucho-more, a deleterious species of mushroom" comprised "entertainments" of Siberian tribal people (Von Kotzebue, 1830: 6). Beginning in the late nineteenth century, it became traditional (almost formulaic) to portray Baba Yaga with fly agarics (history and illustrations in Dugan, 2017), including Baba Yaga collecting these mushrooms (Fig. 1). This trend has continued into the present, enabling a plethora of 'Baba Yaga cum *Amanita*' depictions on coffee mugs, lacquer ware, tapestries, coasters and numerous other commercial items available online at Amazon, eBay, Zazzle and other venues.

Fig. 1: Baba Yaga collecting mushrooms (and a prospective dinner). Boris Zworykin's Baba Yaga (1917, in V.A. Smirnov, *The Witch and the Strawberries*, public domain). Her basket is laden with fly agarics.

Baba Yaga of the North?

It is frequently conjectured that Baba Yaga's formulaic greeting to her visitors (who sometimes become her dinner) implies that she is not Russian. She nearly always states that she smells a Russian, and that it has been a long time since a Russian has come to her (Forrester et al., 2013; Johns, 2004). Her domicile (the rotating hut, on chicken legs, Fig. 2) bears strong resemblance to huts erected off the ground by Finno-Ugric peoples for storage (Goscilo, 2013; Hatto, 2017), for sacrificial objects (Kodolány, 1968), or in some instances in the Baltic region, for mortuary functions (Mägi, 2005). Mansi (Vogul) folklore "features monstrous supernatural beings who live … in a revolving dwelling" (Johns, 2004: 164) and this motif persists in Hungarian lore (in the same language family as Ugric) (Johns, 2004).

Baba Yaga, although quite plausibly of non-Russian origin (at least in part) is now regarded as the quintessential Russian fairy tale witch (Dugan, 2017; Forrester et al., 2013; Johns, 2004). Russian language and culture are now strongly ascendant over regional Finno-Ugric and other Uralic peoples (Khanty, Mansi, Selkup and others) of the far north. But Finno-Ugric pride in their unique traditions is also manifest. For instance, *Λoŋ-vertĭ-imi* ("woman who ties blood vessels") is a figure in Khanty folklore. *Λoŋ-vertĭ-imi*, "of course eats people and lives in the forest, but this is the only thing that connects her to Baba-Yaga" (interview with Nikolai Nikitich Nakhrachëv, in Siikala & Ulyashev, 2011: 110). Khanty people may use the term 'Baba Yaga' to indicate 'witch,' but at least some Khanty refer to *Λoŋ-vertĭ-imi*, their own cannibal witch, as distinct. Nonetheless, the Russian term co-exists with other words for witch in the lexi-

Fig. 2: Baba Yaga's hut on chicken legs, by Ivan Bilibin (1899). Note fly agarics in foreground. Wikimedia Commons, public domain.

cons of various Finno-Ugric peoples. For example, Limerov (2005), in discussing pre-Christian and Christian forest peoples of the Russian north, notes an entry in Lytkin and Guliaev's (1999) dictionary of the Komi language in which 'baba yaga' is synonymous with several other terms designating witch or sorceress.

There are other instances: Among the Mordvins (another Finno-Ugric group) there is a figure intriguingly similar to Baba Yaga. "In initiation rites Vir'ava ['Forest Mother'] tends to have similar functions as the Russian Baba Yaga; in general, they both look similar, being repulsive, not pleasant. Vir'ava has extremely long legs [as does Baba Yaga], or only one leg [an attribute of Baba Yaga in some sources] ... her teeth are big and sharp" (Yurchenkova, 2011: 179). Vir'ava is, like Baba Yaga, known for abduction of children, but is also a midwife (Yurchenkova, 2011). Birds are prominent among her animal familiars, as they are with Baba Yaga (Dugan, 2017; Yurchenkova, 2011). The theme of one-leggedness connects Baba Yaga with Vir'ava, and also with *A. muscaria*; Ivanova (2013) discusses at length the sources and significance of Baba Yaga's sometimes one-legged condition, but without reference to *A. muscaria*, whereas Wasson (1971b) related one-leggedness to properties of *A. muscaria*, but without mention of Baba Yaga.

According to Limerov (2005), most Finno-Ugric peoples long had forest-based economies, and came late to agriculture. Their female deities, although eventually acquiring associations with fertility, were less likely to represent agriculture, but were "the female embodiment of the forest" or "the guardian of the forest," with "supremacy over all fauna" (Limerov, 2005, pp. 110, 115). As for Baba Yaga, "Animals venerate her, and she protects the forest" (Zipes, 2013: viii). This aspect of Baba Yaga as "the mistress of the woods, birds and animals" may imply pagan associations (Ivanova, 2013: 1858), significant in the light of the rather late conversion of these northern peoples to Christianity (Limerov, 2005).

Folkloric Anastomoses

Thus, we can find Finno-Ugric witches with Baba Yaga-like attributes, and Baba Yaga has distinct Finno-Ugric or Uralic attributes. Although Baba Yaga lives in a Finno-Ugric-like legged hut, it is interesting that Uralic folktales of cannibal witches include Selkup witches living in dugouts, as is the case with the witch Tom-nänka (Tuchkova et al., 2007). The dugout was the traditional home of the pagan Chudes (another Finnic people of northwest Russia), as opposed to the log house of the increasingly Christianized Komi (Limerov, 2005). Perhaps it should not surprise, with all this merging of traditions, that we find in a contemporary animated Mansi folktale, *Pumasipa*, a Mansi cannibal witch (Fig. 3) who lives in a dugout (filled with bones of her victims), and whose regalia include (as if you couldn't guess), a fine specimen of *Amanita muscaria*!

Some caveats are essential. The concept of a cannibal witch is hardly unique to Baba Yaga (think Hänsel and Gretel of the famed Brothers Grimm tale), and the motif of cannibal shaman is wide-spread in Siberian traditions (Stépanoff, 2009). How widely the motif of a hideous cannibal witch may have diffused, and from where, is highlighted by Karazhanova's (2016) analysis of the witch Zhalmauyz Kempir in Kazakh folklore: "The comparison with Baba-Iaga was an immediate reaction of many Russian scholars and travelers who recorded tales of Zhalmauyz Kempir" (p. 18). This brings to mind the possibility of even broader and deeper context.

Fig. 3: Mansi cannibal witch with fly agaric, in *Pumasipa*. She lives in a dugout (not a hut on chicken legs). Her 'iron nose' recalls the "Iron-Nose witch" in Hungarian folklore, and ritual idols of the Mansi and Khanty peoples may have iron noses, as may Baba Yaga (Johns, 2004). Originally from Studio Pilot (USSR's first privately-owned animation studio, closed in 2010), the video is still available on a blog (www.metafilter.com/109069/17-Hours-of-Russian-Animation) or can be found by searching for "*Pumasipa*" on YouTube.

How Global? How Ancient?

The use of *Amanita muscaria* as a psychotropic agent was widespread, including (putatively) North America among the Algonquian (Ojibway) Native Americans (Navet, 1988; Wasson, 1979). Also, tales of cannibalistic witches were early documented in western North America, e.g., the hideous Kwakiutl witch Tsonoqua, who lived on a diet of little children (Feldman, 2008, summarizing reports of the ethnographer Franz Boas). Most intriguingly, highly specific cannibalistic witch motifs are common to Siberia and Native American folklore (Berezkin, 2006, citing over twenty-five ethnographic records). These motifs involve two women who go to collect grass. One who is "witchy" kills the other and intends to eat the other's children (as in the example of Tomnänka above), but the children escape, typically with instances of "obstacle flight" (where obstacles are created to obstruct the pursuing witch). Another such example, with obstacle flight, is from Samoyed tradition (collected 1853, source in Coxwell, 1925). The Samoyed version in Coxwell (1925) never specifies the witch by name, but a girl's escape from Baba Yaga via obstacle flight comprises part of the tale 'Baba Yaga' in Aleksandr Afanas'ev's famed *Russian Fairy Tales* (Afanas'ev, 1945, the Russian counterpart of Brothers Grimm). In North America, these motifs are distributed from the west to the Great Lakes (Salish to Menomini) (Berezkin, 2006). It must be noted that the occurrence of the above motifs in both Siberia and North America is on firmer ground than the use of *A. muscaria* by the Ojibway, the latter depending on the testimony of famed ethnomycologist R. Gordon Wasson's associate, Keewaydinoquay, whose reliability and reputation have been both criticized (Millman, 2015) and defended (Moses, 2016).

There is a report of religious use of the related and psychoactive/poisonous *A. pantherina* by the Ajumawi near Mt. Shasta in California (Buckskin and Benson, 2005), and the authors cite "indications" of use of *A. muscaria* by Athabascan people in northwest Canada. Also pertinent is an unpublished report of the Na-cho Nyack Dun (Big River People) of the northern Yukon previously using *A. muscaria* in a manner analogous to some Siberian usages; ingesting the mushroom and/or urine of those who had, and communing with animal spirits (Lawrence Millman, personal communication of an interview with a Big River People elder). Much less contested, of course, is the use of numerous other psychotropic plants or mushrooms throughout the Americas (e.g., the classic study of Ott, 1976a).

Assuming the accounts of Keewaydinoquay are accepted, the instances from Buckskin and Benson (2005) pertinent, and Millman's unpublished interview credible, these circumstances elicit speculation about possible diffusion of a millennia-old, circum-Arctic shaman/witch-*Amanita* motif throughout Arctic regions and southwards. Contemporary fly agaric enthusiasms, grounded in arguments philological (Allegro, 1970), philological and anthropological (Wasson, 1968), and including the highly imaginative (e.g., Ruck et al., 2007, 2011), all essentially in Eurasian contexts, might be expanded to Northern Hemisphere proportions.

Fig. 4: Baba Yaga Dancing, watercolor on card (1908) by Ivan Bilibin. Original in Russian State Museum, St. Petersburg, public domain. Note fly agarics lower right.

There are few limits once this extended path is chosen, and the true believer might find enjoyable the vistas in Puharich (1959), wherein persons in contemporary North America channel conversations with ancient Egyptian *Amanita* users ("Key to the Door of Eternity"). The chief danger of intense attachment to an *Amanita*-centric view is challenge by an equally fervent belief in the dominant role of *Psilocybe* species, mushrooms quite well documented in ethnographic contexts (e.g., Dugan, 2011). For those inclined to such perspectives, an ecumenical synthesis (McKenna, 1988) accommodates both camps and elevates psychoactive mushrooms to the driver, not just of history, but of human evolution. However, the winner, if there is one in this weighing of fungal impacts, is probably not an agaric at all, but more likely the yeast *Saccharomyces cerevisiae*, the primary agent of alcohol production worldwide (e.g., Dugan, 2009). One can be quite philosophical about all this if pondering the topic over a cold pint of pale ale.

Amanita muscaria is a striking a mushroom, and the cannibal witch is pervasive in world folklore. The melding of the two in now classic illustrations of Baba Yaga (e.g., Fig. 4) has a resonance that begs for explication. Finno-Ugric shamanistic practices and female figures in Finno-Ugric folklore represent very plausible foundations for our standard image of the Russian Baba Yaga. Other threads, deeper and less visible but worthy of further investigation, include ancient mushroom sculptures and bird-women from the Balkan Chalcolithic, Bronze Age Indo-European forest goddesses, and the pan-European association of herbs and fungi with witches (sources in Dugan, 2011, 2017). Readers are encouraged to a skeptical review of evidence, but can decide for themselves if Uralic lore on mushrooms and witches is only one facet of an all-encompassing Amanitan universe.

Acknowledgements: The author thanks Lawrence Millman and Monica Stenzel for constructive comments on the manuscript.

Chapter 8

Magical Potions:
Entheogenic Themes in Scandinavian Mythology
Steven Leto

Early on in my studies at the University of Lund in Sweden, I became fascinated by the degree to which the known forms of pre-Christian Scandinavian religion have become shrouded in mystery. As an American Swede, I was particularly intrigued by the traces of early "pagan" cultures that were reflected in Scandinavian petroglyphs, stone monuments, place names and mythological literature. In the little town where I lived, you still see streets named after Thor and Odin and shops selling jewelry from old Viking designs. But in spite of this tangible, historical presence, when my professors lectured about the great religious sites and monuments left by early Scandinavians, they openly admitted that the religious practices and concepts associated with them remain largely mysterious.

On an Autumn day, driven by my curiosity about these ancient sites, I made a trip to a nearby place known as Tofta Högar, in Northwest Skåne. As is true of many other archeological sites, its original purpose is now unclear. Recent archeological work has revealed low stone walls surrounding large open areas, possibly used for ceremonial gatherings, as well as a number of boat-shaped stone grave sites, likely built during the Bronze Age. With a light drizzle falling, I parked the car and entered the site alone, trying to open myself to new impressions that might hint at its original function.

Apart from it being a beautiful place, sheltered from the wind and with breath-taking views of the sea, I could see nothing unusual about the site. The whole area was now overgrown with grass and wildflowers, and most of the grave sites and walls were barely visible. As I walked around the site, I came upon a large, flat-topped stoned. Etched into it were several small, cup-shaped petroglyphs, known as *skålgruppristningar* (cup hole carvings). Pondering the significance of these round, concave indentations – most about an inch and a half in diameter – I took a seat on the stone and looked out over the site. A few cows grazed calmly in the field to the east.

My attention was drawn to the grassy ground where all around me stood healthy groups of *Psilocybe semilanceata,* a powerful entheogen. I picked several of the mushrooms and, on a whim, placed one, cap down, into a cup hole. The caps of the mushrooms seemed to generally correspond with the sizes and shapes of the cup holes. This serendipitous discovery prompted me to wonder if psychedelic mushrooms might have played a role in pre-Christian Scandinavian religions.

As a rule, the Swedes are mushroom lovers, and it is something of a national pastime to venture out into the forests and fields in spring and fall to harvest the great variety of fungal treasures available there. The Swedes have few taboos about mushrooms, with one notable exception, *Amanita muscaria*. Although the red-and-white-capped *A. muscaria* is well known in Scandinavia, there is a powerful and curious ambivalence regarding it in Sweden.

Known colloquially in Sweden as *flugsvamp* (literally, "fly mushroom"), the red-and-white mushroom is considered to be quite poisonous. The scientific name, *Amanita muscaria,* was coined by the renowned Swedish naturalist, Carl von Linné (*Carolus Linnæus*). In *Flora Svecica*, Linné states that he named this mushroom after the word, *musca*, Latin for "fly," because during his childhood in southern Sweden, the mushroom, chopped and mixed with milk, had been used to kill flies.[2] Although Swedish children are taught to avoid eating and even touching it, I have heard stories of it being used as a food during difficult times, with special care taken to salt and cook out its "poisons."

Despite its reputation as a dangerous species, *A. muscaria* remains very dear to the hearts of the Swedish people. Its image appears everywhere, especially during the Christmas holiday season when figurines of the red-and-white mushrooms are used to decorate Christmas trees, store windows and flower arrangements. It is also a common image at other times of the year, appearing frequently in paintings, greeting cards, children's books and elsewhere.

The reason for the ambivalence regarding *A. muscaria* may stem back to the prehistory of Eurasia. In *Soma, Divine Mushroom of Immortality*, a book that traces the origins of the Vedic entheogen, Soma, ethnomycologist R. Gordon Wasson

(1971a) provides both linguistic and historic evidence that suggests *A. muscaria* was widely known and used by the ancient Finn-Ugrian cultures. Based on European taboos against eating *A. muscaria*, Wasson suggests that it may have once been used as an entheogen by European shamans, but that its use by commoners was probably prohibited (Wasson, 1971a: 191). The fact that *A. muscaria* is one of the best-known mushrooms in Scandinavia yet is treated as poisonous certainly fits a typical taboo pattern.

In contrast to *flugsvamp*, the *Psilocybes* have not, to my knowledge, been associated with any folk traditions in Scandinavia, apart from their reputation among youths in the 1970s as an inexpensive way of becoming intoxicated. Nevertheless, my serendipitous discovery of the correspondence in size between the *P. semilanceata* caps and the still mysterious cup hole carvings at Tofta Högar continued to intrigue me.

One of the first questions I needed to answer was whether botanical conditions two or three thousand years ago could have supported the appearance of either or both of these entheogens, so I sought out Sigvard Svensson, the mycological expert at the Botanical Institution of the University of Lund, and I questioned him about the evolution of the Swedish biota. I specifically wanted to know if there was any reason to assume that the *Psilocybes* were recent arrivals in Sweden, or if they would have been native to the area during the Scandinavian Bronze Age.

Svensson assured me that climatic conditions conducive to both types of mushrooms had existed in Sweden during the Bronze Age and even long before. He also indicated that more than a half-dozen varieties of *Psilocybe* mushrooms are currently found in Sweden, with *P. semilanceata* being one of the most common. Interestingly, he pointed out that *Psilocybes,* but not *Amanitas,* are also found in Iceland.[3] Apparently, during the Viking period, Iceland's birch forests were entirely decimated, thereby removing the primary symbiotic habitat for the *Amanita*.

Having established that climatic conditions favorable to *A. muscaria* and *Psilocybes* had existed in early Sweden, I decided to see if I could find any other evidence suggesting the use of entheogenic mushrooms in the region. This chapter presents some of my more interesting discoveries along these lines of inquiry.

The Shamanic Heritage of Scandinavia

Given the prevalence of *A. muscaria* motifs in contemporary Swedish holiday celebration and popular culture, as well as the overwhelming likelihood that psychoactive mushrooms were available in ancient Scandinavia, my next question was whether there is any evidence that the Swedes made use of psychoactive mushrooms or other entheogenic plants in their religious or perhaps shamanic activities. There are very few extant firsthand accounts describing the pre-Christian belief systems of Scandinavia. Most early writers tended to assume these religions were similar to the classical polytheistic religions of Rome and Greece, with a pantheon of gods and goddesses. However, more recent scholars, including Bjorn Collinder and Jörgen Eriksson (1991), and Ronaldo Grambo (*cited* by Hoppál, 1989), have pointed out the strong shamanic character evident in the Scandinavian mythological sources. The two views are not mutually exclusive: shamanism can and in fact, often does coexist with other religious forms.

Even earlier, religious historian Mircea Eliade (1988: 379) gave several examples that seemed to reflect shamanic traits in the Scandinavian literature:

> *To acquire the occult knowledge of runes, Odin spends nine days and nights hanging in a tree. Some Germanists have seen an initiation rite in this; Otto Hefler even compares it to the initiatory tree climbing of Siberian shamans. The tree in which Odin "hanged" himself can only be the Cosmic Tree, Yggdrasil; its name, by the way, means the "steed of Ygg (Odin)." Odin's steed, Sleipner, has eight hooves, and it is he who carries his master, and even other gods (e.g., Herm Udhr), to the underworld. Now, the eight-hoofed horse is the shamanic horse par excellence; it is found among the Siberians, as well as elsewhere (e.g., the Muria), always in connection with the shaman's ecstatic experience. Describing Odin's ability to change shape at will, Snorri writes: "his body lay as though he were asleep or dead, and he then became a bird or a beast, a fish or a dragon, and went in an instant to far-off lands." This ecstatic journey of Odin in animal forms may properly be compared to the transformations of shamans into animals; for, just as the shamans fought one another in the shape of bulls or eagles, Nordic traditions present several combats between magicians in the shape of walruses or other animals; during the combat their bodies remained inanimate, just as Odin's*

did during his ecstasy.

We can see from the above example that vestiges of a shamanic cosmology and perhaps even the forms of some shamanic practices may well have been alluded to in early Scandinavian literature. Folklorist Peter Buchholz (1984: 428) from the University of South Africa writes, "The Scandinavian god Odin... has a number of traits reminiscent of shamanism. His main name, which must designate a decisive aspect of his being, signifies nothing else than the 'master of ecstasy' (O.N. Óðr)."

Along with the similarities between the Scandinavian mythological descriptions and accounts of Eurasian Shamanism in general, the rich tradition of Sami Shamanism in northern Scandinavia is worth mentioning briefly here. In *Sejd*, his study of the ancient *sejd* divination rituals of Scandinavia, Dag Strömbäck (1935) points out that there has been substantial contact between the southern Scandinavians and the northern Lapps (Sami) since ancient times. Strömbäck further argues that a close relationship exists between the Sami Shamanic séance and the *sejd* divination ritual.

Shamanism remained the dominant religious practice among the Sami long after Christians first invaded southern Scandinavia in the 1300s and even after Christianity was adopted as the state religion of Sweden in the 1700s. Missionary reports from the 1600s and 1700s provide detailed descriptions of certain aspects of Sami Shamanism. The divinatory seances of the Sami *noide* (or *noijd*) fit the classical form of ecstatic journey found in Siberian Shamanism, and the *noide*'s drum and costume were considered vital tools of the trade, as in many parts of Eurasia.

While early reports often assumed that drumming was the principal method used by the *noide* to enter into trance states, those reports came from Christian missionaries who tended to be quite hostile toward the shamanic religion of the Sami. These clerics had the power to mete out punishments of torture or death for practicing pagan rituals, so it is quite conceivable that the missionaries weren't told the whole truth about shamanic ceremonies.

In an early description of a Sami shamanic ritual from 1670, the Swedish explorer Nicolai Lundius (1670: 6) relates:

> *Shortly after the Noid had begun beating his drum, he fell dead to the ground and his body was hard as stone. During this time, the others present contin-*

ued to sing a song. After about an hour, the Noid got up and started to sing as well, as he slowly beat the drum. After a time he began to tell how he had been under the earth [author's translation].

The description of the *noide* falling to the earth and remaining inanimate for long periods of time (during which he was not to be disturbed) reveals certain similarities to accounts of shamanic divination "under the cloak" reported from the Viking Age in Sweden, Norway, and Denmark (Adalsteinsson, 1978). In the Viking ritual, the man who was to seek insights relative to a difficult and important matter would draw his cloak over his head and lie under a bench or in a dark place in total seclusion for up to twenty-four hours.

Both the Sami and the middle and southern Scandinavian forms of divination appear to have involved the shaman entering a coma-like altered state of consciousness, a common trait of *A. muscaria* inebriation among Siberian shamans using the mushroom. The fact that coma-like trance states were used in divination does not prove the use of psychoactive means, but it does suggest a context in which entheogens could have been used, as they seem to have been in a directly adjacent cultural pattern. And it is not necessary to suggest that psychoactive mushrooms were the primary ecstatic tool of the Sami *noide*, because this certainly does not appear to be the case. On the other hand, such use has been documented for example by T. I. Itkonen that *A. muscaria* was used by the Inari Sami for this purpose (Wasson, 1971a: 279). Moreover, Åke Hultkrantz has suggested that other Sami may have, in earlier times, used the fly agaric as a means to achieve ecstatic states (Bäckman & Hultkrantz, 1978: 93).

While these reports are suggestive, I have yet to find any conclusive archeological or historical evidence that the pre-Christian Swedes consumed magical mushrooms. More work remains to be done. However, I have discovered several references in the early Scandinavian literature referring to gods and heroes drinking "magical," ecstasy-inducing drinks from magical cauldrons. I began to wonder if the source of the magical nature of these drinks could have been something more than common Mead; perhaps entheogenic mushrooms?

A little later in this chapter, I will present some of what seem to be entheogenic and mushroom themes appearing in three of the oldest Scandinavian texts: the *Gylfaginning* from the *Edda* of Snorri Sturluson (1179-1241), the *Voluspá* from the *Poetic Eddan*, and the story of Lill-volvan from the *Eiriks Saga Rautha*. The written versions of these legends have been dated to around 1100-1300 A.D., but most scholars agree that they were likely based on oral traditions that would date back several hundred years earlier.

These texts were originally recorded in early Icelandic, which is significantly different from contemporary Swedish. In preparing this paper I had to rely heavily on available Swedish translations of the Old Icelandic. It is worth noting that most translators, when confronted with a range of possible meanings for obscure or complex words, naturally tend to choose those meanings that fit their own world views. Such choices are unavoidable, but they can inadvertently obscure a great deal of important information.[4]

An interesting example appears in *Lokasenna*, verse 24, from the *Poetic Eddan*, which Strömbäck (1935) discusses in *Sejd*. In the verse, Loke, describing Odin's ability to perform various feats of shamanic power, uses the phrase "*oc draptv á vétt sem valor.*" Concentrating on *draptv á vétt*, or *drepa á vétt* in present tense, we learn that *drepa á* means "is involved with, devoted to, or having an affection for," and *vétt* is often translated to mean "magic potion" or "magical means." So, *drepa á vétt* can be understood to mean, "devoted to a magical potion (or means)." Strömbäck goes on to say that the phrase refers to something of central importance to the magical act of *seið* or *sejd*, an early Scandinavian ritual considered by many scholars to have been closely related to shamanic seeing. He points out that the older Western Norwegian meaning of *vétt*, "A convex or raised lid," is the most appropriate translation. He also suggests that this raised, convex form could refer to the shaman's drum, a well-known "magical means" to ecstasy used by Sami *noides,* and he consequently translates *drepa á vétt* as, "devoted to the magical means of the drum."

It seems plausible that Strömbäck had to extrapolate a bit in order to link a convex form with a round, flat drum. Still, he may have been on the right track when he identified *vétt* as a convex form of central importance in shamanic work. I would like to suggest that *vétt* could refer to another convex magical means, the *A. Muscaria* used in various Siberian and Finno-Ugric shamanic cultures. If Strömbäck and other translators had been familiar with the use of psychoactive mushrooms as shamanic potions, I think this information might have influenced their translations. Instead of identifying *vétt* as a raised or convex form related to ecstasy and soul travel, scholars have variously translated the line from verse 24 as, "you made use of superstitions," or, "in magic you were like a female magician," (Strömbäck, 1935). As I hope to show, a more accurate translation might be, "devoted to the magical mushroom."

Gordon Wasson's Research on Soma

Before discussing my own thoughts on other mushroom-like themes in Scandinavian mythological sources, I think it may be helpful to review Gordon Wasson's (1971a) theories on the Vedic Soma tradition described in the Rig-Veda and to identify some of the poetic metaphors and associations that persuaded Wasson to propose the most likely candidate for Soma as A. *muscaria*. His research is an important precedent for this type of investigation, and it may provide ideas that could prove useful in our understanding of the ancient link between ecstatic practices and religion in early Scandinavia as well.

Soma is described in the Rig-Veda as being a plant and a drink made from that plant, as well as one of several gods surrounding Indra. When Soma is ingested, it gives rise to extraordinary feelings of "Divine Unity," and it is glowingly described as the giver of inspiration, song and poetry (Wasson, 1971a: 4). Out of a total of 1,028 hymns in the Rig-Veda, most are associated with Vedic *Agni* (Fire) and Soma rituals, and over a hundred hymns are devoted to describing and praising Soma (p. 4-5). Based on the descriptions in the Rig-Veda, it is generally assumed that Soma was a powerful and highly revered psychoactive.

The word Soma comes from the Sanskrit word *su*, which means, "to press" (1971a: 62). Along with the descriptions of the Soma ritual, the Rig-Veda texts describe the plant's ritual preparation, which involves washing and pressing it between two stones, then passing its juices through filters and into a bowl. The Soma is then combined with milk and, sometimes, honey (p. 63). The pressing of the juices creates a liquid that is at first golden or tawny yellow, but becomes white with the addition of milk. There are other texts that describe the Soma plant being chewed and then being passed on to another to swallow.

Poetic metaphors regarding Soma abound in the Rig-Veda. The sun and the colors red and golden yellow are recurring images. Soma is repeatedly associated with cattle and, in particular, with the udder. The "head" of Soma is specifically mentioned. Other very interesting metaphors include the "single eye" of Soma and the "support of the sky." Indra, as the "great Soma-drinker," frequently appears in the Soma hymns. Other names for Soma include *amrita,* (ambrosia, the liquor of immortality), "clear flowing," "the bright drop," *máda* (inebriation). And *mádhu* (honey) (1971a: 63).

Until a more convincing argument appears to suggest otherwise, I am inclined to accept Wasson's proposition that the Vedic Soma was probably *A. muscaria*. Whether *A. muscaria* or some other mushroom was used in Sweden remains to be seen. However, it would seem that Wasson's research is important when considering this possibility, because in the process of presenting his case that *A. muscaria* was the original Soma, he provides considerable linguistic evidence and ethnological reports documenting the widespread use of the psychoactive mushroom among various Eurasian tribal groups. Most importantly, Wasson's research helped establish that the effects of the *A. muscaria* are enhanced by drying or heating it, and that drinking the urine of one who has ingested the mushroom creates much the same effect as consuming them (1971a: 153).

Alternatives to Wasson's Soma Theory

Although Wasson's thesis that the Vedic Soma *A. muscaria* is persuasive, it should be remembered that other scholars have suggested various plant candidates for Soma – from *Peganum harmala* to *Cannabis sativa* (Wasson, 1971a: 114). It is also well known that various substitutes for Soma have been used in the Soma ritual since very early times. It is even possible that substitutes were already in use when the Rig-Veda was first recorded. Wendy Doniger O'Flaherty makes an interesting comment, "The statement that Soma was 'intoxicating' as it appears in various discussion and in the *Rig Veda* itself does not really exclude any plant capable of producing a state of exhilaration, including narcotic or psychotropic plants" (p. 144).

The historic use of Soma substitutes raises several questions regarding the interpretation of Soma descriptions and attributes. Even if we accept Wasson's persuasive proposal that *A. muscaria* was the original form of Soma, is it possible that the Vedic drink could have included other plant additives, such as *Psilocybe* mushrooms? Wasson cites reports that mushrooms other than *A. muscaria* were used in Siberia for shamanic purposes (pp. 305, 308-9). When the Rig-Veda describes the addition of milk and honey, could the Vedic poets be metaphorically implying the use of other psychoactive admixtures?

In one counterproposal to Wasson's theory, Terence McKenna (1992: 110-120) suggests that the most obvious candidate for Soma is another mushroom, *Stropharia cubensis* (also known as *Psilocybe cubensis*). Although McKenna devotes more time to attacking Wasson's proposal than to supporting his own, his thesis is intriguing. He essentially endorses Wasson's views regarding the prevalence of mushroom-like motifs in the Rig-Veda, but he suggests that the frequent literary links between bulls and Soma in the text may be best explained by the unique symbiotic relationship between cow manure and *S. cubensis*. He charges that Wasson ignores Vedic evidence pointing toward *S. cubensis*, while placing undue emphasis on the reports of extant *A. muscaria* usage in distant Siberia.

McKenna's views on the Soma candidates seem to have been informed by two less-than-ecstatic experiences with *A. muscaria,* contrasted to his many favorable visionary experiences with *S. cubensis*. Based on his experiences and some of Wasson's comments, McKenna argues that Wasson's Soma theory was inspired by his ecstatic experiences in Mexico with *S. cubensis*, and that he never proved that *A. muscaria* was capable of producing ecstatic visions. McKenna (1992: 110) consequently concludes that, "the rapturous visionary ecstasy that inspired the Vedas… could not possibly have been caused by *Amanita muscaria*."

Compared to the powerful visionary effects of *S. cubensis*, *A. muscaria* may not provide as dramatic an experience. However, it does not follow that *A. muscaria* cannot produce ecstatic states. There is the clear example of Wasson's (1971a: 75; also *cited* by McKenna 1992: 108-9) associate Imazeki, who had a marvelous visionary experience after consuming dried *A. muscaria*. Moreover, as we shall soon see, other psychedelic explorers claim to have experienced ecstatic visions and dreams under the influence of dried *A. muscaria* (Festi & Bianchi, 1992; Heinrich, 2002). These explorers have established that the potency of *A. muscaria* must be enhanced through drying the mushroom, and that its effects can be further enhanced by preparatory fasting and by drinking one's urine, which recycles the ibotenic acid, a psychoactive alkaloid found in the mushroom that is passed unchanged into the urine.[5]

We also know that shamanic cultures are often quite adept at enhancing entheogenic effects through highly sophisticated plant combinations. Wasson (1971a: 248) mentions a report from Georg Langsdorf that some Siberian peoples add the juice of a particular berry to *A. muscaria*, which is thought to intensify the mushroom's effect. Wasson also cites a report by A. Kannisto that mushrooms other than *A. muscaria* were sometimes used by the Voguls for shamanic purposes (pp. 308-9). Considering the latter account and McKenna's *S. cubensis* theory, it would seem logical that both ancient Vedic and Scandinavian seers and priests could have used *Psilocybes* as substitutes for, or in combination with, *A. muscaria*.

Wasson places considerable emphasis on statements in the Rig-Veda that Soma was consumed in two (or more) forms; he holds that one form involved a drink made from Soma, water, and milk, and that the other form was Soma urinated by the *Maruts* (Vedic deities). He is correct that the drinking of urine is uniquely associated with the consumption of *A. muscaria*. As Wasson noted, this practice is well documented in Siberia, and it is hinted at in the Rig-Veda.

I think Wasson (1971a: 25) may have been wrong to assume that, "the fly-agaric is unique among the psychotropic plants in one of its properties: it is an inebriant in two forms." Several other entheogens, including *Cannabis*, *Datura*, and *Tabernanthe*, to name a few, can also be ingested in multiple forms, such as smoking, eating, drinking, snorting and so forth (Schultes & Hoffman, 1992: 66-79). Moreover, Wasson's (1971a: 26) proposal does not adequately explain the passage from the Rig-Veda stating that the two forms, "stand facing us." In simple terms, the image of two forms "standing" would fit most comfortably with the idea of two mushrooms, and it seems less appropriate if the two forms were juice and urine.

While the preponderance of evidence from the Rig-Veda points to *A. muscaria*, some Soma metaphors could easily refer to mushrooms from the *Psilocybe* group. For example, the color of Soma in the Rig-Veda ranges from red to tawny yellow. *Psilocybes* can appear tawny brown or yellowish, which would certainly allow for them to be included in this description. In the Rig-Veda, the bull is a very common metaphor for Soma (Wasson, 1971a: 42), and *Psilocybes* are known to thrive in association with cattle manure, whereas *A. muscaria* typically does not.

Based on the above, I would be hard pressed to argue whether *Amanita muscaria* or the *Psilocybes* were the original or substitute form of Soma. However, in light of the reference to the two forms of Soma that, "stand facing us," I am not ready to dismiss the possibility that more than one psychoactive mushroom may have been used by the Indo-Europeans.

Identifying Magic Mushroom Motifs

Knowing that Scandinavian languages and myths reveal early Indo-European influences, and that, according to Wasson and others, psychoactive mushrooms may have inspired various Indo-European myths, I decided to follow the examples of Wasson and McKenna and to look for specific motifs and other circumstantial evidence in the early myths and sagas that might point to the use of either *A. muscaria* or *Psilocybes*.

Before attempting a discussion of metaphorical motifs, it may be useful to have a clear idea of the practical realities of known practices. For this reason, I will review a few descriptions of experiences with *A. muscaria* and *Psilocybes*, in order to establish profiles of the distinguishing characteristics and signature traits of each. Then, based on these profiles, we can see if Scandinavian legends dealing with magical drinks contain motifs that are typical of these entheogens.

As previously noted, most Western European cultures have long-standing taboos against eating *A. muscaria*, so, until recently, most of our knowledge about its effects came largely from reports of Siberian usage. Building on Wasson's research, Francesco Festi recently compiled a general profile of *A. muscaria* usage based on the reports of explorers and scientists who have studied and observed its use among Siberian populations. In addition to pointing out that drying the mushroom enhances its psychoactive potency and reduces unpleasant effects, Festi explains that typical doses can range from one to more than ten mushrooms, possibly reflecting regional differences in potency, as well as personal and cultural preferences. The first effects, which appear fifteen minutes to an hour after ingestion, include feelings of vertigo and nausea, accompanied by tremors and involuntary movements of the limbs or head and sometimes interspersed with strong soporific states. The second stage can include perceived increases or decreases in strength and agility, and, sometimes, attacks of frantic movement followed by a state of calm or depression. Festi states, "Some auditory or visual hallucinations, the latter consisting more in a splitting of objects and alterations of profiles than in colored vision, have been observed" (Festi & Bianchi, 1992: 80). In many shamanic cultures, persons who eat these mushrooms report hearing and seeing the spirits of the fungus. The third state involves a deep sleep, often including vivid lucid dreams, followed by feelings of nausea, headache, or depression upon waking. Festi emphasizes that this third, "hallucination-dream stage" may be a natural consequence of, "the precocious sleep-inducing effect of Amanita muscaria" (p. 81).

In a second part of the same article, Antonio Bianchi offers personal observations based on his firsthand experiences with *A. muscaria* (presumably dried mushrooms) (Festi & Bianchi, 1992: 82). Bianchi stresses that *A. muscaria* experiences seem to be influenced by a variety of factors, including the dosage, the relative potency of specimens, (most influenced by seasonal factors), and the practice of fasting beforehand. He also observes that, as with other psychedelics, *A. muscaria* experiences seem to be heavily influenced by personal expectations, prior experiences and the setting in which the mushroom is used.

Bianchi also identifies three stages of *A. muscaria* intoxication. The first stage involves nausea and, more rarely, vomiting. The second stage involves different kinds of narcotic effects, such as sleep with "colorful bright dreams with a particular sense of 'lucidity'" (Festi & Bianchi, 1992: 82). He notes a third stage, typical of intense inebriation, that includes dizziness, disorientation, and motor difficulties, but also evokes lucid dreamlike states involving "a peculiar kind of imagination where the thoughts [are] immediately transformed into images" (p. 82). Bianchi concludes that the most powerful quality of *A. muscaria* experience is the sense of silently talking to someone inside oneself: "It's a kind of internal dialogue where a person has the feeling of important revelation about his life…" (p. 82).

Based on personal experimentation, Clark Heinrich (2002) includes detailed descriptions of his remarkable experiences with *A. muscaria* in his book *Magic Mushrooms in Religion and Alchemy*. Heinrich argues that properly drying the mushrooms, fasting beforehand, and, perhaps most significantly, drinking one's own urine after ingesting the mushroom may be essential to producing an ecstatic experience with *A. muscaria*. Heinrich (2002: 201) relates that, by following his prescription, he and his friend Michael were both able to experience significant peak ecstasies:

> *I became aware of tremendous energy at my feet that rose up through my body in wave after wave. "Feeling good" was rapidly changing into the most blissful feelings I had ever experienced. I looked at Michael and he was radiant, truly radiant. We started laughing and exclaiming in disbelief as the bliss kept increasing. My mind and entire body were in the throes of a kind of meta-orgasm that wouldn't stop –*

not that I wanted it to.

Heinrich adds that he subsequently felt a spontaneous understanding of the "hidden wisdom" of certain Biblical texts, and he experienced a kind of wordless, psychic communication between himself and his friend. The increasing feeling of bliss culminated in sleep and a dream-vision in which Heinrich (2002: 203) experienced himself being filled with light from above that eventually absorbed and lifted him into the "unspeakable Godhead."

In contrast to that heavenly experience, Heinrich also describes an encounter with *A. muscaria* that verged on being hellish. Soon after ingesting the dried mushrooms, he began to sweat and salivate profusely, and he became convinced that he would literally drown if he did not keep swallowing quickly. He also felt intense cold, on a very hot day, and he experienced a crushing nausea that ended in vomiting and immobility. He felt that he had died and was being pressed into the ground by the weight of the universe. He heard his name being spoken, then returned to the waking world by way of a hallucination of giant ropes reaching into the sky, (which turned out to be small threads in his blanket). The rest of the afternoon, he experienced what he describes as repeatedly dying and coming back to life, interspersed with nausea and vomiting.

Heinrich's accounts confirm the profile of traits compiled by Festi and Bianchi, but it must also be recognized that some veteran psychonauts, including Terence McKenna, have had less-than-ecstatic experiences after consuming substantial quantities of dried *A. muscaria,* experiencing little more than nausea and disorientation. If Bianchi and Heinrich are correct, fasting beforehand and drinking one's urine may be vital to achieving ecstatic experiences with *A. muscaria*; such practices could explain how the Vedic priests and Siberian shamans could have achieved ecstatic states with the mushroom, while many modern psychonauts have had such bad luck with it.

In contrast to the scarcity of contemporary accounts involving *A. muscaria* experiences, descriptions of visionary experiences involving the consumption of *Psilocybe* mushrooms are almost commonplace in contemporary psychedelic literature. The literature on these mushrooms shows that, unlike *A. muscaria, Psilocybes* may be ingested either fresh or dried, without any apparent change in potency. Moreover, *Psilocybe* alkaloids are easily absorbed and consistently popular with modern users.

For the sake of comparison, I present a typical experience related by Paul Stamets (1996) in his book *Psilocybin Mushrooms of the World*. Incidentally, Stamets stresses the influence of "set and setting" when taking *Psilocybe* mushrooms, and he also recommends fasting or dieting lightly beforehand. In this experience, after ingesting over thirty fresh *Psilocybe stuntzii* mushrooms, he found the first hour to be somewhat unsettling. Then, he began to see unfolding geometric patterns surging towards him "in wave after wave of beauty and complexity" (p. 6). For the next two or three hours, until he eventually drifted off to sleep, he found himself contemplating the great matters of existence, while continuing to view grids of geometric patterns. Before waking in the morning, Stamets had a significant premonitory dream, which was later revealed to have been accurate. Stamets makes these observations about *Psilocybe* experiences in general: The senses come alive, elevated to a level far above ordinary consciousness. Vision becomes more acute. Hearing is enhanced… Many have deep religious revelations, or feel that they have been forever changed for the better" (p. 34). He suggests that at a lower dosage (one gram of dried *P. semilanceata*), people experience colorful geometric patterns and changes in auditory perception. At higher dosages (two to three grams of dried *P. semilanceata*), they experience a visual wave phenomenon in which the air appears to be in a liquid state (p. 41).

McKenna has also published a large number of descriptions of his experiences with *Psilocybes*. In *True Hallucinations*, he describes an experience he had after ingesting five grams of dried *S. cubensis* (McKenna, 1993: 217-218). After a variety of "hallucinations from behind closed eyelids," he began to see images of future events in his own life, and he experienced an unusual synchronicity between his internal state and events in the world around him. He also heard a distinct inner voice that seemed to speak from a position of greater knowledge than normally experienced. Then, he abandoned himself "to the waves of visions and vistas," before eventually drifting off to sleep.

The above descriptions are not meant to be viewed as exhaustive profiles of these mushrooms, but they do highlight some key effects associated with each. These traits can be used to identify and interpret motifs appearing in ancient literary sources. For example, both *A. muscaria* and the *Psilocybes* seem to be capable of catalyzing ecstatic experiences, with feelings of unity and bliss, but the content and quality of the experiences may be significantly different. A prevalence of literary motifs and themes involving intense salivation, significant changes in muscular-motor control, dramatic size distortions in visual perception, and a tendency to fall asleep and have lucid dream-vision states may hint at *A. muscaria* usage. In contrast, motifs of strong visual distortions and web-like hallucinations, accompanied by pronounced experiences of intuition and precognition, might be indicative of *Psilocybe* mushrooms.

Considering the above might still support *A. muscaria* as the most likely candidate for the Vedic Soma. However, in the early Scandinavian legends and sagas, there are some intriguing passages that suggest the use of both *A. muscaria* and *Psilocybes* in early Scandinavian forms of divination.

It is my own suspicion that the literary record suggests a possible transition from one entheogenic mushroom to the other.

The Bubbling Cauldron under Yggdrasil

In the *Gylfaginning*, King Gylfi is amazed by the sudden disappearance of a large tract of land in the middle of the kingdom, and he decides to make his way to Val-hall in the land of the gods to find out more about the nature of reality. Once there, he is able to ask a series of questions to which he is given fairly elaborate answers. He is told that even before the earth was created, there was a spring in *Niflheim* from which flowed the great rivers lying near the entrance to the underworld, *Hel*. It is noteworthy that this spring, the source of the rivers of *Niflheim*, is called *Hvergelmir*, which can be understood to mean "the bubbling cauldron."

After a while, the great regions of fire and ice met, producing a giant called *Ymir*, the father of all giants. Above *Ymir*, an extraordinary gigantic cow appeared, from whose udder flowed rivers of milk, which nourished *Ymir*. It is noteworthy that the motifs of the cow, the udder, and the milk flowing like rivers are also prominent in Rig-Veda references to Soma. In any case, this primeval cow created the first man, (or male god), who mated with some unnamed female relative of *Ymir*, thereby creating another male god. This male god married another of *Ymir's* female offspring, who gave birth to Odin, ("the inspired or intoxicated one"), and his two brothers, *Vili* and *Ve*. So, it can be said that this primeval cow was actually the *de facto* mother of the gods in Scandinavian mythology.

When trouble arose, as might be expected, among the original gods, Odin and his brothers killed Ymir and used his body to create the world. One aspect of this creation process was the spontaneous appearance of dwarfs from the giant's carcass. When Odin and his brothers used the skull of the giant to create the heavens, the dwarfs served handily as four pillars to hold it up. As we shall see, dwarfs played a pivotal role in the story. A little later, after the gods had managed to establish the earth realm and things were going really well, three women from giant-land made a strange appearance, an event that mysteriously managed to upset everyone. The nature of this upset in the god's camp will become clearer when we look at the *Voluspá*. But, somehow, the arrival of the women stimulated speculation among the gods as to the origin of the dwarfs.

Already, at this early stage in the creation of things, dwarfs leaped into the spotlight with such importance that sixty-five of them are mentioned by name. Significantly, the first two dwarfs named were *Muóðsognir* and *Durinn*. The name *Muóðsognir* may mean "*som suger mod i sig,*" (literally, "one who guzzles courage"). Also, the part of the name that refers to courage, *Muóð*, is related to "mead," the drink of poetry. *Durinn* probably comes from *durr*, meaning "sleep." As already noted, cycles of sleeping and waking are typical of *A. muscaria* inebriation.

It is not difficult to observe striking similarities between the shamanic themes in this Scandinavian myth and the entheogenic mushroom metaphors of Siberian shamanism. In the *Gylfaginning*, it is said that the dwarfs originally appeared spontaneously, "like maggots from the flesh of Ymir," and then lived within the earth (Sturluson, 1987b: 16). This association of maggots and Ymir brings to mind the link between flies, maggots, and *A. muscaria*.

In Scandinavian mythology, dwarfs are considered to be chthonic creatures that dwell in the earth, (or in mountains), and it may not be by chance that they pop up mysteriously and spontaneously in the stories, much like mushrooms do from time to time. In reports from the Siberian Chukchee, tiny mushroom men frequently appear in *A. muscaria* visions, often guiding the Chukchee shamans to the underworld (Wasson, 1971a: 159, 276). The Koryak speak of "little folk" who appear as small girls, only a few centimeters high, in their mushroom visions (p. 160).

Dwarfs, in fact, make recurring appearances in the traditions of many entheogenic mushroom cultures worldwide. Chthonic dwarfs are even prominent in *Psilocybe*-based shamanic traditions. In the *Popol Vuh,* one of the few pre-Columbian texts preserved from the Quiche Maya of Guatemala, there are references to dwarfs in association with entheogenic mushroom use. Wasson (1971a: 163) observes:

> *There is a striking similarity in the imaginative world of the Mexican Indians and the tribesman of Siberia: both have created a community of dwarfs who take over. In Mexico the mushrooms command. They speak through the* curandero *or shaman. He is as though not present. The mushrooms answer the questions put to them about the sick patient, about the future, about the stolen money or the missing donkey. The mushrooms take the form of duendes, to use the Spanish term; dwarfs in English. Similarly the eater of fly-agarics comes under the command of the mushrooms, and they are personified as amanita girls or amanita men, the size of the fly amanita.*

We have already noted that, in the *Gylfaginning*, four dwarfs were used to hold up Ymir's skull, or the dome of the heavens. There are numerous references in the Rig-Veda to Soma as the "mainstay" of the sky. Wasson (1971a: 47-8) lists nine examples of this in *Soma*, and he explains that the Sanskrit word for "mainstay" can be (and often is) translated as "pillar" or "fulcrum." I find it significant that in this story from Scandinavian mythology, we also see four dwarfs functioning as the "mainstay" of the heavens.

Next, King Gylfi asks where the center or holy place of the gods is located. He is told that this center is Yggdrasil (the primeval tree). Yggdrasil is the biggest and best of all trees; it is always green and it covers the whole world and the sky. A serpent lies at the roots of Yggdrasil, and an eagle that is full of knowledge sits in the branches. Four stags feed on the branches, and everywhere there is a sweet "honey-dew" that drips.

We have already seen Eliade's reference to Yggdrasil as the "world tree" of shamanism. Wasson (1971a: 214-215) goes further to suggest that this shamanic archetype, also described in the Rig-Veda, is actually Eurasian in origin:

> *Repeatedly we hear of the Food of Life, the Water of Life, the Lake of Milk that lies, ready to be tapped, near the roots of the Tree of Life. There where the Tree grows is the Navel of the Earth, The Axis of the World, the Cosmic Tree, the Pillar of the World.*

The Rig-Veda poetically praises Soma as the source of rivers. It states, "He has spilled forth, mainstay of the sky, the offered drink, he flows throughout the world" (Wasson, 1971a: 47, *citing* Rig Veda IX: 86). Another verse says, "Enter into the heart of Indra, receptacle for Soma, like rivers into the ocean thou [O Soma] who pleaseth Mitra, Varun, Vayu, supreme mainstay of heaven!" (p.48, *citing* Rig Veda IX: 108). Again, Soma was seen as the source of rivers, like the wells under Yggdrasil, and as the support of the heavens, as were the chthonic dwarfs in Scandinavian literature.

In the *Gylfaginning*, Yggdrasil has three roots; under one root is *Hvergelmir* (the bubbling cauldron), under another root is *Mimir's* well, and under the third is *Urd's* well. *Mimir*, the master of *Mimir's* well, is full of knowledge, because he drinks from the well. Odin was not allowed to drink from that well until he placed one eye in it. *Urd's* well, where the gods hold court, is so holy that whatever touches it becomes white.

Mimir's well grants knowledge to those who drink from it. *Mimir* can be translated as "wisdom and poetry." In this description of *Yggdrasil*, we see the metaphor of drinking from the waters near the tree of life to gain knowledge, wisdom, and poetry. Other Soma-like metaphors, such as honey, drops, white liquids, and the knowledge-giving drink, are also plentiful.

Odin places one eye in Mimir's well, an act most commonly understood as Odin pawning his eye for a drink, and thus he and the well both become "single-eyed" – a motif associated with Soma (Wasson, 1971a). Interestingly, the Nordic word *veði*, translated as "pawn," is also related to *vaða*, "to venture forth." When the word *vaða* is used in verse 15 to describe how Thor gets to the court of the gods, it is translated as "to wade." So, Odin placing his eye in Mimir's well could be interpreted to mean that he sets out on a journey – possibly a shamanic journey. In any case, both Mimir, the embodiment of wisdom and poetry, and Odin, the inspired and intoxicated one, regularly drink from the well.

The Rig-Veda refers to Soma as "single-eyed." It says, "Quickened by the seven minds, he [Soma] has encouraged the rivers free of grief, which have strengthened his single eye" (Wasson, 1971a: 46, *citing* Rig Veda IX: 9). And finally, Soma imbues the power of poetry. Soma is the "father of poems, Master-Poet never equaled" (p. 53, *citing* Rig Veda IX: 76). Once again, the parallels between the Soma descriptions and the descriptions of the Scandinavian "drink of poetry" are hard to miss.

The Soma-like metaphors continue to flow. Urd's well is white, like the milk mixture of the Vedic Soma. This well is home of the creator goddess – the goddess of "becoming" and the goddess of *skuld* (debt or payment). This well of white liquid is where the gods meet. The root of Yggdrasil located there also mysteriously extends into stunningly beautiful and mystical heavenly realms populated by "light-elves." The description of this heavenly world is highly reminiscent of mushroom visions.

In short, these and other Soma-like motifs appear throughout the *Gylfaginning*. For example, in Val-Hall, there is a magical boar, *Sæhrimnir*, that miraculously regenerates whenever it is eaten. This boar is literally one of the principal "foods of the gods." *Sæhrimnir* contains the word *hrimnir*, which can be understood to mean "drink of poetry." Moreover, a goat named *Heidrun* stands on the roof, with mead flowing from its udder. *Heidrun* is derived from *heiði*, a word associated with magic, which means "clear, blue" (as in sky) as well as the "drink of poetry." As we will discuss more fully later, blue has a special meaning relative to the *Psilocybe* group of mushrooms. In any case, here we understand that the gods subsisted largely on foods that were related to "the drink of poetry."

Skaldskaparmal and the "Magic Mead"

In the *Skaldskaparmal*, Ægir, a man skilled in magic, asks one of the gods, *Bragi*, where the "power of poetry" originated. *Bragi* tells of a dispute between the two groups of gods, the *Vanir* and *Æsir*. As in the *Gylfaginning*, the dispute arose when three mysterious women from giant-land appeared. I will return to this matter later on. In any case, the two groups of gods eventually made a truce that they sealed by spitting into a cauldron. From their combined saliva appeared a being named *Kvasir*. It is said that *Kvasir* was so wise that he could answer any question put to him. He traveled around the world, teaching his knowledge.

The name *Kvasir* comes from the proto-Slavic word *kvase*, which has a number of meanings. It can mean "the drink of poetry," and in some Russian languages, it is associated with the production of intoxicating beverages. Curiously, it can also refer to pressing, as does Soma. Most interestingly, it can also mean "rapid breathing," which is an occasional symptom of *A. muscaria* intoxication.

I consider it even more noteworthy that *Kvasir* was born from the saliva of the gods. *A. muscaria* inebriation is often associated with excessive salivation. As we saw earlier, Soma was sometimes prepared by chewing, as was *A. muscaria* among the Koryak. Wasson notes that in several Siberian legends, the mushroom was created from the saliva of a god. According to Koryak legend, "The Supreme Being spat upon the earth, and out of this saliva the agaric appeared" (Wasson, 1971a: 268). Here "agaric" refers to the fly-agaric, or *A. muscaria*. According to the Vasyugan people, "The power of the mushroom derives from the fact that is was created from the spittle of the God of Heaven" (p. 281).

There are other interesting *A. muscaria* motifs in the story of *Kvasir* recorded in the *Skaldskaparmal*. Two dwarfs, *Fialar* and *Galar*, kill *Kvasir*. The name *Galar* comes from a word that means "to sing or shout" and that is associated with intoxication and ecstatic trance. The dwarfs pour *Kvasir's* blood into a cauldron and into two vats. They then add honey to the blood and it becomes a "magical mead," which turns anyone who drinks it into poet or a scholar. The cauldron is named *Odreir*, and *oðroerir* literally means "puts the soul in motion" – as in soul travel – or "the one that stimulates to ecstasy." The two vats are called *Son* (*Súnn*), which means "sound," and *Bodn* (*Boðn*), which means "poetry" and "message or messenger." In short, chthonic dwarfs are associated with a process in which the saliva of the gods is transformed into the blood of *Kvasir*, highly suggestive of the blood-red color of *A. muscaria*, and placed into two vessels, which are coincidentally named after shamanic traits associated with the use of entheogenic mushrooms.

The dwarfs then lose the mead to a giant who places it inside a mountain, leaving his daughter *Gunnlod* to guard it. Odin appears and decides to retrieve the mead for the gods. In order to do so, he disguises himself and cleverly overcomes a series of obstacles. First, he drills a hole through the mountain, and then he transforms himself into a snake (a classical shamanic metamorphosis), and enters the chamber where the magical mead is kept.

Over three days and nights, Odin manages to drink all of the mead, and then he transforms himself into an eagle and begins to fly back to *Asgard*, the god's camp. The giant, seeing Odin's flight, also transforms into an eagle, and the chase is on. Just as Odin (in the form of the eagle) nears his home, where he will spit the magical mead, (again the association with saliva), into the containers provided by the gods for this purpose, he sends some of it out "backwards." The phrase, commonly assumed to mean that he defecates, has been variously translated as "the shit-poet's share" or "the rhymester's share." The Icelandic word used for backwards, *aptr* (similar to the English *aft*), does not indicate whether he urinates or defecates. In fact, according to my Icelandic informant, this word could even indicate that he somehow returned to the cave and left some of it back there, since *aptr* means something like "return," as well.

Taking the various interpretations into account, one might imagine that after drinking three huge vessels full of the mead over a three-day period and being stressed by an angry giant chasing him, Odin might well need to urinate a bit. This interpretation is further supported by the statement that this lost portion of magical mead could then be used by "anyone." It is reported that, in some Siberian societies, individuals who could not afford the price of *A. muscaria* would sometimes wait until a person who had ingested the mushrooms urinated, and they would then drink the urine (Wasson, 1971a: 257, 259, 263, 267, 269, 275, 276). It is not known how this unusual practice of recycling the urine evolved, but it has been established that most of the ibotenic acid is passed unchanged into the urine.

According to my reading of the *Skaldskaparmal*, the magical mead of the gods, the mead of poetry and wisdom, appeared from the saliva of the gods and was transformed through the blood of *Kvasir*, placed into cauldrons, and drunk by Odin. Odin transformed himself into an eagle, perhaps urinated some of the psychoactive, and then spit the rest out so that it would once more be available for drinking. This entire legend makes an uncanny metaphor for the preparation, use, and effects of *A. muscaria*.

In Siberian mythology, the birch tree, (the tree under which the *A. muscaria* grows), is the "tree of life," and the "waters of life" lie near the roots of the tree (Wasson, 1971a: 214). Again, in the Scandinavian tradition, the wells that are the source of wisdom lie at the foot of Yggdrasil. Recent archeological work has identified the remains of at least one great birch tree that was apparently used in a ritual manner during the Viking Age.

In the Rig-Veda, we find descriptions of the juice of Soma being golden yellow, then changing to white, as in Urd's well. It is squeezed into a bowl and mixed with milk (and honey), much in the same manner as the drink of poetry is prepared in the Scandinavian literature.

Odin's feats of drilling through the mountain, shapeshifting, and flying are also reminiscent of the great feats of strength associated with *A. muscaria* usage in Siberia. It is used by ordinary people among the Koryak for help in healing, divining the future, and gaining access to the upper and lower worlds, and by the Chukchee for assistance in hunting (Wasson, 1971a: 269, 274).

Most of the Scandinavian themes that we have examined thus far seem to have close parallels to both *A. muscaria* motifs and Wasson's Soma themes. Nevertheless, inspired by my discovery of *P. semilanceata* at Tofta Högar, I continued to wonder if the *Psilocybe* group of mushrooms might have played a role in Sweden, as well.

The *Voluspá* and the Blue Drink of Poetry

While the preponderance of evidence in the *Gylfaginning* and the *Skaldskaparmal* seems to point to the possible use of *A. muscaria*, as I began to look further into other Scandinavian texts, I found that *A. muscaria* themes were sometimes joined by motifs that look remarkably like metaphors for *Psilocybe* mushrooms.

The *Voluspá*, from the *Poetic Eddan* (Sturluson, 1990), opens with a description of the nine abodes of Yggdrasil below ground. In the *Voluspá*, we find many of the same traits of Yggdrasil that are mentioned in other texts, including the dripping "white dews" and Urd's well. Once again, three wise maidens are referred to in association with the "white" well of Urd, and the conflict mentioned briefly in the *Gylfaginning* is brought up again. In this account as well, the conflict that started the gods wondering about the origin of dwarfs begins shortly after the arrival of three women from giant-land, and it leads to a truce that is sealed with spitting into a cauldron, creating *Kvasir*.

There is one potentially significant difference. In the *Voluspá*, immediately after the three wise maidens are mentioned, another cryptic female, *Gullveig,* is named. In this story, *Gullveig* is the source of the conflict. Abrupt changes of subject are not unusual in Eddic poetry, and her introduction may imply nothing at all. However, my reading of the passage suggests that *Gullveig* could be viewed as a substitute or representative of the three women.

In the *Voluspá*, the first "great war" in the world is started when the gods spear *Gullveig* and burn her three times, and yet she remains alive. Who is this mysterious *Gullveig*? The name *Gullveig* is from *gull*, which means "gold," and *veigu*, which means "strong drink." So, she can be seen to be named after a golden, strong drink – an image that bears a striking resemblance to the Rig-Veda's description of Soma.

The character of *Gullveig* would make an apt metaphor for the preparation of *A. muscaria*. She is speared through and burned three times, and yet she remains active. In Siberia, the *A. muscaria* mushrooms are cut, (harvested), and "speared through" with a needle in order to be hung on a string for drying. Then, they are "burned" once by drying in the sun or by a fire, a second time by cooking, in many cases, and a third time in the fires of digestion when they are consumed, and yet the mushroom remains potent. Alternatively, perhaps the story could refer to the idea that *A. muscaria* is harvested, eaten and digested, urinated out and then digested again two or more times, yet it keeps its potency.

As the *Voluspá* continues, *Gullveig* is renamed *Heith* (*heiði*), and she travels from house to house, performing magic. A similar image of a seeress who travels from house to house will be addressed in the forthcoming discussion of Erik the Red's story. As noted previously, the word *heiði* means, "clear blue," as well as "drink of poetry." The color blue immediately brings to mind the well-known "bluing reaction" of *Psilocybes*, in which their stems turn bluish when they are picked or handled.

Scandinavian scholars have suggested that some kind of difficult transition seems to be described in this passage (For example, Ross, 1994), perhaps a conflict between two cultural traditions or even two forms of entheogenic practice. If *Gullveig* can be linked to *A. muscaria*, and *Heith* to the *Psilocybes*, the name change from Gullveing to Heith could explain what all the fuss among the gods was about. In any case, the dilemma here seems to be related to the magical mead.

I can't help but wonder if this mythic conflict may have been caused by a shortage of *A. muscaria* habitat due to the cutting of Iceland's birch forests. Under such conditions, control over access to the means of sacred ecstasy could have become an issue of concern for leaders of the conflicting cultures. It is also possible that the conflict suggests the introduction of a new type of rite, or divinatory practice, based on a new entheogenic mushroom associated with the proliferation of cattle in Scandinavia. If I am correct about the possible confrontation of two psychoactive mushroom traditions in Scandinavia, then it may be significant that the conflict was eventually resolved with the creation of the "magical mead" of the gods and the beginnings of its use in non-ritual contexts via the "mead of poetry." However the events are interpreted, appearances of entheogenic mushroom metaphors seem to abound, suggesting the use of both *A. muscaria* and *Psilocybe* mushrooms.

The Blue Mantle of Lill-Volvan

As with the *Eddan*, it is difficult to say with any certainty whether or not the stories of Erik the Red are part of a much older oral tradition, or whether they might in some way reflect actual events. It is known that Erik was from Norway and that he traveled to Iceland and then Greenland, where this particular story is said to have taken place, probably in the late 980s A.D. (Strömbäck, 1935: 49). Incidentally, it should be remembered that birch trees, a primary host of *A. muscaria*, were wiped out in Iceland during the Viking period.

The fourth chapter of the *Eiriks Saga Rautha* gives a detailed description of a woman named *Lill-volvan*, a "seeress" who travels from farm to farm predicting the future for the landowners (Strömbäck, 1935: 52). She was one of ten sisters, (all but her now dead), who did this work. Unlike earlier Scandinavian shamanistic myths, this account is filled with intriguing mushroom motifs that are highly suggestive of *Psilocybe* mushroom metaphors. The seeress is dressed in a very special way. She wears a blue cloak, jewels, and a headpiece of black lamb decorated with white cat skins, and she carries a staff. I would suggest not only that her outfit appears to be a shamanic ritual costume, but also that it serves as an entheogenic metaphor.

To begin with, Lill-volvan wears a black-and-white fur cap. It may be merely coincidental, but the cap of *P. semilanceata*, the most common and most potent of *Psilocybes* found in Scandinavia, is frequently black and white, as well as "furry" looking (Stamets, 1996: 143). The seeress also wears a blue mantle (or cloak), which immediately brings to mind the tendency of *Psilocybe* stems to turn blue when they are handled. Furthermore, she sits on a cushion of chicken feathers. When a mushroom is picked, one often sees white, downy material at its base, which is part of the mycelium (a network of fungi cells that amass under the ground, out of which the mushroom fruits). So, here we have a woman who is dressed in a manner that can be seen as a *Psilocybe* mushroom metaphor. It is stated that the seeress wears a belt with mushrooms hanging from it (Strömbäck, 1935: 52). Moreover, on her belt hangs a pouch in which she keeps a "magical substance" that she reportedly uses in order to go into trance.

As noted earlier, ecstatic coma-like trances are a hallmark of Siberian and Sami shamanism and *A. muscaria* inebriation, but they are not typically associated with *Psilocybe* experiences. Significantly, there is nothing in the Lill-volvan story that indicates that the seeress lies down during the ceremony, a feature that distinguishes this story from earlier accounts of Scandinavian ecstasy.

The account also mentions that Lill-volvan travels with a troop of fifteen young men and fifteen young women who assist her in her ceremonies. The number fifteen may be coincidental, but it is reported to take about fifteen of the small *P.semilanceata* mushrooms to induce visionary experiences (Stamets, 1996: 37-40). In any case, the account says that these young men and women were trained in certain musical arts to help this woman go into trance.

Training so many individuals to perform such a service would represent a sacrifice of important labor in a relatively small agrarian community. Merely housing and feeding a troop of thirty-one individuals would have involved a significant cost, so we can assume the ritual must have had great importance to Viking culture. Closer examination of this form of practice might help us to understand the scope and importance of the role of entheogenic divinatory practice in pre-Christian Scandinavian society.

Conclusions

We have seen that there are numerous similarities between the descriptions of Soma in the Rig-Veda and the descriptions presented here from Scandinavian mythology. We have also seen that many of the concepts and practices found in *Amanita*-based Siberian shamanism are also present in Scandinavian mythological literature.

The prominent theme of a magical drink that gives rise to ecstasy, poetry, and soul travel is certainly suggestive of psychoactive compounds. A sophisticated knowledge and skill in using botanical resources has been found among shamans and other healers throughout human history. We have no reason to believe that early Scandinavians would have been less intelligent or skillful.

Based on the climate and biota of the period, it is likely that the early Scandinavians had ample access to both types of entheogenic mushrooms discussed here. It would seem that the mere presence of these mushrooms in the environment of an agrarian society would invite some experimental use. In addition, there is solid evidence of a geographically related form of entheogenic shamanism among the Inari Sami, as well as among linguistically related Siberians. Exposure to stories and rumors of the Sami using psychoactive mushrooms to divine the future and to locate lost items could have provided southern Scandinavians with incentive to embrace the use of mushrooms.

Based on the literary evidence presented here, I believe it can be said that Scandinavian mythology reflects the use of entheogenic mushrooms. My studies so far have led me to believe that these stories may represent descriptions of two distinct entheogenic traditions that came into conflict and that eventually found it useful to merge. Most of the early motifs seem to closely fit *A. muscaria* metaphors, but some of the later motifs could also function as *Psilocybe* metaphors.

Naturally, it is difficult, if not impossible, to make concrete judgments based on comparative literary analysis alone. Still, there appear to be many descriptions in Scandinavian literature that are highly reminiscent of entheogenic mushroom usage in other parts of the world. I am continuing my research, following ethnological and archaeological evidence, with the hope that further work can bring us closer to compiling a coherent description of the appearance, forms, and status of entheogenic shamanism in pre-Christian Scandinavian religion.

Notes

1. This chapter was originally published as a featured article in the Winter 2000 issue of *Shaman's Drum* and is republished here for the first time.
2. It was once assumed that *A. muscaria* contained insecticidal properties, but it is now known that its flesh, which can smell similar to rotting meat, attracts and inebriates the flies, causing them to drown in the milk. In fact, flies often lay their eggs in wild specimens, turning them into nesting grounds for maggots. *See* Wasson (1971a: 199) for the evolution of accounts linking *A. muscaria* with flies.
3. While *Amanita muscaria* was unknown in Iceland for a long time, it was first documented occurring there in 1959 (*see* Hoffman & Ruck, Ch. 10). It is popularly known in Iceland as *berserkjasveppur*, or Berserker's Mushroom, presumably named after Samuel Ödman's hypothesis connecting the battle-fury of the Berserkers' to the effects of *Amanita muscaria*.
4. Wherever possible, I have returned to the language of the original texts to find the meanings of specific words and names. After noting the pertinent words in the English translations, I found their Icelandic counterparts in either *Eddu kvædi* (Briem & Briem, 1968) or *Edda* (Sturluson, 1987a). Then, using a cross-referencing method of verification between *Ordbog over det Norsk-Isnadske Skjaldesprog* (Egilsson, 1931) and *Islandsk-Dansk Ordbog* (Blöndal, 1924), I found their meanings and their variations. I was able to then check my versions with an Icelandic doctoral candidate in the Nordic Languages Department at the University of Lund, to be sure that my interpretations were reasonable. I chose meanings that I considered primary and relevant, but I am aware that there is often a wide range of interpretations possible for a given word. Rudolf Simek's (1954) *Dictionary of Northern Mythology* offers other variations.
5. For an overview of contemporary experimentation with *Amanita* species, *see* Ott (1993: 334-339). *See also* The Vaults of Erowid (2018): Psychoactive Amanitas, available at: https://erowid.org/plants/amanitas/amanitas.shtml

An Attempt to Explain the Battle-Fury of the Ancient Berserker Warriors through Natural History

Samuel Ödman (1784)

Many stories from the oldest chronicles of our motherland receive, without the knowledge of natural science, erroneous explanations. The risk of this practice is that scholars will inadvertently relegate these stories and histories to the myths of the dark ages. I do not believe I would be mistaken to include among them the old Nordic sagas about the ancient *Berserkers* and their legendary rage, known as *Berserka-gång (going berserk),* which have been described in such peculiar ways.

It is not my aim, nor is this the place, to comprehensively examine these historic monstrosities. A short description will suffice to lay the foundation for this present study.

These warriors were, according to the thinking of their own time, necessary tools against the advances of intruders. But they were also often feared by the Ruler whom they served, and were often viewed more as wild animals than humans. As soon as they went berserk, you could see them in such rage, like ravenous wolves, fearing neither fire nor iron, and throwing themselves into severe danger. They would rush forward to meet the most superior enemy, bite their own shields and so forth (Figure 1). If no enemies were present, they would direct their rage to lifeless things, ripping out trees, turning over rocks and in this state of ecstasy they could barely tell friend from foe. The Berserkers of King Halfdans are described in Hrólfs saga in the following manner: "*these fighters*", it says, "*were at times so overcome by rage that they could not control themselves. They killed people and animals alike, and anything that came in their way, without regard for themselves. While this rage lasted they did not ever step aside for anything, no matter what. But when the rage left them they were without power, not having even half their strength, and so weak, as if just having survived a disease. The rage would last about a day."* After this, the text continues with another enlightening example from the fighting on Samsö in the Saga of Hervarar.

The respect people had for these heroes was always mixed with a kind of secret hatred – even within paganism – and it is likely that their arrogance gave good reason for this. As the peaceful principles of Christianity began to be preached to and instilled in the barbaric Nordic people, the Berserkers soon lost all respect. Their presumed relationship with the Devil contributed to an increasingly fearful view as people thought of their skills and occupation with horror. Wars did not end, but the change in thought no longer permitted using the help of such dark forces, and consequently their knowledge died with them. From then on, no other explanation of the Berserker battle-fury was sought other than the aid and participation of unclean spirits. Professor Verelius was not the only one to call this a diabolical art.[2] This century, two disputations making the same claim have been published at Uppsala.[3] The likely foundation for this is that people all too literally accept the stories about these warriors, perhaps exaggerated, that have survived into our own time.

I am not of the opinion that these ecstatic states should be seen as exclusively caused by some special ability of the mind, as if they could, through the forces of imagination, assume such extraordinary abilities. For although we do not completely lack examples that support such a supposition, few people could suffer such consequences and still, in between the outbursts, skillfully

Fig. 1: The chess piece at left comes from a 12th century chess set known as the Lewis Chessmen, or Uig Chessmen. The chess pieces, carved from walrus ivory were discovered in 1831 on Lewis Island in the Outer Hebrides, located off the west coast of Scotland. The featured piece, a rook, is believed to represent a "Berserker" as he is depicted with "wild" eyes and is biting down on his shield. The chess pieces are currently held at the British Museum (Photo by Rob Roy, 2005: https://www.flickr.com/photos/robroy/60530196/).

maintain the arrogant pride that was part of the hated character of the Berserkers, even in times of peace. On the other hand, since there are plants providing several different substances that can cause delusions and confusion, as well as excessive rage, I am inclined to believe that the Berserkers used some similar intoxicating substance. They would have used this substance on occasion, and kept it a secret between themselves, so that the people's regard for them would not be lessened by the simplicity of the substance.

The fact that *Opium* can cause these same symptoms is known to everyone. What we can read about the people of the island Celebes [Sulawesi, an Indonesian island east of Borneo], and their use of opium to induce rage when they go to war, correlates closely to the Nordic Berserkers of old. What Kempfer tells of the rage, called *Hamuk*,[4] common in his time on Java, overwhelmingly supports the possibility that opium fueled the Berserkers, such that the testimony of Alpinus can be left out.[5] But, since no ship could supply our forefathers with this substance, and neither Allen nor Dillenius had discovered how to extract opium from the Poppies in Europe, and that it is even less likely that such attempts had happened in Sweden, even though Lindestolpe later did succeed, it is unreasonable to attribute the Berserker-fury to this intoxicating sap.

If *Atropa belladonna* were native here in the north, the example given by Mr. Gmelin the younger could give much illumination on the Berserkers.[6] The same is true about many intoxicating substances from India.[7]

A suitable substance, that would be sufficient to induce this periodic rage, could also be prepared from hemp leaves.[8] It has still not been carefully studied if the hemp that grows in our climate is similar to that of the southern one that gives the Persians, Indians, and Egyptians their *Bangve*. The intoxicating capabilities of hemp leaf have already been pointed out by Galeno, and according to Doctor Russel's observations it is still used by the Turks to mix with and strengthen their tobacco,[9] a practice similar to that of the Scythians, who, according to Herodotus would throw the seeds onto hot rocks and inhale the vapors to induce dizziness of the mind. It is also certain that this East Indian plant was not known here in the north. Because of all of this, it is not possible that it could have been used by the Berserkers.

Among our native plants that we really should consider here, many are notorious for some intoxicating property, but not to such a degree to catch my attention. Among them are *Crambe maritima* [Sea Kale], *Lolium temulentum* [Darnel], and others that would, on the one hand, require a larger dose to induce such violent result, and on the other would have been more likely to make the Berserkers incapable of accomplishing what has been said about them – because they have a numbing effect and cause sleep and unease. *Datura Stramonium* [Thornapple] does still deserve a special mention because its properties, not unknown among our doctors, have been presented as new evidence by the same Mr. Gmelin.[10]

Of all our native Swedish plants, it is my opinion that the Fly Agaric, *Agaricus muscarius*, especially solves the riddle of the Berserkers. The use of it is so commonly known in the north of Asia, that practically all of the nomadic peoples there use it to subdue their feelings and minds, and to enjoy the beast-like pleasure of not being tormented by the beneficial control of reason. Ostyaks, Samoyeds, Yukaghirs, and others, use it daily, and the Tungusic people, whose cold ice-climate cannot bring forth this mushroom, trade their reindeer – their most precious resource – for it. The dosage of this poisonous substance is from 1 to 4 mushrooms, according to size. The Ostyaks can only stand the use of one, or an infusion of three. The people of the Kamchatka Peninsula drink it with an infusion of *Epilobium*. Those who use this mushroom first become merry, and sing or shout out, and so on. Soon the brain functions are affected, and they feel like they have become big and strong. The rage increases, and along with it, uncommon strength and convulsive motions. Often it is necessary for those who are present – and sober – to guard them, so that they don't become violent towards themselves or others. The rage lasts for 12 hours, more or less, then faintness takes over and eventually complete powerlessness and sleep. Steller adds the peculiar fact that the urine of a person affected by this mushroom has the same intoxicating property, and that the Tungusic Shamans, as they use their ceremonial drums, take a big gulp of this urine to fall into the epileptic or ecstatic state belonging to their ritual. This, Mr. Georgii has reported, along with many related notes, in his *Description of the Peoples belonging to the Russian Imperial Government* (T. II pages 329, 336).

One thing in particular, that in my opinion speaks for the use of Fly Agaric in this context, is that its use is a custom coming from the part of Asia from which Odin, with his Æsir, made the famed move to our Nordic home. For even though the art of distilling subsequently invented a shortcut to this dishonoring abuse of man, and that in turn led to the mushrooms no longer being used around Danube, it

Fig 2: Woodcut reproduction of one of the Torslunda plates found on the Swedish island Öland in 1870. The plate, dated to between the 6th and 8th century CE is believed to depict Odin (left) and one of his Berserker warriors (right) (Academy Publishers, 1872: 90).

spread from there with those that moved further north and still made use of it. I find that the history of the Berserkers begins when Odin came (Figure 2), not only because it is assumed by those who have put forth all kinds of probabilities from the dark legends of old, but also because it fits well with the intentions of an intruder that with twelve furious men [the typical number of men in a Berserker war party, *see* Kershaw (2000)] could make himself feared and safe amidst a foreign people. The honor that other warriors could earn by killing a Berserker, that they viewed as a malefactor, also reveals that the custom was not native. Because this mushroom, as well as other similar substances, causes early emaciation and makes the body stiff and without skills, it would have probably strengthened Odin's principle of an early voluntary death, geronticide, walking to Valhalla, so that the honor of the proven heroes would not be obscured. At that time the wisdom of the land also demanded that a Berserker, as the foremost defender of the land, was considered invincible.

Notes

1. Originally published as "Försök at utur Naturens Historia Förklara de Nordiska gamla Kämpars Berseka-gång," in *Kungliga Vetenskaps Akademien*, Vol. 5. Stockholm. 1784. Pp. 240-247.
2. Epist. Dedicatoria Hist. Præfixa.
3. The first by Mr. Hamnell 1709. De *Magia Hyperboreorum*, states, on page 42, that it is likely that *going Berserk is caused by the devil*; the second by Mr Ramelius 1725 de *Furore Bersekio*, where he on page 24 laments that *he, no matter how much he wishes to acquit the Berserkers from involvement with the Devil, does not dare to defend them in this matter.*
4. Am. Exot. Fasc. 3. p. 649. *Opii deglutiunt bolum, quo intentionis idea exasperatur, turbatur ratio, et infrœnus, reddîtur animus, adeo ut stricto pugione, inſtar tigridum rabidarum in publicum excurrant, obvios qu?vis, sive amicos, sive inimicos trucidaturi.*
5. De Med. Aeg. p. 121.
6. Russian Travel. Vol. 3 p. 361. 15 grains (928 mg) in wine made a Persian soldier dizzy.
7. Diss. Linn Inebriantia. §. 3.
8. Alpin. l.c. p. 121. Kemph. l.c. p 645.
9. Nat. Hist. Of Aleppo. p. 83.
10. Gmelin. l.c. Tom. I. p. 43. A man that collected seeds from *Datura* at Voronezh [a city in Russia] was asked how they were used and answered that they were added to beer to double the intoxication.

Chapter 10

The Berserkers:
Odin's Warriors & the Mead of Inspiration

Mark A. Hoffman & Carl A. P. Ruck

The Berserks howled,
Gods were in their minds,
the werewolves howled,
and bit iron
                         Hornklofi
                         (early 10th century poet)

Many legendary traditions underlie the berserkers of the Vikings and ancient Germans, the warriors of Odin, who for the battle have assimilated the power of the bear or wolf and have become fierce like them. At least at times, this transformation was catalyzed by the ritual ingestion of *Amanita muscaria* mushrooms (Ödman, *see* Ch. 9). As the English phrase 'to go berserk' indicates, they are understood to have 'gone out of their minds.' Odin himself is named for this battle rage, and was sometimes seen to materialize as a bear, fighting alongside the king. Etymologically, the berserker is a "warrior with the skin (*serk*) of the bear (*bjorn*)" (Jones, 1961: 313), and they can easily be likened to other secret and sacred male warrior brotherhoods, such as the Greco-Roman Mithraists or the Mannerbünde, the initiatory societies of the ancient Germans (Kershaw, 2000). The shamanic role of the *Amanita* in Scandinavia is documented as early as the 3rd or 4th millennium in rock paintings, and it is intricately involved in the metaphoric aspects of Yggdrasill (Kaplan, 1975; Nichols, 2000).

The transformation of humans into 'spirit animals' is a profound and universal shamanic theme. Setting aside the crass interpretation of this phenomenon – the idea that humans literally transform into animals – we point out that when one 'changes' into an animal familiar, it signifies that one has ceased to be a human being. Though often for a specific purpose such as warfare, such *nahualism* (animal transformation) always carries with it distinct spiritual significance. It also signifies that one has been converted into a kind of god, since, in the context of elemental religious experience, the beast of prey represents a superior mode of existence. This bestial transmogrification is an initiatory experience that involves a profound alteration of consciousness.

In fairytales, the wolf and the bear are largely interchangeable. Both represent the 'beasts' required for the shamanic transformation and ritual. The sanctity of the wolf and bear can be sensed from the fact that in many localities, they have no name, for fear of conjuring them forth, but are called by descriptive epithets, such as 'bear' in English for its 'brown' color, or in Russian where it is simply the 'honey-eater.'

In the *Ynglinga Saga*, Sturluson described the berserkers, Odin's men, who "went without a coat of mail and were crazy like dogs or wolves, biting on their shields; they were strong like bears or bulls; they killed the people, but neither fire nor iron harmed them" (Sturluson, 1911). Because of their great strength and fearsome appearance, they are often called either giants or little trolls in the sagas. The descriptions of the onset of their madness, including the terrible sweating, the drastic increase in physical strength, as well as the inevitable exhaustion that ensues, correspond to the unique and well-documented effects of fly agaric (Heinrich, 2002; Ott, 1993). When the fit had left them, they were often so debilitated that they then became easy prey. Sturluson recorded a myth of the founding of the berserker cult in the *Heimskringla*, written prior to 1241. This preeminent authority records that, Odin, "when he sent his men to a battle or other errands...first laid his hands upon their heads and gave them *bjànak*, then they believed that all would go well" (Sturlu-

Fig. 1: A mushroom shaped memorial stone depicting the journey to the Other World. From Lärbro St. Hammers, Gotland (Davidson, 1969: 44.)

son, 1911). In his notes, Wasson astutely discovered that "*bjànak*: is commonly interpreted as *benedictio*, but it is no doubt the Scottish "bannock", from Gaelic *banagh*, an oat-cake (Cleasby, Vigfússon & Craigie, 1957); and he adds: "It could possibly be the dough of the shelf fungus"[2] (Wasson Archives). Sturluson is explicit in establishing Odin's warriors as *berserksgangr*, thus there can be no doubt that the laying on of Odin's divine hands, the sacred meal, and the sense of well-being or confidence described in this passage represent key elements of the mythological and ritual complex of the historical berserkers.

However, as with the 'long dream' of the Greek shamans, medieval Europe believed that the soul wandered abroad during sleep; that the conscious self had a double, like the Egyptian *ba,* that stayed with the body until death, often turning into an animal like a wolf or a bear, during which state, accessed via the entheogen, the possessed person might perform acts that manifested themselves in conscious reality, or that another person similarly dreaming might witness prophetically, in effect, each having the other's dream, since in the dreaming state one may access another's zoomorphic double (Lecouteux, 1999). This suggests that the berserker rage and similar episodes of lycanthropy as well as encounters with the whole host of fairies and the like were experienced in something more complex than ordinary reality (Dumézil & Hiltebeitel, 1970). Hence, it may not be appropriate to question whether the effect of fly agaric is physically enervating or stimulating, especially since King Erik officially banned the cults in 1015 and most accounts of the berserker rage are legendary or mythologized history.

Mead

The heroic drink of the Vikings and the north European Indo-Europeans was mead, a fermented honey water. Though mead could easily reach a modest alcoholic content when prepared properly, and this content could theoretically reach upward of 40% alcohol via a process of freeze-distillation, the intoxication produced by alcohol alone is not conducive to the kind of battle-ready *nahualism* for which the berserkers were known. Odin, himself a symbol of wisdom, sacrificed an eye for this mead, as it was the source of all knowledge. We can safely assume that this was no mere alcoholic beverage.

The mead of the skaldic (oral Scandinavian epic) tradition, as described in the *Prose Edda*, was made from the blood of the wise giant Kvasir, who was spawned from the saliva of the rivaling Vanir and Aesir, to which two dwarfs added the honey, clearly indicating that the honey was the sweetener for the already potent 'blood of wisdom,' and also that its origin involves the macropsia/micropsia phenomenon of the big-little people. Kvasir lent his name to the Norwegian *kvase* (Russian *kvass* 'sour drink'), which are both alcoholic,[3] and both sometimes made with fermented milk.

Fig. 2: Odin riding his eight-legged horse and carrying a drinking horn. From a memorial at Alskog Tjangvide, Gotland (Davidson, 1969: 45).

The derivation of mead from Kvasir's blood fits the identification of the main psychoactive component of this inspirational mead as *A. muscaria*. Obviously, the drink was red (not the honey yellow of mundane mead) like the tawny reddish liquid that is pressed from the rehydrated red skin of the mushroom cap; it is only Kvasir's blood, not his body, that is used. Preparing the mushroom for safe consumption so as to eliminate the adverse toxic reaction requires the thorough drying of the fresh mushroom to maximize the conversion of the ibotenic acid to the visionary muscimol, and then rehydrating it as a simple water infusion. Hot water increases the detoxification and decarboxylation initially achieved through the drying process. The 'blood' is 'pressed' from the cap and the solid body of the giant is discarded.

The involvement of the saliva of the elfin Vanir and the Aesir gods in the creation of the giant Kvasir also suggests the identification of mead as a preparation of *A. muscaria*. The mushroom was 'seeded' by droplets of saliva and blood, the same two ingredients, from the mouth of Odin/Wotan's horse as they flew through the sky in the wild hunt on the Winter solstice (Figure 2), with its obvious connotations of the ecstatic shamanic flight. In addition to the honey, as in the *Prose Edda*, milk was added to the 'bloody' potion of mead, a perpetuation of the venerable mixture that characterizes the preparation of Soma and Haoma.[4] This is, of course, the drink constantly praised for bringing divine inspiration to the Vedic poet-priests.

Another suggestive example of sacramental or magical blood in Nordic mythology can be found in the story of Sigurd, where Sigurd touches dragon's blood to his tongue, and is thereafter able to understand the language of the birds and, because of their warning, avoid treachery (Gelling & Davidson, 1969: 161). The symbol of the World Serpent and the winged fire-dragon were both established by the Viking Age, and the Scandinavian dragon, a slithering serpent, was thought to "guard over treasure in the earth" (p. 155). The 'fiery' and 'winged' associations of the *Amanita muscaria* need not be revisited here, other than to say that the metaphors are apt in the myth of Sigurd, especially when the conspicuous consumption of 'dragon's blood' is joined with the symbols of the twin birds atop a (world) tree.

Odin, like the mushroom, carried the epithet of "One-eye" because he had sacrificed an eye in exchange for Knowledge, dropping it into Mimir's well-spring of Wisdom at the base of the World Tree, Yggdrasill. This lost eye established Odin's shamanic pathways between the realms, and it was reputed to have great mystical powers. It was placed beneath his palace in Asgard as its metaphysical fundament. Also, like other anthro- and dio-pomorphisms of the *muscaria* sacrament, Odin wore a wide-brimmed cap. Nor should we forget that Odin brought to his citadel a psychoactive Drink of Immortality, the magical mead, made from the spittle of the gods and the blood of the knowledgeable giant, Kvasir (Figure 3).

Similarly, Odin metamorphosed himself into a serpent in order to slip into Suttung's mountain to steal away the giant's mead. Odin stole the three cauldrons of mead by drinking all three (three 'cauldron' mushroom-caps being the traditional dosage reported from Siberia, *see* Wasson, 1968), but as he fled, to elude his pursuer he had to release the contents of one of them by urinating, to lessen his load.

Fig. 3: An eagle and woman with drinking horn. The eagle is thought to be Odin, either returning with or being offered the Mead of Inspiration. From a memorial at Lärbro St. Hammers, Gotland (Davidson, 1969: 45).

This metabolite in the urine was considered the source of bad poetry, hence still psychoactive and an indication that the intoxicant mead is involved in the symbolism of the fly agaric.

Wasson and the Berserker Controversy

Writing in 1965, Wasson addressed the Berserker controversy thus:

> I should be only too delighted to accept the belief [that berserker rage was brought on by *A. muscaria*], as it supports my theory of the importance of mushrooms in Eurasian pre-historic cultures. Furthermore, I am unable to prove that the Vikings did not eat *A. muscaria*. But since mushrooms are nowhere mentioned in the sagas, and since there is no mention of an unidentified plant that caused the Vikings to rage, and since in the remotest valleys of Norway I can find no tradition of a mysterious mushroom, I am reluctantly obliged to abandon the story as a latter-day yarn (letter to Miss Fay Hall, editor's secretary, *The Times Literary Supplement*, July 19, 1965).

Wasson never found adequate reason to change his above expressed opinion. In *Persephone's Quest*, published the year of his death, Wasson (1986: 74) wrote: "There is not a shred of historical evidence that the 'rages' were provoked by Soma." He continues: "Nowhere in the Soma world, neither in the New nor the Old World, is there support for the notion that it made warriors better fighters. On the contrary, the evidence shows that it is a pacifying agent." He concludes, "Ödman must have been a mycophobe – all his compatriots were – and he knew nothing about mushrooms."

Though many bioassays and much research has since been published that disproves these rather sweeping statements about the effects of his beloved mushroom, Wasson himself had collected in his files a surprising amount of detailed research that might have satisfied other scholars that Ödman's original conclusion was more than tenable. What follows is from an anonymous, five-page typed manuscript in the Wasson Archives (Harvard Botanical Museums), written by a resident of Iceland after 1959, when *A. muscaria* was finally documented as occurring there:

> A pair of unidentified mushrooms known as the

> *reiðikúl*a ('red- or rage-ball,' 'red' being spelled with a y, though the alternative spelling is equally likely; perhaps the ambiguity was even intended) and *bleikjukúla* ('pink-ball') may be the Icelandic common names – of few recorded in this fungus-indifferent country – that hold the key to demonstrating a connection between *A. muscaria* and the berserkers. In the 1630's Bishop Gísli Oddson identified the later as 'less edible' and 'possibly poisonous,' it's name (*'bleikur'*) indicating that it is yellowish or reddish. At the end of the 17th Century, Björn Halldórssen seems to have misidentified both, claiming that the *bleikjukúl*a was *Boletus pollidus*, a choice culinary and aromatic variety. More significantly, Halldórssen clearly misidentified the red- or 'rage-ball' as *Boletus lutes*, which is not red (as both he and it's *reiðikúla* folk name claim) but yellow, and which only grows with imported conifers.

Thus, these two intriguing mushrooms remain unidentified, though the all too conspicuous – and present – fly agaric remained without a designation. The *Amanita muscaria* is now known in Icelandic as the 'berserk-mushroom' (*berserkjasveppur*), although this may derive from the debate spurred by Ödman's claim.

Wasson had also collected passages from the old Sagas that explicitly describe the sacramental cakes given by Odin to his berserkers before battle, several occasions of lycanthropy, and the berserker's entheogenic condition (i.e. that 'gods were in their minds') – all of which indicate a sacrament beyond simple mead, a theory that Wasson never espoused. Indeed, contemporary scholars including Robert Graves, Dr. Howard Fabing (1956), and others followed the insightful early work of Ödman (1784) and Schubeler (1786), accepting and contributing to the theory (letter to Mr. Lars O. Berglund, October 18, 1974, Wasson Archives).

Wasson and other critics had little or no knowledge of the specific preparations that would minimize the toxic effects of the fly agaric, effects that were often cited as evidence that Ödman must have been mistaken in his identification. In addition, the frenzy of 'be-mushroomed' Siberians was conveniently overlooked, as was, more excusably, the often-determinative role of subjective experience and expectation upon entheogenic inebriation. Wasson also did not consider the initiatory role of entheogens in cementing the bonds of battle brotherhoods, as was the case, for example, with hashish among the Assassins, and with the *muscaria* itself in Mithraism (Ruck, Hoffman & Celdrán, 2011).

The fact that Wasson found no tradition of a 'mysterious mushroom' begs the question as to whether the mushroom in question is not mysterious at all, but a celebrated and (in)famous icon – the fly agaric. As in Germany, the knowledge of this mushroom's inebriating effects and former religious role was all but lost to the popular imagination, though traditions and images of the mushroom have survived and thrived in Scandinavia, being a motif that is "seen everywhere" (letter from G.L. Phillips, December 28, 1959; *see also* Arizona Daily, 1960) – a fact directly traceable to its once unique and august place in the natural – and entheogenic – world. Wasson should have perhaps asked himself if the sweeping acceptance of Ödman's theory by the Scandinavians didn't itself reflect some evidence that the intoxicating, stimulating or 'poisonous' properties of the fly agaric were widely known, if somewhat unconsciously by this Germanic population.

Notes

1. This chapter is a condensed version of "Hunting the Berserkers", an unpublished article that has been in limited electronic circulation since 2002.
2. Wasson abbreviated S.F., which probably refers to the shelf fungus (*Fomes formentarius*), used as tinder for fire, hence called 'tinder polypore'; in folklore, it was seen as a seat for the little people, hence also called 'dryad's saddle,' for the wee folk of the sacred oak tree or *drys*, and the metaphor of 'saddle' obviously associates it with the 'steed' for the shamanic journey. It is part of the metaphoric complex of the Cosmic Tree and the firestick and the making of fire. Although not psychoactive, its pulpy texture makes it a kind of magical bread.
3. Kvass is alcoholic beer-like home brew with an acidic taste, made by pouring warm water over a variety of grains and allowing it to ferment.
4. Additional scholarship has convincingly argued for the identification of *A. muscaria* as the Celtic/Germanic 'Soma' plant, uncovering the close Indo-European mythological connections between the pantheons and sacraments of India and Europe (Wilson, 1999).

Chapter 11

Speckled Snake, Brother of Birch:
Amanita Muscaria Motifs in Celtic Legends

Erynn Rowan Laurie & Timothy White

References to magical brews and foods abound in Celtic legends dealing with journeys to *Tir Tairngire* (Land of Promise) or into the *sidhe* (faery mounds). In the Welsh *Hanes Taliesin*, the young Gwion Bach imbibes three drops of magical brew simmering in Cerridwen's cauldron; he is immediately gifted with inspiration, and then he is launched on a magical journey that entails shapeshifting into various animal forms, being eaten and rebirthed by Cerridwen, and then being set adrift in a dark skin bag on an endless sea for forty years. In the Irish *Adventures of Cormac*, Manannan, king of the Land of Promise, gives Cormac a magical, sleep-inducing silver branch with three golden apples and, before long, Cormac travels to the otherworld where he discovers a marvelous fountain containing salmon, hazelnuts, and the waters of knowledge. Considering that the old Celtic legends of Ireland and Wales are filled with motifs of sleep-inducing apples, berries of immortality, and hazelnuts of wisdom, it is remarkable that Celtic scholars have largely ignored the possible shamanic use of psychoactives and entheogens in the British Isles.[2]

There are several sound reasons why Celtic scholars have feared to tread where amateurs now dare to venture. First, due to the prohibition on writing that surrounded the ancient Celtic druids and Irish *filidh* (poet-seers), we know few specifics about the religious practices of the ancient Celts. Second, there are no direct references in the early histories to the Celts using psychoactives other than meads and wines in their ceremonial rituals and practices. Third, there is no irrefutable archaeological evidence – such as the discovery of an archaic medicine bag filled with psychoactive mushrooms – to prove the Celts actually used psychotropic substances capable of inducing ecstatic, visionary experiences.

Nevertheless, the abundance of Celtic legends about crimson foods which induce mystical experiences, inspire extraordinary knowledge, and impart the gift of prophecy, is highly suggestive. To our knowledge, no one has adequately explained why apples, berries, hazelnuts, and salmon were selected by the *filidh* as magical foods, or why they were associated with otherworldly journeys, and with the training of poets. None of these foods are inherently psychotropic.[3]

Even if one assumes that the frequent Celtic literary references to magical brews of knowledge indicate that the Celts utilized some type of psychotropic substance, several questions remain – most notably, exactly what was used and how was it used? Given the paucity of reliable information on Celtic religious practices, the answers to these questions may remain forever speculative. However, the absence of direct evidence is not proof that evidence is nonexistent.

The Celtic druids and bards had a definite penchant for poetic metaphors – for always speaking in "riddles and dark sayings," as the Roman historian Diogenes Laertius observed (Piggot, 1987: 117). It can be assumed that if the druids and *filidh* did use a psychotropic substance to access knowledge, healing, and wisdom, they would have carefully protected its identity from Roman invaders and Christian missionaries. We contend that the motifs of magical foods can best be explained as metaphoric references to *Amanita muscaria*, the highly valued, red-capped mushroom that was once used shamanically throughout much of northern Eurasia (see Saar, 1991a; and Wasson, 1967).

Wasson's Findings on Celtic Toadstools

One reason the possible role of a psychoactive mushroom in Celtic mythology has been overlooked is that *A. muscaria* is difficult to find in Ireland today. *A. muscaria* grows only in a symbiotic, mycorrhizal relationship with the roots of birch, spruce, and some conifers – and Ireland has been almost totally deforested over the last thousand years. However, there were once great forests of birch and pine in Ireland, so the red-capped mushroom could easily have grown there, as it still does in the forests of England and Scotland, and on the Isle of Man (located between Ireland and England).[4] Furthermore, even if *A. muscaria* never grew in Ireland, the *filidh* could have easily obtained supplies of dried mushrooms from their Celtic neighbors.

The mere availability of *A. muscaria* does not prove its use, however. Even ethnomycologist R. Gordon Wasson – the most enthusiastic proponent of the theory that *A. muscaria* was used by the ancient Indo-European peoples – once admitted that, in all his research, he had found little evidence suggesting the shamanic use of fly-agaric (*A. muscaria*) among the Celts, Germans, or Anglo-Saxons. He stated explicitly that he could find no direct evidence that psychoactive mushrooms had been used either by the "shadowy Druids," or medieval witches (see Wasson, 1968: 172-203).

Despite the lack of hard evidence, Wasson never totally dismissed the possibility of *A. muscaria* use in Europe. Based on his studies into why most European languages are filled with mycophobic references toward mushrooms in general and fly-agaric in particular, Wasson arrived at a very interesting conclusion:

> *I suggest that the 'toadstool' was originally the fly-agaric in the Celtic world; that the 'toadstool' in its shamanic role had aroused such awe and fear and adoration that it came under a powerful tabu, perhaps like the Vogul tabu where the shamans and their apprentices alone could eat it and others did so only under pain of death...This tabu was a pagan injunction belonging to the Celtic world. The shamanic use of the fly-agaric disappeared in time, perhaps long before the Christian dispensation. But in any case the fly-agaric could expect no quarter from the missionaries, for whom toad and toadstool were alike the Enemy.* (Wasson 1968: 191)

The absence of evidence led Wasson to conclude that Indo-European usage of the sacred mushroom may have disappeared early during their migrations into Europe. He hypothesized that, as the Indo-Aryans migrated into warm, dry climates, they were forced to adopt various local psychoactive plants as substitutes for *A. muscaria*. Although historical evidence in India, Turkey, and the Mediterranean may support his theory, the proto-Celts would not have needed to find substitutes for *A. muscaria* in their new homelands – the red-capped mushroom flourished throughout much of northwestern Europe.

Did the Irish Practice a *Soma* Cult?

Peter Lamborn Wilson suggests that – in light of the "well-known affinity between Celtic and Vedic cultures," and the fact that "entheogenic cults can thrive under the very nose of 'civilization' and not be noticed" – it should be considered whether the Irish may have once had a "*soma* cult" (Wilson, 1995: 43). Although Wilson seems reluctant to draw definitive conclusions, he argues that if the Irish did use soma, the evidence should be encoded in early Irish literature and folklore. "I think we can take for granted," he states, "that whatever we find in Ireland that looks like soma, and smells like soma, so to speak, might very well be soma, although we may never be able to prove the identity."

Although Wilson does not conclusively identify the Vedic soma, he seems to accept R. Gordon Wasson's theory that it was probably *A. muscaria* or – if not that – another psychoactive mushroom.[5] Whatever its source, soma was clearly an ecstasy-inducing drink once used by the ancestors of India's Vedic priests, who recorded hundreds of hymns praising its miraculous powers in the 3,500-year-old *Rig Veda*, the oldest extant Indo-European text. Utilizing Wasson's research on soma and *A. muscaria*, Wilson focuses primarily on identifying soma motifs – such as one-eyed, one-legged beings – that also appear in Celtic mythology.[6] Wilson suggests that the Greek legends of one-eyed, one-legged Hyperboreans may be connected to the Irish legends of the Fomorians (the mythic primordial inhabitants of Ireland), who are sometimes depicted as one-eyed, one-legged giants.

The theme of one eye, arm, and leg certainly appears prominently in several Celtic legends about the Fomorians, but the most fascinating reference occurs in the *Second Battle of Magh Tuired*, when the Irish sky god Lugh performs a curious shamanic ritual. During the battle, Lugh adopts a strange posture, standing on one leg, one arm behind his back, and closing an eye in order to cast spells on his opponents, the Fomorians. Working magic in this posture is called *corrguinecht* or "crane sorcery," and Lugh's practice of corrguinecht is a clear indication of his shamanic associations.[7] Lugh is well renowned as a shamanic magician who used his magical weapons and spells to win battles. As a deity associated with thunderbolts and magic healing as well, Lugh may also qualify as a god of *A. muscaria* – suggesting a possible link between shamanism and *A. muscaria* in early Irish legends.[8]

As Wilson ultimately admits, the mere existence of soma motifs in Celtic literature does not prove the use of soma by the insular Celts. It is possible, given their conservative nature, that they preserved soma motifs in their myths without actually continuing the use of soma – just as Christians still cherish many ancient pagan religious symbols, such as Yule logs and decorated trees at Christmas, and fertility bunnies and eggs at Easter, without understanding their original pagan context.

While we believe that Wilson is essentially correct in his identification of soma-like motifs in Celtic literature, our quest into the roots of Celtic religion has further convinced us that Celtic legends dealing with foods of knowledge point directly to the use of *A. muscaria* in Celtic shamanism. Of course, even if we can demonstrate the presence of psychoactive mushroom metaphors and motifs in Celtic legends, that still does not prove that the druids or *filidh* used the red-capped mushroom. As in the case of Wilson's soma motifs, the veiled references to *A. muscaria* could theoretically be faded memories of earlier pre-migration Indo-European practices, preserved in oral legends passed down from generation to generation.

Nevertheless, the Celtic legends contain certain innovative occult symbols for *A. muscaria* that are fairly unique to the insular Celts. We contend that these new symbols would not have been needed unless a sacred mushroom cult was still being practiced there and the druids wanted to communicate teachings about the sacrament's use to initiates, while maintaining a protective veil of secrecy around its identification.

Dreams of Paradise

The first hint that the Celts may have used *A. muscaria* can be found in the Irish descriptions of the beautiful, magical Land of Promise and the *sidhe* realms of the Tuatha de Danaan, the old Celtic gods of Ireland. Celtic otherworlds are almost always exquisitely beautiful places endowed with many attributes typical of psychotropic experiences. Brilliant colors abound, and humans and animals shift from shape to shape. Time and space are typically distorted, faery music is often heard on the wind, and foods tend to taste particularly delicious. Some of these otherworld motifs could theoretically have been inspired by various psychotropic plants, by other forms of spiritual journeying, or even by hunger-induced hallucinations. However, when considered as a whole, the Celtic legends paint pictures that look remarkably similar to dream-visions experienced under the influence of *A. muscaria*.

Consider the following description of the Land of Promise in *The Adventures of Art MacConn*. In the middle of the story, the father, Conn, embarks in a magical, oarless coracle (skin boat) that takes him wandering over the sea for a month and a fortnight until he comes to a fair, strange isle:

> *And it was thus the island was: having fair fragrant apple-trees, and many wells of wine most beautiful, and a fair bright wood adorned with clustering hazel trees surrounding those wells, with lovely golden-yellow nuts, and little bees ever beautiful humming over the fruits, which were dropping their blossoms and their leaves into the wells. Then he saw nearby a shapely hostel thatched with bird's wings, white, and yellow, and blue. And he went up to the hostel. 'Tis thus it was: with doorposts of bronze and doors of crystal, and a few generous inhabitants within. He saw the queen with her large eyes, whose name was Rigru Rosclethan, daughter of Lodan from the Land of Promise...* (Cross & Slover, 1988)

Now compare the above scene to a description of an *A. muscaria* experience translated by Wasson (1968) from the journal of Joseph Kopec, a Polish brigadier who tried the mushrooms while visiting Russia's Kamchatka Peninsula in 1797. Once, while very ill with a fever, Kopec sought medical help from a local Russian Orthodox priest, who recommended that he take some "miraculous mushrooms." Because Kopec's description of his dream-visions is fairly typical of accounts of *A. muscaria* experiences, it is worth quoting here:

> *I ate half my medicine and at once stretched out, for a deep sleep overtook me. Dreams came one after the other. I found myself as though magnetized by the most attractive gardens where only pleasure and beauty seemed to rule. Flowers of different colors and shapes and odors appeared before my eyes; a group of most beautiful women dressed in white going to and fro seemed to be occupied with the hospitality of this earthly paradise. As if pleased with my coming, they offered me different fruits, berries, and flowers. This delight lasted during my whole sleep, which was a couple of hours longer than my usual rest. After having awakened from such a sweet dream, I discovered that this delight was an illusion.* (Wasson, 1968: 244)

Delighted by the results of his first experience, Kopec took an additional dose of dried mushrooms and had a series of new visions, which he unfortunately did not describe. He did, however, volunteer some intriguing observations about their nature:

> *I can only mention that from the period when I was first aware of the notions of life, all that I had seen in front of me from my fifth or sixth year, all objects and people that I knew as time went on, and with whom I had some relations, all my games, occupations, actions, one following the other, day after day, year after year, in one word the picture of my whole past became present in my sight. Concerning the future, different pictures followed each other which will not occupy a special place here since they are dreams. I should add only that as if inspired by magnetism I came across some blunders of my evangelist* [the priest] *and I warned him to improve in those matters, and I noticed that he took these warnings almost as the voice of Revelation.* (Wasson, 1968: 245)

The parallels between Kopec's *A. muscaria* dream-visions and the chronicles of Celtic journeys to the Land of Promise are noteworthy. Kopec visits a land "where only pleasure and beauty seemed to rule," encounters beautiful women dressed in white, and comes back with visionary insights – not unlike the gift of inspired sight found frequently in Irish myths. By themselves, such parallels might seem to be coincidental and inconsequential. After all, beautiful people and magical objects are the building blocks of many myths and legends. However, as we shall soon show, Celtic myths of the otherworld are filled with motifs of magical, wisdom-inducing foods and brews that closely parallel what we know about the use of red-capped mushrooms in Siberian shamanism. But first, let us see if there is any historical evidence that could have involved the use of *A. muscaria*.

Traces of Celtic Shamanism

So little is known about the spiritual practices of the druids that some scholars have questioned whether it is appropriate to even speak of Celtic shamanism per se. However, based on comments scattered throughout the early records of Roman historians as well as later accounts recorded by Christian monks, we can conclude that the druids performed shamanic functions comparable to those performed by

Siberian shamans.[9] Celtic legends mention that the druids practiced battle magic, invoked storms, conducted healings, used enchantments to put crowds of people to sleep, and performed oracles to predict the future.[10] We also know that the *filidh* were not only inspired poets but also visionary prophets, healers, and workers of magic.[11]

Working knowledge of druidic shamanic practices may have vanished with the druids, but Irish histories and commentaries have preserved many short descriptions and notes about the divinatory practices of the *filidh*. Through statements made in the tenth-century book *Cormac's Glossary* and elsewhere, we know that the pre-Christian druids and *filidh* practiced three oracles, at least one of which could be considered shamanic: *imbas forosnai*, which can be translated as "manifestation that enlightens" or "kindling of poetic frenzy;" *teinm laída*, or "illumination of song;" and *dichetal do chennaib*, or "extempore incantation."

Nora Chadwick (1935: 97-135) has compiled an informative study of the many historical references to these three methods of divination. Unfortunately, the extant historical notes are usually brief and occasionally contradictory, and they deal primarily with the external forms of the oracles, so we can only speculate on how these divinatory practices actually worked. Nevertheless, because documentation is available on these oracles, any evidence linking them to the use of *A. muscaria* would add historical flesh and bones to the *A. muscaria* metaphors found in Gaelic legends.

According to *Cormac's Glossary*, the *imbas forosnai* ritual involved chewing a substance described as the "red flesh" of a pig, cat, or dog; chanting incantations; and invoking and making offerings to idols of the gods.[12] After this the *fili* (singular of *filidh*) covered his cheeks with his palms or went to sleep in a dark place for a three- or nine-day period of incubatory sleep called a *nómaide*. During that time, several other *filidh* usually stood watch to make sure that the sleeping *fili* was not disturbed and did not move. The seer was expected to experience visions of the gods and the future, and to receive answers to questions being asked. This oracle would qualify as a shamanic ritual under the most stringent definitions of shamanism.

None of the extant accounts of *imbas forosnai* adequately explain how the divinatory visions were induced, but they all indicate that the ritual involved eating "red" flesh and being confined in darkness. Perhaps the *filidh* were natural psychics or lucid dreamers, and chewing the red flesh was merely incidental to the ritual. However, if they were chewing on pieces of dried red-capped mushrooms, that would explain how the ritual induced prophetic dreams. As Wasson (1968: 244) and Saar (1991a) note, *A. muscaria* is often used in Siberian shamanism for the incubation of prophetic dreams. The idea that the red flesh used in the *imbas forosnai* ritual could be a veiled reference to *A. muscaria* may seem farfetched at this point, but it should make sense after we examine other motifs of magical crimson foods found in Celtic legends.

Accounts of the other two divinatory traditions – *teinm laída* and *dichetal do chennaib* – are less consistent, perhaps because those practices were less formal and could be conducted extemporaneously, without specific ceremony.[13] *Dichetal do chennaib* has been translated variously as "extempore recital," "incantation from the ends (of the fingers)," and "inspired incantation." It appears to have involved the recitation of *dicetla* (spells) or verses in order to find the answers to the questions posed. This was the one form of divination that Saint Patrick tolerated, reportedly because it did not involve the invocation of pagan deities.

The varied accounts of *teinm laída* suggest it involved the chanting of intuitive images received through the psychometric reading of objects. In one of the Fionn stories, the hero Fionn is asked to identify a headless body. Fionn puts his thumb into his mouth and uses a repetitive chant – referred to as *teinm laida* – divine that the body belongs to Lomna, his fool (Figure 1). Interestingly, Fionn's ability to achieve poetic insight by sucking or chewing on his thumb harks back to his childhood consumption of a magical red and white speckled salmon, and as we will show later, the salmon may be a metaphor for *A. muscaria*.

Fig. 1: Fionn-like figures with thumbs in mouths and clerics with ceremonial staves, on church stele from Drumhallagh, County Donegal, Ireland, circa 10th century AD. (T. White).

Other references also suggest metaphorical links between *teinm laida* and *A. muscaria*. As Joseph Nagy (1985: 137) points out, the word *teinm* means "cracking or chewing of the pith," and this word is found in the phrase *teinm cnó*, to crack open a nut; thus *teinm laida* can be translated literally as "the chewing (or breaking open) of the pith (or nut)." Chewing the nut could conceivably refer to mulling over poetic images, but if crimson hazelnuts are *A. muscaria* metaphors (as we hope to show), then the *teinm laida* could have been inspired by chewing the red-and-white mushroom.

Vague references to chewing red meat or nuts are hardly conclusive evidence of an underground Irish mushroom cult, but they do suggest that the Irish seers were chewing something "red." In light of the many Celtic legends about magical red foods – red berries, crimson nuts, and apples – which inspired the gift of insight and induce prophetic visions, we do not think the red flesh used in the *imbas forosnai* was incidental. We also do not think it is coincidental that all these red foods happen to exhibit traits reminiscent of *A. muscaria*.

Assuming that the red-capped mushrooms were used in the *imbas forosnai* ritual to induce prophetic dreams, the purpose of covering the eyes and retreating into a dark environment could easily be explained. *A. muscaria* intoxication can cause such a pronounced visual sensitivity to light that the light of a single candle can hurt the eyes. Since the shamanic use of *A. muscaria* has tended to rely on dream-visions rather than waking journeys, the darkness would also have helped secure the trance-sleep necessary to gain prophetic visions.

The Red Berries of Immortality

Various legends mention that the Tuatha de Danann ate magical red rowan berries which had the properties of preserving immortality, returning youth, and offering the gift of healing to those who consumed them.[14] In the medieval Irish tale "The Pursuit of Diarmuid and Grainne," we are told that one of the magical berries fell from the table at one of these feasts and grew into a tree. This tree is guarded by a giant who refuses to allow any mortal access to the berries (could this be a literary relic of a Celtic taboo against commoners using *A. muscaria*?).

Diarmuid's description certainly supports the *A. muscaria* metaphor: "In all the berries that grow upon that tree there are many virtues, that is, there is in every berry of them the exhilaration of wine and the satisfying of old mead; and whoever should eat *three* berries of that tree, had he completed a hundred years he would return to the age of thirty years" (Cross & Slover, 1988). Although no known psychoactive reverses aging so dramatically, many Siberian tribes, such as the Koryak, consider *A. muscaria* brews to be rejuvenating, and they say that in moderate dosages – approximately *three* mushrooms – *A. muscaria* produces a mild inebriation comparable to drinking wine or beer.

The Voyage of Maelduin provides a description of some other magical berries found on a tree on an otherworldly island. They are as large as an apple but have a tough rind. When their juice is squeezed out and consumed by Maelduin, he falls into a deep intoxication and sleeps for an entire day. His companions cannot tell whether he is dead or alive, but when he awakens he tells them to gather as much of the fruit as they can, for the intoxication that it produces is wondrous.

Maelduin's magical berries sound suspiciously like *soma* and the red-capped *A. muscaria*. The *Rig Veda* describes *soma* as being pressed out as a juice, and *A. muscaria*, when consumed in large doses, can result in an intoxicating sleep associated with wondrous visions. Assuming Celtic druids wanted to maintain a veil of secrecy around their use of *A. muscaria*, the image of large, red rowan berries would make a fair substitute for red-capped mushrooms. There are other reasons besides their shape and color why rowan berries would make a useful metaphor for *A. muscaria*.

Throughout northern Europe, the red rowan (*Sorbus aucuparia*) commonly grows in association with the birch (*Betula sp.*), one of the primary hosts of *A. muscaria* mycelia. In many parts of the Eurasian tundra – the primary habitat of *A. muscaria* – the fruiting season of the mushroom follows the early fall rains and coincides fairly closely with the peak of the berry season. Thus, forests in which red rowan berries grow would make excellent places to look for red-capped mushrooms.

There is another less obvious yet vital connection between berries and *A. muscaria*. Many Siberian cultures drink psychoactive brews made from *A. muscaria* mixed with berries. The Khanti believe that mixing *A. muscaria* with bog bilberry (*Vacciunum uliginosum*) strengthens the effect of the mushroom.[15] If adding acidic berries helps make the *A. muscaria* brew not only more palatable but more psychoactive, then large red rowan berries could have made an instructive metaphor for the sacred mushroom.

Journeys to the Land of Apples

Diarmuid's descriptions of magical red berries "as large as apples" may help explain the frequent association between magical apples and the Celtic otherworld. Apples are so commonly associated with the Land of Promise, ruled over by Manannan Mac Lir, lord of the mists, that his kingdom is sometimes called *Emain Ablach*, the Land of Apples.

Ross (1967: 269) quotes a passage in *The Sickbed of Cú Chulainn*, translated by Myles Dillon, where the motif of magical apple trees appears: "*There are at the great eastern door / Three trees of crimson crystal, / From which sings the bird-flock enduring, gentle / To the youth from out the royal rath.*"

In the twelfth-century Irish collection of tales known as *Táin Bó Cualigne*, a figure variously identified as Lugh, Eochaid Bairche, or Manannán gives a wheel and a magical apple to Cú Chulainn when the young warrior goes to seek out the martial school of Scathach in the Otherworld. The wheel and apple miraculously guide Cú Chulainn on his quest to the gates of the woman warrior's domain.

In one *immrama* (vision voyage), when Teigue MacCian reached the shores of the otherworld, he noticed "a wide-spreading apple-tree that bore both blossoms and fruit at once." When Teigue meets a fair youth holding a fragrant golden apple, he asks, "what is that apple tree yonder?" the answer he receives is revealing: "That apple tree's fruit it is that for meat shall serve the congregation which is to be in this mansion…" (O'Grady, 1892: 394).

Magical red apples would make a good visual metaphor for the fresh red-capped mushroom. Furthermore, one well-known trait of *A. muscaria* is that the dried mushrooms are more psychoactive than the fresh red caps, and dried *A. muscaria* caps tend to look a bit like dried red-brown apples. But how does one explain the fairly frequent references to golden apples in Celtic legends? Some varieties of the mushroom turn a metallic golden color when dried.[16]

It is not hard to see that red rowan berries and apples might make effective visual substitutes for *A. muscaria*. But why would the Celts have used hazelnuts and salmon as magical foods of knowledge? Could they be less obvious metaphors for the red-and-white mushroom? As we shall soon demonstrate, there is ample circumstantial evidence linking those foods to *A. muscaria*.

The Crimson Nuts of Wisdom

At first glance, it might appear difficult to explain the many references in Celtic legends to hazelnuts that impart instantaneous knowledge and foresight.

There is no evidence that eating hazelnuts is likely to induce visions, wisdom, or precognition. However, there is evidence that the nuts could have served as useful metaphors for *A. muscaria*. First, let us examine some of the linguistic evidence linking hazelnuts and mushrooms.

In Celtic legends, hazelnuts are variously called *cuill crimaind*, the hazels of knowledge; *bolg fis*, bubbles of wisdom; *bolg gréine*, sun bubbles; and *imbas gréine*, sun of inspiration. These terms refer not only to the nuts but also to the bubbles caused by the nuts falling into the waters of the well of wisdom. Significantly, *bolg* is a word frequently found in both Irish and Scots Gaelic names of mushrooms. Wasson and Wasson (1957: 93) offer this analysis on the use of the word *bolg*:

> *In Irish there are two words for a bag or a pouch,* bolg, *which is related to the Latin* bulga, *and* púca, *which was probably borrowed between AD 800 and 1050 from Scandinavian sources…In Irish one way to refer to a wild fungus is* bolg losgainn, *literally "frog's pouch," and another ways is* púca beireach, *"heifer's pouch." If the "heifer's pouch" refers to the udder, as we suppose, the same figure of speech that in Albanian means "toad" turns up in Irish meaning "toadstool." In Irish,* bolg seidete, *"blown-up bag," is a term for the puffball. It is easy to see why the fungi figure in all these metaphors; puffballs, toadstools, all the wild fungi of the forest and field, impress the visual sense as creatures that quickly swell up.*

Colloquial Gaelic preserves other links between mushrooms and the traditional hazelnuts of wisdom, even today. In Irish, we find the phrase *caochóg cnó*, literally a "blind nut" – which means a nut without a kernel. Scottish Gaelic, which often preserves older uses of the language than does Irish Gaelic, gives us the words *caochag*, which means either a nut without a kernel or a mushroom, and *caochagach*, the state of being full of nuts without kernels or full of mushrooms. Colloquially, the word *caochóg* is also used in phrases referring to shyness or to winking. To wink is to close one eye and, as we have seen, Celtic sorcery is often performed one-eyed. The linguistic links between nuts, mushrooms, and the one-eyed winking state are intriguing but there is even more definitive evidence.

In the seventh-century Old Irish text known as "The Cauldron of Poesy," there is a direct statement that the *filidh* found inspiration by chewing on the hazelnuts of wisdom. The text explains that poetic inspiration and the gift of poetry originate in three full cauldrons found within the body of the *fili*. Poetry is said to arise from the experience of sorrow and joy, and one of the divisions of joy experienced by the *fili* – which leads to *imbas*, the gift of prophetic vision – is the "joy of fitting poetic frenzy from grinding away at the fair nuts of the nine hazels on the Well of Segais in the *sidhe* realm" (For a more detailed discussion see Laurie, 1996).

The Middle Irish gloss on this passage offers a very tantalizing item – the phrase *bolcc imba fuilgne*, "the bubble which sustains or supports *imbas*." One possible translation of the gloss on this phrase is: "The bubble that sustains *imbas* is formed by the sun among the plants, and whoever consumes them will have poetry." In short, we have a direct statement that consuming *bolcc* produces the gift of prophetic vision.

In light of the linguistic link between *bolcc*, *bolg*, and mushrooms, it is not hard to read these phrases as statements that some type of mushroom was chewed to sustain poetic inspiration. The fact that *imbas*, the poetic inspiration or poetic frenzy of the Irish, is frequently described as a "fire in the head" also suggests that the most likely mushroom would be *A. muscaria*. "Fire in the head" is an excellent term for a signature symptom of *A. muscaria* inebriation – a pronounced heating of the head, apparently caused by blood rushing to that area.[17]

The Sacred Trees of Knowledge

From numerous sources, including the ogham alphabet, we know that the Celts, like other Indo-European cultures, once venerated the birch. According to Celtic myths, the ogham was given to druids and *filidh* by the god Ogma as a secret language for the preservation of their wisdom. The ogham alphabet begins with *beith*, the letter for "birch." This tree is given primacy of place in the alphabet because it was the first letter created by Ogma, who carved seven *beith* strokes on a birch branch. This reference alone suggests that the Celts held great reverence for the birch tree.

As Wasson points out in his book *Soma: Divine Mushroom of Immortality*, most cultures that use *A. muscaria* as a sacred psychoactive have adopted the birch as the sacred world tree. He attributes this association to the symbiotic relationship between the birch and the mushroom. As noted before, the mycelia of *A. muscaria* can grow only in symbiotic relationship with the roots of a few types of trees – primarily the birch, the spruce, and certain conifers.

Wasson (1968: 215) suggests that the motif of sacred world trees found in many Mediterranean cultures originated in the birch forests of Eurasia at a time when Indo-European cultures were still using *A. muscaria*:

> *The Peoples who emigrated from the forest belt to the southern latitudes took with them vivid memories of the Herb of Immortality and the Tree of Life spread also by word of mouth far and wide, and in the South where the birch and the fly-agaric were little more than cherished tales generations and a thousand miles removed from the source of inspiration, the concepts were still stirring the imaginations of poets, story-tellers, and sages. In these alien lands far from the birch forests of Siberia, botanical substitutions were made for Herb and Tree.*

Wasson's theory may accurately describe what took place in the drier lands of Iran, India, and the Mediterranean. However, the birch has always grown well throughout northwestern Europe, so the insular Celts would not have needed to substitute the hazel for the birch, (the scarcity of birch in many parts of Ireland and Scotland today is due not to warming trends but the relentless overgrazing of sheep and cattle).

Assuming that the Celts never abandoned their use of *A. muscaria*, why would they have bypassed the sacred birch and chose the hazel as their tree of knowledge? As far as we know, *A. muscaria* does not grow under hazel trees. However, the hazel (*Corylus avellana*) is a member of the *Betulaceae* family, and its leaves and catkin flowers resemble those of the birch. Thus, the image of crimson nuts appearing under the hazel tree would make an apt occult metaphor for the birch and the *A. muscaria* that grows at its roots. It would also make a good teaching tool for reminding initiates how to identify the tree beneath which the crimson nuts of knowledge can be found.

The Red-Speckled Brother of Birch

The ogham alphabet may contain another most interesting veiled reference to *A. muscaria*. The ogham alphabet is not unlike the Norse runic alphabet in that the letters are named after various objects starting with the particular letter in question. Although the original names of most ogham letters have known meanings – such as *h-uath* (terror), *tinne* (a bar of metal), and *sraiph* (sulfur) – one letter, *edad*, is a nonsense word with no known meaning.[18] Fortunately, each letter in the ogham has a color, bird, tree, and other objects, as well as kenning phrases called "word oghams," associated with it.

According to Damian McManus (1991: 43) and Howard Meroney (1949), the color for *edad* is *erc*, or "red-speckled," and its word oghams are "discerning tree" and "brother of birch." This association of "red-speckled" and "brother of birch" is very suggestive, and the often white-speckled, red-capped *A. muscaria* grows best at the roots of the birch tree. Is it possible that *edad* was one of the names of the mushroom? We may never know, but the notion is intriguing.

The association of "brother of birch" with *erc* or "red-speckled" provides a vital clue linking several potential *A. muscaria* motifs. According to the *Dictionary of the Irish Language*, the word *erc* can refer to speckled fish, particularly the salmon and the trout, and to speckled or red-eared cattle (Quin, 1990: 278). Significantly, *erc* can also refer to "a reptile of some kind" – for instance, a "viper." In fact, the phrase "red-speckled" is often used in Celtic legends to identify liminal objects and creatures that come from the otherworld.

Red-Speckled Salmon

There are a number of Celtic legends in which eating salmon instantaneously imparts miraculous powers of knowledge. In "The Boyhood Exploits of Finn," a boy named Demne went to learn poetry from Finn Éices, or Finn the Poet (Nagy, 1985: 214). Finn Éices had spent seven years watching for the salmon of Féc's pool, because it had been prophesied that nothing would remain unknown to whomever ate the salmon of Féc. When the salmon was found, Finn Éices told Demne to cook it but not to eat any of it. While turning the fish over in the pan, the lad accidentally burnt his thumb on the fish and, without thinking, stuck his thumb into his mouth. Honoring this quirk of fate, Finn Éices renamed the lad Finn and gave him the salmon of knowledge to eat. From then on, whenever Finn "put his thumb in his mouth, and sang through *teinm laida*, that which he did not know would be revealed to him" (Figure 2).

Is there any reason the Celts might have selected salmon as a source of instantaneous wisdom? In some of the stories, we are told that salmon gain their miraculous powers by nibbling on the bubbles of knowledge or the hazelnuts floating in the fountains of knowledge, but as far as we know, consuming salmon – even those that have eaten hazelnuts – does not lead to instantaneous enlightenment or poetic insight. However, assuming that the druids used occult metaphors for their miraculous inebriant, the salmon – which is silver and white, speckled with red spots – would make a fair metaphor for *A. muscaria*. Pieces of dried *A. muscaria* could be discretely referred to as pieces of dried salmon – and, in some cases, students may have inadvertently eaten them as salmon. Suddenly the legends about eating the salmon of knowledge begin to make more sense.

The Speckled Snake

Alexei Kondratiev, former president of the Celtic League American Branch, has encountered folk references to *A. muscaria* being called *an náthair bhreac*, the speckled snake, in Scotland and Ireland.[19] This direct linguistic link between speckled snakes and *A. muscaria* is certain-

Fig. 2: Fionn-like figure with thumb in mouth and serpent headdress motifs on gold bracteate from Lellinge, Denmark, date unknown (T. White).

ly intriguing, particularly in light of folk legends that Saint Patrick exiled all serpents from Ireland. As we have already noted, *Cormac's Glossary* claims that Saint Patrick banned the oracular practices of *teinm laida* and *imbas forosnai* because those rituals invoked pagan gods. Is it possible that Saint Patrick was waging war against a sacred mushroom cult that involved the invocation of snake deities? If so, post-Christian *filidh* might have had good reasons to adopt the speckled salmon, crimson hazelnuts, or red rowan berries as new metaphors for the most potent *A. muscaria* metaphor – the speckled serpent.

The serpent is certainly an important figure in Celtic mythology, particularly on the continent but also on the islands. Serpents are associated with a number of Celtic deities – especially Brigid, the goddess of poets, and Cernunnos, the god of shamans (Figure 3).

The De Danaan goddess Brigid – whose mythology was later transferred to the fifth-century Saint Brigid (Saint Bride) of Kildare – was originally a triple-aspected goddess, the patroness of smithcraft, medicine, and poetry. Brigid, whose name means "high or exalted one," was associated with the sun and fire, and she was invoked as a guardian of the home and the hearth fire. As a goddess of healing, Brigid was also closely linked with healing springs. However, Brigid is best known as the goddess of poets, and *Cormac's Glossary* praises her foremost as the archetypal female sage and woman of wisdom. Given the tradition of poets seeking inspiration at wells of wisdom, Brigid may have been one of the pagan deities invoked in the shamanic divinatory ritual of *imbas forosnai*.

Fig. 3: Serpents devouring heads on solar cross from Gallen Priory, County Offaly, Ireland, circa 9th century A.D. (T. White).

In "The Life of Brigid," recorded in the fifteenth-century *Book of Lismore*, Saint Brigid is portrayed as a dutiful virgin of Christ, but her metaphoric links to the liminal otherworld are well preserved, even there. Saint Brigid was born at sunrise on the threshold of a house – neither within nor without the house. She was raised in the house of a "wizard," who instructed her to drink only the milk from a white, red-eared cow. Once, when her nurse was ill, Brigid went to a well and fetched some water, which tasted like ale and also healed the nurse. Later, as a nun, she reportedly turned water into milk, which she used to heal one of her sister nuns. Fiery pillars often appeared over Brigid's head, and in one story, sun rays supported and dried her wet cloak (MacDonald, 1992).

Brigid's numerous associations with possible *A. muscaria* themes – red-eared cows, fiery pillars over her head, and healing waters – are highly suggestive. Her ability to transform water into healing wine and milk could hark back to the *New Testament* miracle, but it could also refer to the healing, rejuvenating powers of *A. muscaria*. Ultimately, it is Saint Brigid's close link to snakes that reveals her essentially pagan nature. Long after Patrick reportedly banned snakes in Ireland, the Scots continued to believe serpents came out of their lairs on Saint Brigid's Day, originally a pagan holy day called Imbolc or Oimelc, celebrated in early February

(Figure 4). Significantly, Ó Catháin (1995: 160) notes that the feast day of Brigid is associated with two colors – "the Eve of Brigit's Feast, speckled – the Day of Brigid's Feast, white" (For more on Brigid *see* Ch. 13).

While the cult of Brigid was tolerated in subverted form under the Celtic Church, the cult of the antlered god was forced underground – or into the forests. In Gaul, the antlered lord of the animals – known there as Cernunnos – was closely associated with ram-horned serpents of wisdom, and with the world tree through his connection with the stag.[20] It doesn't take a theologian to explain why Christians incorporated the shamanic horned god with his serpents into their imagery of the devil.

Fig. 4: Entwined serpents and skeletal motif on solar cross from Kilamery, County Kilkenny, Ireland, circa 10th century AD. (T. White).

One of the earliest European images of the stag-antlered deity may be the so-called "dancing sorcerer" painting, at Les Trois Frères, in France. Some of the most interesting petroglyphs depicting the antlered lord of shamans are found in the Valcamonica (Valley Comonica) of northern Italy – an area inhabited by early proto-Celtic and Celtic cultures (see Figure 5). Michael Ripinski-Naxon (1993: 155-157) points out that the rock carvings in the Valcamonica depict several classic Indo-European shamanic motifs, including stylized sun disks and representations of antlered "shamans," and "shamans" dancing around or over small trees. Noting that a prehistoric rock carving of a naturalistic spotted mushroom has been found near Monte Bego (*see* Ch. 17, Fig. 4) – also in northern Italy – Ripinski-Naxon cautiously concludes that Indo-Europeans in the region were familiar with *A. muscaria*. If more evidence can be found in those regions directly linking antlered shamans to the use of spotted mushrooms, it would support our speculations about the use of *A. muscaria* in Celtic Ireland. For now, we must wait.

Probably the most familiar image of the stag-antlered figure is that depicted on the Gundestrup Cauldron (Figure 6), a marvelous work of mythic art undoubtedly inspired by Celts, though most likely crafted by Thracian artists (Taylor, 1992). One panel shows a man, crowned with seven-tined stag antlers which branch like tree limbs, who is holding a speckled, ram-horned serpent. Other panels on the cauldron may portray other Celtic gods or scenes from Celtic legends. Although there is no evidence how the cauldron was used, numerous scholars have assumed that it probably had ritual significance. Could the cauldron have been used in a shamanic cult of the speckled serpent? Could it have been used to serve a brew of *A. muscaria*? The idea is certainly provocative.

Fig. 5: Antlered deity figure, one of many shamanistic motifs depicted by early Celts at Valcamonica, Italy, circa 4th century BC. (T. White).

The Red-Peaked One

An old Scottish folk tale, *Fionn and the Man in the Tree*, directly links the primary magical food motifs of the shamanic lord of the animals and to the world tree. In this story, Fionn is hunting in the forest for Derg Corra, a version of the lord of the animals. The epithet *Derg Corra* means "red-peaked" or "red-pointed," and "red-peaked" would make an excellent metaphor for the red-capped *A. muscaria*. Coincidentally, Derg Corra – like several other heroes of the Celtic otherworld – is famed for his power of leaping, a trait which is associated in Siberia with *A. muscaria*.[21]

While searching through the forest, Fionn beholds a strange sight – a man perched at the top of a tree. A blackbird sits on the man's right shoulder. In his left hand is a bronze vessel in which a salmon leaps. Below him, at the base of the tree, is a stag. This peculiar figure, who is really Derg Corra in disguise, cracks a nut (*teinm cnó*), giving half to the blackbird and eating half himself. He splits an apple in two and shares it with the stag at the base of the tree. Then, he shares the water of wisdom with the salmon, the blackbird, and the stag.

The story is highly significant because we know that druidic apprentices gathered in sacred groves hidden deep in the forests in order to learn their poetic arts. We can infer from other sources that the ranks of the *filidh* were viewed as positions on the world tree.[22] Minor or student *filidh* were the roots, experienced *filidh* were the trunk, and the most powerful *filidh* were the uppermost branches. In this story, Fionn is considered to be a powerful *fili* – having consumed the red-speckled salmon of wisdom in his youth – but now he stands at the root of the tree below the figure of Derg Corra, who holds and eats not one but several wisdom-inducing substances.

In short, this legend weaves together several *A. muscaria* metaphors into an unmistakable tapestry. Fionn has come to study with Derg Corra – the red-peaked lord of the animals, who in many cultures

Fig. 6: Antlered deity with ram-horned, speckled serpent, detail on Gundestrup Cauldron from Jutland, Denmark, circa 2nd century BC. (T. White).

is the lord of shamans (Figure 7). Each of the magical items – the apples, waters of wisdom, and sacred hazelnuts – shared with the animals by Derg Corra is closely linked in Celtic literature with the otherworld and with the imparting of poetic inspiration and prophetic knowledge. Based on this story – and the other evidence we have presented – we will climb out onto the limb of that sacred tree and suggest that the use of *A. muscaria* could have played a prominent teaching or initiatory role in the ancient druidic sanctuaries.

Kindling Poetic Inspiration

Some Celtic scholars may rise up in arms at our suggestion that the druids and *filidh* used *A. muscaria* as a source of poetic inspiration. However, given what we know about *soma*'s role in inspiring the *Rig Veda*, it is certainly conceivable that *A. muscaria* could have inspired the visions and verses of Celtic poets. In Siberian shamanism, *A. muscaria* is definitely associated not only with ecstatic visions, but also with the inspiration of poetry and songs.[23] We know from historical accounts that the *filidh* had to memorize great tracts of legends and poems, but they were much more than just versifiers. In addition to being epic poets, they were inspired philosophers and powerful enchanters. If we are correct that crimson hazelnuts and spotted salmon are metaphors for *A. muscaria*, then we can assume that the ancient *filidh* used the red-and-white mushrooms as a significant source of poetic inspiration and prophecy.

In the Old Irish poem known as "The Song of Amergin," the primal *fili* of the Milesian invaders announces himself as he sets foot in Ireland: "I am a wind of the sea, I am a wave of the sea, I am a sound of the sea, I am a stag of seven tines, I am a hawk on a cliff, I am a tear of the sun, I am fair among flowers, I am a salmon in a pool, I am a lake on a plain, I am a hill of poetry, I am a god who gives inspiration (literally: forms fire for a head)" (Matthews & Matthews, 1994: 11).

Fig. 7: "Lord of the animals" figure holding bird-headed serpents, on gold bracteate from Riseley, Kent, England, date unknown. (T. White).

Could the god who gives inspiration – who forms "fire in the head" – refer to a god connected with *A. muscaria*, or to the mushroom itself? We have already noted that Brigid, the goddess of poets, was associated with fires around her head and that *A. muscaria* intoxication can produce a pronounced heating of the head. Moreover, we know from many sources that *imbas forosnai* and *teinm laida* were both associated with light and the fire of illumination itself.[24]

In the twelfth-century *Hanes Taliesin* (*Romance of Taliesin*), the Welsh poet Gwion announces himself with a similar list of associations, plus a few additions: "I have been a fierce bull and a yellow buck. I have been a boat on the sea…I have been a blue salmon. I have been a spotted snake on a hill…I have been a wave breaking on the beach. On a boundless sea I was set adrift" (Graves, 1966: 211).

A commoner without a broad background in Celtic myth would have found it hard to catch the many metaphors woven into these poems. On one level, the phrases undoubtedly refer to key themes of Celtic legends. However, it is intriguing that many of the phrases could refer to the motifs that we have linked to *A. musca-*

ria: the stag of seven tines to the stag at the base of Derg Corra's tree; the salmon in pools to the salmon that eat the hazelnuts of wisdom; the boat on the sea to the magical coracles that carry poets to and from the Land of Promise; and the spotted snake on the hill directly to the red-and-white mushroom that grows on the *sidhe* mounds.

Judging from the manner in which poems similar to "The Song of Amergin" are repeated by various Irish and Welsh poets, these poems may have served as the oral calling cards of the *filidh* entrusted with Celtic legends. In the text known as "*Immacallam in do Thuarad*" (*Colloquy of the Two Sages*), two bards meet and politely test each other's knowledge (Matthews & Matthews, 1994: 203-218). In the opening salvo, the aged Ferchertne asks: "A question, wise lad, whence have you come?" The young Nede answers:

> *Not hard: from the heel of a sage,*
> *From a confluence of wisdom,*
> *From perfection of goodness,*
> *From the nine hazels of poetic art,*
> *From the splendid circuits in a land*
> *Where truth is measured by excellence,*
> *Where there is no falsehood,*
> *Where there are many colors,*
> *Where poets are refreshed.*

In short, Nede appears to be saying that he has learned the poetic arts by eating the hazelnuts of wisdom. After a lengthy exchange of poetic metaphors, Ferchertne and Nede acknowledge and honor each other's poetic wisdom. Then, Nede asks: "And you, O aged one, have you tidings?" Ferchertne launches into a long list of oracular predictions for the future.

If hazels are metaphors for *A. muscaria*, then we can assume that Nede and other *filidh* probably chewed on inspirational mushrooms during their training, perhaps in a ritual form similar to *imbas forosnai*.[25] Ferchertne's long list of predictions certainly suggests that he also could have been engaging in some type of prophetic divination practice on a regular basis.

While the early historic references to the schools of the druids and *filidh* offer few details about their training practices – beyond mentioning the long lists of verses that had to be memorized and the subjects of grammar, law, mathematics, and natural philosophy – we know from the legends and poems that the training ultimately culminated in the kindling of poetic inspiration. Descriptions of eighteenth-century bardic schools indicate that their initiates – the descendants of the *filidh* – spent long hours practicing and composing poetry in small, dimly lit cells.[26] Although sensory deprivation may be conducive to the practice of poetry and the kindling of prophetic insight, the darkness may have served a more immediate purpose.

If *A. muscaria* was used in the training of the *filidh*, its tendency to increase sensitivity to light would have necessitated the use of dark environments. This pronounced sensitivity to light could also explain the initiatory account of Taliesin being sown into a skin and set adrift on the seas. Based on the many references linking *A. muscaria* metaphors to poetic inspiration, we hypothesize that the training of *filidh* probably included visionary journeys undertaken in darkened rooms or inside bull skins, under the watchful eyes of trained *filidh* – and probably under the influence of *A. muscaria*.

Ethnographic studies of other shamanic cultures show that folk stories are often used to inculcate esoteric shamanic teachings, particularly in oral cultures.[27] If we are correct in identifying the presence of *A. muscaria* motifs in Celtic legends and literature, we should fully expect to find traditional teaching about its use woven into those same legends. Indeed, as we have already seen, Celtic stories may include useful information and teachings relevant to the identification and use of *A. muscaria*.

Is it possible that many of the legends of magical voyages into the otherworld were teaching stories based on the *A. muscaria* experiences of previous generations, refined and polished through the art of poetry? If so, that would explain why new initiates started their training by learning the legends of the past: the legends would provide metaphoric maps and teachings for novices preparing to embark on journeys into the waters of inspiration.

The Wells of Inspiration

Celtic legends are filled with heroes drinking from wells of wisdom or from streams that flow from those wells, and the well of wisdom is often referred to as the ultimate source of the *fili*'s art. One of the highest grades of *fili* is even called *ansruith*, or "great stream," referring directly to this flowing of watery wisdom. Although the motif of cosmic wells of wisdom is found in many parts of the world, the Celtic legends contain numerous elements linking these wells fairly directly to the shamanic use of *A. muscaria*.

Cormac's Adventures in the Land of Promise, also called "Cormac's Cup," mentions several key *A. muscaria* motifs in direct association with the well of knowledge. For brevity, we will offer only a very abridged version of the story here.

One day at dawn, Cormac encounters a grey-haired warrior carrying a silver branch with three golden apples. Cormac is fascinated by the branch, which makes such wonderful music that when it is shaken it puts to sleep sore-wounded men, women in childbed, and folk in sickness. The mysterious warrior – who is Manannán in disguise – explains that he comes "from a land wherein there is naught save truth, and there is neither age nor decay nor gloom nor sadness nor envy nor jealousy nor hatred nor haughtiness" (Cross & Slover, 1988: 503). The two men agree to make an alliance, and Cormac asks for the branch to seal the deal. The warrior agrees, but he asks in return for three unnamed boons to be granted later. Each year afterward, Manannán returns and asks for one of his three boons – first Cormac's daughter, then his son, and then his wife. A man of his word, Cormac grants the boons, but after the third request, he follows the mysterious warrior into a great mist – the *ceo-druidechta* (druid's fog) that appears around him on the plain, and soon he finds himself in a strange fortress in the Land of Promise.

Cormac is shown a silver house, half-thatched with the wings of white birds. Then, he sees a man kindling a voracious fire. Finally, he enters another fortress with a palace that has bronze beams and silver wattling and that is also thatched with the wings of white birds. He sees people drinking from a marvelous fountain with five streams flowing from it. Nine hazel trees grow over the well, dropping their nuts into the water, and "five salmon open the nuts and send their husks floating down the streams."

Cormac is shown Manannán's magical pig, which is cooked in a cauldron by telling four truths. After his hosts tell three truths, cooking three quarters of the pig, Cormac reveals his truth – that he is saddened by the loss of his daughter, son, and wife. When his hosts give him a portion of the pig to eat, Cormac declines, saying he never eats without fifty in his company. Manannán puts Cormac to sleep, and when he awakens he is accompanied by fifty warriors – and by his wife, son, and daughter. During the banquet, Cormac becomes intrigued by a marvelously crafted gold cup. Manannán explains that whenever three falsehoods are spoken under it, the cup breaks into three, and that the only way to restore the cup is to speak three truths under it. Demonstrating how the cup works, Manannán tells three falsehoods, breaking it, and then three truths to fix it again. Manannán then offers Cormac the magical cup as a gift, promises to let him return home with his family, and interpret the strange visions that Cormac has seen.

Manannán functions as a teacher and guide who helps interpret the meaning of the visions for Cormac – and for any students being told the story. According to Manannán, the fountain is the well of knowledge, and the streams are the five senses through which knowledge is obtained. Manannán explains: "No one will have knowledge who drinketh not a draught out of the fountain itself and out of the streams. The folk of many arts are those who drink of them both."

Drowning in the Waters of Wisdom

By studying the lessons preserved in Celtic legends in the context of what is known about *A. muscaria* and about the training and divinatory practices of the *filidh*, a diligent student could conceivably be able to resurrect what appears to be one of the most viable and best documented Celtic shamanic practices. However, before imbibing large quantities of *A. muscaria* in the hopes of becoming instantly enlightened, enthusiastic students would do well to read and heed the warnings found in some of the Irish legends. For example, new initiates interested in exploring the marvelous otherworlds of *A. muscaria* visions would do well to remember that during the *imbas forosnai*, the *fili* seeking the vision was watched over by other *filidh*, who presumably had experience in dealing with the effects of consuming the mushrooms.

In the Irish *dinnshenchas* (land-name tale) of Siannan, the goddess of the Shannon River, we find a teaching that the wisdom from the nine hazel trees can be overwhelming. In this story, the goddess travels to Conla's well, which is said to be the origin of the Shannon, the Boyne, and a number of other rivers. Siannan goes there seeking wisdom, and the waters of wisdom overwhelm her. She flees before them along the course of the river, drowning in wisdom at the mouth of the Shannon. Significantly, the hazel trees around Conla's well miraculously produce their leaves, flowers, and nuts in the space of a single hour – much like mushrooms appearing suddenly after a rain.

The story provides a warning that eating the hazelnuts at the well of wisdom can involve certain dangers. For one thing, if one is not properly prepared, the intensity of the experience can be overwhelming. It is perhaps possible that Siannan's death could also be a warning that eating the wrong *Amanita* – the highly poisonous, green-white *Amanita phalloides*, appropriately known as "Death-Caps" – can be fatal.

If Celtic legends such as this served as shamanic teaching stories, we should expect some of them to include metaphoric maps aimed at helping new initiates navigate the potentially overwhelming waters of wisdom. Let us examine a *dinnshencha* which describes the creation of the sacred Boyne River from the waters of Nechtán's well. In this story, the goddess Bóann, whose name means "the white cow," is married to the god Nechtán. While the hazelnuts of wisdom are not mentioned directly in this tale, we can assume they also inspired Nechtán's waters of wisdom. Nechtán – whose name means "clean, pure" or "white, bright" – is a symbol of fire in the water and of the power of poetry that comes from the well of wisdom.

The legend states that Nechtán has three cup-bearers who must accompany anyone seeking wisdom at the well or dire consequences will result. Ignoring the traditional warnings, Bóann went to the well alone and challenged its power by walking counter-clockwise around it three times. In response to this action, the waters of the well rose up and ripped Bóann's right eye, arm, and leg from her. She fled before the stream of water, running down to the sea, only to be drowned at the mouth. The waters became the Boyne River, named after her.

It is noteworthy that when Bóann approached the well of wisdom without assistance and challenged its power, she lost an eye, arm, and leg, and then was drowned in its waters. This legend not only links the motif of one-eyed, one-legged beings with the well of wisdom, but also provides a warning about the inherent dangers of using *A. muscaria* without appropriate guidance. A one-eyed person lacks perspective, and a one-legged person lacks balance – two conditions that can overwhelm a novice.[28]

Many teachings can be conveyed in a single legend. The names of the three cup-bearers at Nechtán's well – *Flesc* (a wand or stave engraved with ogham letters), *Lesc* (which can mean lazy, sluggish, or still), and *Luam* (a steersman or guide) – may convey hidden information about the rituals performed at the well of knowledge. *Flesc* may refer to the *bunsach comairce* or "rod of safe passage" that *filidh* carried during their travels. In a mundane social sense, this rod of safe passage referred to the right of the *filidh* to travel unmolested between tribal territories, but it may have also represented a *fili*'s right to partake of the well of wisdom and to commune with the gods and spirits.

Lesc, as stillness, could be a veiled reference to the period of prophetic incubatory sleep experience during the rites of *imbas forosnai* and *tarbhfeis* and to the fact that ingestion of *A. muscaria* sometimes produces a somnambulant state. *Lesc*, as sluggish, could also refer to the fact that heavy usage of *A. muscaria* – perhaps in intensive initiation rites – can result in post-use depression.

Luam, the guide or steersman, could refer to someone who keeps watch, as during the ritual of *imbas forosnai*. If so, this ritualist, most likely an experienced *fili*, would have made sure that the vision-seeker did not turn over during the ritual sleep. The watcher would have also kept vigil to prevent disturbances and to monitor the experience between the seeker and the spirits.

The one-eyed, one-legged goddess Bóann is associated with another sacred triad. She is the mother of the three sacred harp strains: *Goltraige*, the sorrow strain; *Gentraige*, the laughter strain, and *Suantraige*, the sleep strain. Sorrow and joy are the two emotions that turn the internal cauldrons discussed in the "Cauldron of Poesy" text, while the sleep strain seems to describe the state necessary to access the internal cauldron of wisdom through the trance rites of *imbas forosnai* and *tarbhfeis*. Knowing that sorrow, laughter, and sleep are potential side effects of *A. muscaria* consumption might help to keep vision-seeking poets from being overwhelmed by the waters of inspiration.

It is not hard to identify Nechtán's well with that of Manannán and also with the well of Segais mentioned in the "Cauldron of Poesy" text. They are all the well of knowledge found beneath the sea at the center of the world and at the base of the world tree – the source of all the rivers of the earth, as well as the source of the five senses. Although there are many methods for crossing the thresholds between the worlds and dipping into the well of knowledge, the ritual ingestion of *A. muscaria* is certainly one ancient and honored way. The warnings and teachings woven into Celtic legends about the wells of knowledge suggest that the Celts treated their magical crimson foods as the sacred food of the gods – to be approached with care and respect. The wisdom of these teachings is still applicable today – if approached with care, entheogens can be enlightening, but, if abused, they can be deadly.

In Search of the Land of Promise

The ultimate question remains: Did the Celtic druids and *filidh* use *A. muscaria* in shamanic rituals? Circumstantial evidence suggests that the druids and *filidh* engaged in shamanic oracles, quite possibly involving the use of a vision-producing substance. Based on the complex of legends linking magical red foods with journeys to the otherworld, we believe that the pre-Christian Celts once used one or more vision-producing substances. We think it is significant that all the red substances happen to look a bit like *A. muscaria*, happen to inspire ecstatic poetry as does *A. muscaria*, and happen to induce prophetic visions as does *A. muscaria*. The ongoing use of *A. muscaria* metaphors in the Celtic corpus, in close association with references to the development of poetic wisdom, suggests that the use of the red-speckled brother of birch could have survived underground, perhaps well into the Christian era.

While hard evidence proving the shamanic use of *A. muscaria* in Ireland and Scotland may have disappeared forever – like the Fomorians and de Dananns into the *sidhe* realms – we believe that the Celtic legends are filled with many veiled and shadowy traces of an ancient secret tradition. Perhaps enough remains to inspire the recreation of a viable, fairly authentic Celtic shamanic practice. For those brave souls ready to explore the otherworlds of the red-speckled brother of birch we offer this advice, once given to Bran as he left on a journey in search of the Land of Women:

> *Do not fall into sleeping stillness*
> *nor let your intoxication overcome you*
> *but begin a voyage over the pure, bright sea...*
> (Translation by Erynn Rowan Laurie)

Notes

1. This chapter first appeared as a featured article in *Shaman's Drum* issue 44, Mar/May 1997.
2. Except for the early linguistic inquiries of R. Gordon Wasson (1968), and the mythopoeic studies of Robert Graves (1966), there was, until recently, little scholastic interest in the historic use of psychoactive mushrooms in Celtic Europe.

   Since the 1960s, numerous cross-cultural studies have established that shamanic use of psychoactive substances to induce altered states of consciousness is much more prevalent in all parts of the world than once thought. See for example Furst (1972); Harner (1973); Ripinski-Naxon (1993); Schultes & Hofmann (1992); and Wasson (1980).

   Although there is still no direct evidence indicating *A. muscaria* usage among the Irish Celts, there is growing evidence suggesting its possible use in other parts of Europe. Ripinski-Naxon (1993:154-165) summarizes evidence for the prehistoric use of *A. muscaria* in Europe and includes the interesting research of Italian enthobotanist Giorgio Samorini. Based on the appearance of mushroom-like motifs in the rock carvings of Valcamonica, and one naturalistic image of a spotted mushroom (*A. muscaria*?) depicted in association with an individual or effigy, it is conceivable that the Celts, who once inhabited the region, might have practiced a ritual mushroom complex.
3. It is likely that the apples, berries, and hazelnuts were sometimes made into alcoholic brews, for there are references to hazel mead in Celtic texts. However, alcoholic inebriation tends to produce drunken fools, not instantaneous wise men.
4. Contemporary nature guides indicate that *A. muscaria* is still relatively common in the woods of England and Scotland. In 1995, the Isle of Man's Philatelic Bureau issued a 20-penny stamp depicting *A. muscaria*, indicating it is still found there.

   British naturalist Oliver Rackham (1986: 112) notes that renowned birch forests were once found throughout the British Isles. It is quite probable that *A. muscaria* was once abundant in Ireland, at least until the Irish forests were cleared. Interestingly, as Rackham observes, the last well-wooded remnants of forests in Ireland were typically found on islands or on the ancient earth mounds known as *raths* – the very places said to hold the underground homes of fairies.
5. The hymns of the *Rig Veda* never directly identify the main ingredient of soma, but they are filled with numerous descriptive references to the mysterious plant. Based on these descriptions, Vedic scholars have proposed various psychoactive plants – including *Peganum harmana* and even *Cannabis sativa* – as possible candidates for soma. A few scholars still favor alternate candidates, but many scholars now endorse Wasson's persuasive proposal that soma was probably *A. muscaria*.
6. Although Wasson was primarily interested in proving that *A. muscaria* was the most promising candidate for soma, many of his observations and comments may be extended to support links between *A. muscaria* and the wondrous Celtic brews of knowledge. For example, Wasson (1986: 60-67) argues that Vedic references to soma as "single-eyed" and "not-born single-foot" may have the same roots as the one-eyed and one-legged mushroom beings of Siberia. Wilson (1995) examines

soma motifs found in the saga of Dermat and Grania from the Fionn Cycle, and he suggests that Sharvan the Surly – the one-eyed Fomorian giant who guards the magical scarlet rowan berries of the Tuatha de Danaan – could be a mushroom being. In addition, based on the observation in Wasson (1968: 44) that soma was pounded to extract a divine inebriant, Wilson suggests that the act of Dermat clubbing Sharvan to death could be a symbolic reference to the ritual sacrifice and pressing of soma.

7. The crane is often associated in Celtic legends with druidic magic, and the magical objects of both the otherworld god Manannan and the hero Fionn MacCumhail were stored in crane-skin bags.

8. As Wasson (1980) points out, soma is associated with thunderbolts, and many cultures associate thunderbolts with *A. muscaria*, which typically fruits after the first thunderstorms of fall. Lugh's association with thunderbolts could link him to *A. muscaria*.

9. The earliest written records of Druidic and Celtic religion were, for the most part, based on secondhand reports, which were then filtered through the personal prejudices of the Roman authors. For a discussion on the information and misinformation contained in Roman histories, see Piggot (1987: 91- 120).

10. The Old Gaelic legends – preserved orally for centuries by Irish *filidh* and bards – often mention pagan and druidic practices in passing. Although the written versions of the Gaelic legends were recorded much later by Christian clerics, these legends contain remarkably sympathetic, insightful comments about druidic practices. Unfortunately, the stories seldom provide detailed descriptions of the rituals, so the best we can do is piece together Celtic practices form various sources. For a sampling of passages dealing with the shamanistic practices of the Celts, see Matthews (1991); and Matthews & Matthews (1994).

11. The functions of the *filidh* are documented in great detail throughout the literature of the early Irish, although no one source compiles all this information into one place. As Joseph Falasky Nagy (1985) notes, Fionn MacCumhail is an archetypal shamanic *fili*, composing poetry, healing by offering a draught of water from his hands, journeying into the *sidhe* mounds, and conducting rituals of divination.

12. *Cormac's Glossary* is a tenth-century Old Irish text found in the *Yellow Book of Lecan*. Because the glossator was commenting from a time several centuries after the practice of *imbas forosnai* had been outlawed, he was presumably speaking from historical tradition rather than direct experience. However, it is conceivable that the practice of *imbas forosnai* continued underground.

We know that Scottish descendants of the *filidh* practices two oracles – the *tarbhfeis* and *taghairm* – which bear certain similarities to *imbas forosnai*. Seventeenth-century Scots practiced a rite called the *tarbhfeis*, or "bull-feast," which also involved chewing or consuming red flesh. According to one account by the historian Geoffrey Keating, a bull was sacrificed and then the seer consumed some of its flesh and broth, wrapped himself in the fresh hide of a bull, and waited for a dream or vision (Matthew & Matthews, 1994: 243). In an eighteenth-century book, *Description of the Western Isles of Scotland*, Marin Martin describes another divination ritual, known there as *taghairnu*. "A party of men, who first retired to solitary places, remote from any house…singled out one of their number, and wrapp'd him in a big cow's hide, which they folded about him, his whole body was covered with it except his head, and so left in this posture all night until…[he gave] the proper answer to the question at hand" (Matthews & Matthews, 1994: 334). No mention is made in Martin's account of chewing on red flesh, but such a detail could easily have been overlooked.

13. Matthews (1991: 184) quotes a passage from the *Senchus Mor*, a collection of law texts from various periods, that appears to confuse the practice of *teinm laída* with that of *dichetal do chennaib*: "When the *fili* sees the person or thing before him, he makes a verse at once with the ends of his fingers, or in his mind without studying, and he composes and repeats at the same time…" However, the passage gives a good description of how, "before Patrick's time," *teinm laída* was done differently: "The poet placed his staff upon the person's body or upon his head, and found out his name, and the name of his father and mother, and discovered every unknown thing that was proposed to him, in a minute or two or three, and this Teinm Laegha [sic] or Imus Forosna [sic], for the same thing used to be revealed by means of them; but they were performed after a different manner, i.e. a different kind of offering was made at each."

14. Although Siberians don't associate *A. muscaria* directly with immortality, Salzman et al. (1996) report that the Koryak of Siberia make a tonic of blueberries and *A. muscaria*, which is drunk for health and longevity. Again, we see parallels to the Vedic soma – just as immortality was promised to Vedic priests who drank soma, immortality is promised to those who eat the foods and drink of the rowan berries of the de Danaan.

15. Wasson (1968: 246, 324) cites statements made by Georg H. Langsdorf, in 1809, and Carl Hartwich, in 1911, that it was fairly common for Siberians to consume a drink made from *A. muscaria* mixed with bog bilberries (*Vaccinium ulignosum*) or the leaves of the narrow-leafed willow (*Epilobium augustifolium*). Like Wasson, Saar (1991a: 168) also cites Langsdorf as stating the Khanty believe berries strengthen the brew. Neither source provides any pharmacological explanation for the belief.

    Wasson (1968: 153-155) notes that there is clear consensus among the Siberians that it is vital to dry the mushrooms before use. Some Siberian tribes say that eating fresh mushrooms is dangerous; others say that fresh mushrooms are more nauseating. Wasson describes how he and his friends discovered that toasting the mushrooms enhances their psychoactive strength (p. 155).

    Ott (1993: 339) cites pharmacological studies by Repke that drying the mushrooms causes decarboxylation of ibotenic acid into the more potent psychoactive muscimol. Ott also suggests that stomach acid may convert ibotenic acid into muscimol (p. 328). The mixing of acidic berries and *A. muscaria* may catalyze a similar synergistic biochemical interaction. On the basis of personal use, White has observed a synergistic relationship between blueberries and *A. muscaria*.

16. The references in Celtic literature to golden apples remained a riddle until, while

working on this paper, White came across a photo of a dried *A. muscaria* specimen with a metallic white sheen. In light of this discovery, it should be considered whether the Greek and Indo-European legends involving sacred golden apples may also be teaching stories about the use of dried *A. muscaria* – designed to remind initiates of useful esoteric knowledge.

17. Wasson (1968: 248) quotes an observation by Langsdorf: "The face becomes red, bloated, and full of blood, and the intoxicated person begins to do and say many things involuntarily." Waser (1967: 435) notes that the warming and flushing of the face is caused by ibotenic acid. While the reddened face is most visible to observers, users experience a pronounced heating of the head.

18. Popular understanding of the ogham makes the names of the letters out to be solely names of trees. The tree ogham is but one of some 150 different ogham lists. According to McManus (1991) the strong identification of the ogham with trees was popularized by fourteenth-century antiquarians who were working with a tradition several centuries old and only partially understood. McManus suggests that the original names of the ogham letters included words, varied objects, and concepts. McManus, and Meroney (1949) before him, offer translations of the known ogham letter names, but a few of the names are obscure and not amenable to translation. *Edad* is one of these.

19. Kondratiev notes that the use of the name *an náthair bhreac* in widely separated locations suggests that the name is not a new invention. The use of the term "speckled snake" in thirteenth century Welsh pseudo-Taliesin poetry could be significant.

    Wasson and Wasson (1957) demonstrated the Indo-European link between the serpent and the mushroom but noted that "the snake-mushroom association of Greece and the Indic world, with all its baggage of associations, becomes the toad and toadstool glyph of the West." In the face of "the speckled serpent" associations, however, it appears that the conservative Irish may have preserved the ancient association of the snake with the mushroom.

20. The symbology and functions of the stags, trees, and ram-horned serpents of Cernunnos are discussed in some detail in Green (1989: 86-96) and (1992: 39-61).

    On some Gaulish coins, the stag bears the quartered circle of the solar wheel between its antlers, clearly linking the Gaulish Cernunnos with solar attributes – a trait he shares with the Gaulish Belenos, who is related to the Welsh Beli and the Irish ancestor god Bilé, whose name means "a great sacred tree."

21. Wasson (1968: 249, 273-4) cites statements by Langsdorf and Bogoraz that *A. muscaria* users are sometimes prone to leaping and may exhibit unusual physical stamina. See also Saar (1991a) and Salzman et al. (1996).

22. Laurie (1996) notes that the "Cauldron of Poesy" text describes *imbas* as a tree that is "climbed through diligence," implying that the more training and inspiration the *fili* has, the higher up the tree he or she advances. Nagy (1985: 281) notes the word *taman* "trunk of a tree, stock, stem" is used in some Irish bardic and legal texts to describe a lower order of poet. One of the higher orders of *fili* is called the *druimclí*, the top of the ridgepole or roof-tree of knowledge. O'Curry (1878: 9) says, "the man who was a *druimclí* was supposed to have climbed the pillar or tree

or learning to its very ridge or top." Taken together, these nuggets of information suggest a picture of *filidh* inhabiting levels of the world tree according to their rank and station.

23. Saar (1991a: 164) cites an observation by Langsdorf that the heroic epic singers of the Khanty used to consume several mushrooms and then sing inspired songs all night long. Salzman et al (1996: 42) indicate that *mukhomor* (*A. muscaria*) continues to inspire songs and singing among the Koryak of Kamchatka.

24. Chadwick (1935) records several stories in which *teinm laída* is associated with songs being chanted by severed heads placed near fires. Chadwick observes that the word *teinm* is generally regarded as being derived from the word *tep-* ("heat").

25. According to early Irish law texts, the rituals of *imbas forosnai*, *teinm laída*, and *dichetal do chennaib* were taught during the eighth year of study, after the *fili* was already considered an *ollamh*, the highest rank of poet (Calder 1997: xxi). If our contention that *A. muscaria* was the "red flesh" chewed during the *imbas forosnai* ritual is correct, the students may have begun their training with the mushrooms at some earlier point in their course of study so that they would be prepared to take on the role of diviner during this ritual.

26. Matthews (1991: 123) quotes an interesting account of bardic training found in the *Memoirs of the Marquis of Clanricarde*, a text written in 1722. The text tells how the training of bards – at that time – took place in "a snug, low hut" located somewhere in a solitary setting. Each student had a small, windowless apartment, or cell, without much furniture beyond a bed, and they spent days and nights in the dark practicing their art on subjects assigned by a professor.

27. Franz Boaz, Ronald and Catherine Berndt, Dennis Tedlock, and numerous other ethnographers have emphasized the role of myth in carrying the teachings of indigenous cultures. Nootka shaman and storyteller Jonny Moses often speaks about how traditional stories are used both as teaching tales and as transformative shamanic healing tools. Daniel Merkur (2014) provides an in-depth study of how esoteric information about Inuit shamanic practices is woven into their songs, stories, and legends through the use of archaic symbols and other circumlocutions.

28. The single-eyed motif may have multiple symbolic meanings. In early Norse mythology, Mimir's well of wisdom is located at the roots of the world tree Yggdrasill. At that well, Odhinn sacrificed one of his eyes to the giant Mimir in return for wisdom. "Single-eyed" may indicate that both Odhinn and Bóann have one eye in this world and one eye in the otherworld. The single arm and leg could easily be metaphors for the ability to move and act both in this world and in the otherworld realms.

Fly Agaric Motifs in the Cú Chulaind Myth Cycle
Thomas J. Riedlinger

The ancient myths of Ireland include many fabulous tales regarding Cú Chulaind, a warrior of the Ulaid clan who lived in the province of Ulster. Some of these portray Cú Chulaind's legendary mood swings: at one extreme, a battle-fury so intense that it terrified even his family, friends and fellow warriors; at the other, a torpor with vivid, prophetic dreams in which he languished for a year. It strikes me that aspects of both of these states, as described in the tales, resemble the effects of ingesting *Amanita muscaria*, the psychoactive mushroom with a bright red cap and white "speckles" also known as the fly-agaric.

This connection is especially significant in light of recent theories which suggest that many ancient Irish myths contain thinly-veiled allusions to *Amanita muscaria*. Peter Lamborn Wilson was among the first to point this out in his 1995 article "Irish Soma." According to him it is likely that the Celts who settled Ireland either used it there in rituals or remembered that it had been used by their ancestors prior to migrating westward from an "Indo-European heartland" near the southwestern edge of Siberia. Wilson calls it "Irish Soma" because *A. muscaria* also is considered a strong candidate for Soma, the unknown plant substance praised as a god in the Hindu *Rig Veda*. It is a curious but well-established fact that there are numerous strong similarities between the culture of the Irish Celts and that of the Aryan peoples who migrated southward to India from the same heartland, bringing with them an oral tradition of hymns that became the *Rig Veda* (Olmstead, 1994). Some even think the name "Ireland" derives from a Celtic word, "Erin," which is based upon or shares a common root with the Sanskrit "Arya" (Schulberg, 1968: 39). In Wilson's view, both cultures almost certainly first learned about the entheogenic effects of *A. muscaria* from Siberian shamans in the Indo-European heartland.

Support for Wilson's theory has been offered by Erynn Rowan Laurie and Timothy White (1997) in their *Shaman's Drum* article, "Speckled Snake, Brother of Birch: *Amanita Muscaria* Motifs in Celtic Legends" (*see also* Chapter 11), which contends that motifs of magical foods in Irish Celtic lore are best explained as "metaphoric references to *Amanita muscaria*, the highly valued psychotropic, red-capped mushroom that was once used shamanically throughout much of northern Eurasia" (p. 53). While agreeing that these references "could theoretically be faded memories of earlier pre-migration Indo-European practices, preserved in oral legends passed down from generation to generation," they think it more likely that Irish Celts actually used this mushroom in religious rituals (p. 54). Their reason is that Celtic legends seem, in their opinion, to contain "certain innovative occult symbols for *A. muscaria* that are fairly unique to the insular Celts" (p. 54). They maintain that such symbols would not have been needed unless a sacred mushroom cult was still being practiced in Ireland by Celtic Druid priests "who wanted to communicate teachings about the sacrament's use to initiates, while maintaining a protective veil of secrecy around its identification" (p. 54).

Considered together, these two articles and Wilson's (1999) book on Irish Soma build a very strong case for the general theory that the myths of ancient Ireland reflect at least a lingering awareness of the psychoactive properties of *Amanita muscaria*. But none of them links *A. muscaria* directly to Cú Chulaind. For reasons discussed in the following pages, I think it plausible that he personifies this mushroom or at least its capricious effects. We will start by reviewing descriptions of Cú Chulaind's so-called "warp spasms" and "wasting sickness." Next we will look at reported effects of *A. muscaria* inebriation and compare them with these maladies. Finally, I will discuss the possibility that myths about Cú Chulaind may derive from a shamanic source in prehistoric times that also inspired the Soma cult of ancient India.

Cú Chulaind's Warp Spasms

The first of Cú Chulaind's two maladies is variously translated as "battle-fury" (Rutherford, 1987: 60), "battle frenzy," "battle ardor" (Mac Cana, 1985: 102) or "warp spasm" (Sharkey, 1975: 10). It is always accompanied by a temporary physical distortion called a *ríastarthae* (Gantz, 1981: 136) or *riastradh* (Mac Cana, 1985: 102) that usually includes an aura-like glow around Cú Chulaind's head, described by various writers as the "hero's light" (Cowan, 1993: 37) or the "warrior's light" (Mac Cana, 1985: 102). He first exhibits this proclivity in childhood, when, as related in *Boyhood Deeds of Cú Chulaind* (*Macghnímhartha*), some other boys provoke him with a barrage of toy javelins, balls and "hurleys" (slingshots) – 150 of each. After knocking these missiles aside without much trouble, Cú Chulaind becomes a bit flustered:

> *You would have thought that every hair was being driven into his head. You would have thought that a spark of fire was on every hair. He closed one eye until it was no wider than the eye of a needle; he opened the other until it was as big as a wooden bowl. He bared his teeth from jaw to ear, and he opened his mouth until the gullet was visible.* (Gantz, 1981: 136)

At this point, understandably, the other boys decide that they have somewhere else to go. But Cú Chulaind, now out of control, begins striking them down. He thrashes fifty before he is stopped by his patron and maybe father, King Conchubur of Emuin Machae, who gently points out that he should have secured the protection of the other boys before he tried to play with them. In other words, he should not have presumed to barge into their games without asking permission. Cú Chulaind resolves instead that it is he who will protect the other boys, and all return to the playing field. After that he always wins no matter what they play, including a primitive version of golf, wrestling and "mutual stripping." (The rules of the latter are lost in obscurity, other than what is deducible from the report that Cú Chulaind "stripped them all so they were stark naked, while they could not take so much as the brooch from his mantle" [Gantz, 1981: 139].)

Another episode in *Boyhood Deeds* relates that when Cú Chulaind was seven years old he went for a ride in a chariot driven by King Conchubur's charioteer. Upon encountering three fierce warriors from another clan, Cú Chulaind kills them and cuts off their heads to keep as trophies. Next he leaps from the chariot onto an antlered deer and subdues it. Then he captures twenty swans by knocking them out of the sky with well-aimed stones. The deer is tethered to the chariot and trots along behind; the swans, likewise bound to the chariot, fly overhead; the heads presumably are swinging to and fro inside the chariot. So much excitement in one afternoon proves too much for Cú Chulaind, who finds himself having a warp spasm just as the chariot approaches home. According to the tale:

> *When they arrived at Emuin, the watchman said "A man in a chariot is approaching, and he will shed the blood of every person here unless naked women are sent to meet him." Cú Chulaind...said "I swear by the god the Ulaid swear by, unless a man is found to fight me, I will shed the blood of everyone in the fort." "Naked women to meet him!" shouted Conchubur. The women of Emuin went to meet Cú Chulaind gathered round Mugain, Conchubur's wife, and they bared their breasts before him. "These are the warriors who will meet you today!" said Mugain. Cú Chulaind hid his face, whereupon the warriors of Ulaid seized him and thrust him into a vat of cold water. This vat burst, but the second vat into which he was thrust boiled up with fist-sized bubbles, and the third he merely heated to a moderate warmth.* (Gantz, 1981: 146)

It is related in another of the tales, *The Wasting Sickness of Cú Chulaind* (*Serglighe Con Culainn*), that all of the Ulaid women who loved him "blinded one eye in his likeness," because "it was Cú Chulaind's gift, when he was angry, that he could withdraw one eye so far into his head that a heron could not reach it, whereas the other eye he could protrude until it was as large as a cauldron for a yearling calf" (Gantz, 1981: 156). Also in this tale, which describes the hero's deeds as a young man, a third warp spasm comes upon him not long after he wakes from his year-long wasting sickness, described in the following section. This time he loses his cool (literally) during a furious battle in the land of Eithne Ingubai. Having just slaughtered thirty-four warriors, he finds it hard to stop even after the rest of the enemy army retreats. One of his own countrymen then says:

> *"I fear that the man will turn his anger against us, for he has not yet had his fill of fighting. Have three vats of cold water brought, that his rage might be extinguished." The first vat that Cú Chulaind entered boiled over, and the second became so hot that no one could endure it, but the third grew only moderately warm.* (Gantz, 1981: 171)

In the tale of *Briancu's Feast* (*Fledh Bhricrenn*), Cú Chulaind is challenged to "straighten" a house when, after lifting one side of it, he sets it down off-kilter. When he tries to comply but at first is unable to do so, the hero gets angry.

> *Then his ríastarthae came over him: a drop of blood appeared at the tip of each hair, and he drew his hair into his head, so that, from above, his jet black locks appeared to have been cropped with scissors; he turned like a mill wheel and he stretched himself out until a warrior's foot could fit between each pair of ribs. His power and energy returned to him, and he lifted the house and reset it so that it was as straight as it had been before.* (Gantz, 1981: 230)

Elsewhere in the Irish myths, Cú Chulaind's warp spasms are said to include: a booming heartbeat (hÓgáin, 1991: 131); the ability to revolve within his own skin and thus fight in multiple directions at the same time (Mac Cana, 1985: 102); the aforementioned "warrior's light" which rose from the crown of his head; the "warrior's moon," a projection "as thick as a whetstone" from his forehead; and a "stream of black blood [that] geysered from his skull as tall as a ship's mainmast" (Cowan, 1993: 37-38).

Cú Chulaind's Wasting Sickness

The title event of *The Wasting Sickness* is triggered when Cú Chulaind tries to capture a pair of enchanted birds flying over a lake. The song of these birds, who are joined by a red-gold chain, causes people to sleep. Though King Conchubur's wife warns him that "those birds possess some kind of power," Cú Chulaind twice casts stones at them – and misses! "Now I am doomed," he says, "for since the day I took up arms I have never missed my target" (Cowan, 1993: 157). He then throws his javelin at them and pierces a wing, but both birds manage to escape. After walking a while, the angry Cú Chulaind sits down with his back to a stone and there falls asleep. The story continues:

> *While sleeping he saw two women approach: one wore a green cloak and the other a crimson cloak folded five times, and the one in green smiled at him and began to beat him with a horsewhip. The other woman then came and smiled and also struck him in the same fashion, and they beat him for such a long*

time that there was scarcely any life left in him. Then they left. (Cowan, 1993: 157)

When the Ulaid try in vain to rouse Cú Chulaind, Fergus, another hero, tells them: "No! Do not disturb him--it is a vision." At this point Cú Chulaind awakes, but cannot speak. For a year he remains in this state, confined to a sickbed, then slowly recovers. On hearing of his vision, King Conchubur counsels Cú Chulaind to return to the stone where he first fell asleep. There Cú Chulaind finds the green cloaked woman, who explains that she seeks his assistance as a warrior to fight on behalf of her people who dwell in the Sídh, a Celtic "other world." As noted by Mac Cana, the Sídh is a place of apparent contradictions whose people, though immortal, are not "permanently invulnerable nor exempt from violent death." The relativity of time and space is fluid there: "perspectives are reversed and brevity becomes length and length becomes brevity as one crosses the tenuous border between the natural and supernatural." Control of magic is the thing that most distinguishes lords of the Sídh from mortal kings and heroes (Mac Cana, 1985: 64), as Cú Chulaind learns firsthand from his wasting sickness. When he tells the green-cloaked woman that he is "not fit to fight men," she replies: "That is soon remedied: you will be healed, and your full strength will be restored" (Gatz, 1981: 159). This comes to pass when an Ulaid woman later tells him: "Throw off sleep, the peace that follows drink, throw it off with great energy." Cú Chulaind then rises and, passing his hand over his face, at last conquers "all weariness and sluggishness" (p. 165). It is during a subsequent battle that Cú Chulaind fights for the green-cloaked woman that he slaughters the thirty-four warriors while in warp spasm.

Effects of *Amanita muscaria*

Before proceeding it will be useful to review the main symptoms of Cú Chulaind's two maladies. Those exhibited during his warp spasm are: agitation (manic behavior, bristling hair); visual distortions (one eye protrudes, the other recedes); facial distortions (he grimaces and gapes, a thick "warrior's moon" projects from his forehead); tachycardia (booming heartbeat); blood rushes to his head (a drop of blood at the tip of each hair, a geyser of it spurts from his crown); light rushes to his head (a spark of fire appears at the tip of each hair, the "warrior's light" from his crown); he becomes very strong and very limber. He also becomes very hot, needing three successive dousings in vats of cold water before his temperature returns to normal. The symptoms of his wasting sickness are: sleepiness, vivid dreaming and prolonged lassitude.

Next let us review some eyewitness accounts of the effects of *Amanita muscaria* among Koryaks and other indigenous peoples of Siberia:

> [From an 1809 paper which describes the effects of fly-agaric mushroom eaten by Koryaks in Kamchatka:] *The narcotic effect begins to manifest itself a half hour after eating, in a pulling and jerking of the muscles or a so-called tendon jump...The face becomes red, bloated, and full of blood, and...the head and neck muscles are also in a constantly convulsive state...According to their own statement, people who are slightly intoxicated feel extraordinarily light on their feet and are then exceedingly skillful in body movement and physical exercise. The nerves are highly stimulated, and in this state the slightest effort of will produces very powerful effects. Consequently, if one wishes to step over a small stick or straw, he steps and jumps as though the obstacles were tree trunks....[T]hese persons exert muscle efforts of which they would be completely incapable at other times; for example, they have carried heavy burdens with the greatest of ease, and eye-witnesses have confirmed to me the fact that a person in a state of fly-agaric ecstasy carried a 120-pound sack of flour a distance of 10 miles, although at any other time he would scarcely have been able to lift such a load easily.* (Wasson, 1968: 248-249)

> [From an 1893 book describing travels in Eastern Siberia from 1861-71:] *When a Koryak consumes the fly-agaric...the mushroom seems to produce a peculiar effect on his optic nerves which makes him see everything on a greatly enlarged scale. For this reason it is a common joke among the people to induce such an intoxicated man to walk and then to place some small obstacle, such as a stick, in his way. He will stop, examine the little stick with a probing eye, and finally jump over it with a mighty bound. Another effect of the mushroom is said to be that the pupils become much enlarged and then contract to a very small size; this process is said to be repeated several*

times. (Wasson, 1968: 254-255)

[From a 1903 book based on two years of field research with Siberian tribes:] *The effect...became evident by the time the men had swallowed the fourth mushroom. Their eyes took on a wild look...with a positively blinding gleam, and their hands began to tremble nervously....After a few minutes a deep lethargy overcame them, and they began quietly singing monotonous improvised songs...They suddenly sprang raving from their seats and began loudly and wildly calling for drums....And now began an indescribable dancing and singing, a deafening drumming and a wild running about...during which the men threw everything about recklessly, until they were completely exhausted. Suddenly they collapsed like dead men and promptly fell into deep sleep... It is this sleep that provides the greatest enjoyment; the drunken man has the most beautiful fantastic dreams.* (Wasson, 1968: 262)

[From the diary of a Polish army officer, published in 1863, who was fed fly-agaric as medicine during a visit to Kamchatka in 1796 or 1797:] *I ate half my medicine and at once stretched out, for a deep sleep overtook me. Dreams came one after the other. I found myself as though magnetized* [mesmerized] *by the most attractive gardens where ...a group of the most beautiful women dressed in white going to and fro seemed to be occupied with the hospitality of the earthly paradise...* [Later, upon taking more of the mushroom] *I fell asleep anew and did not wake up for twenty-four hours. It is difficult, almost impossible, to describe the visions I had in such a long sleep...What I noticed in these visions and what I passed through are things that I felt I had seen or experienced some time before, and also things that I would never imagine even in my thoughts.* (Wasson, 1968: 244-245)

These are only a few of numerous such accounts collected by R. Gordon Wasson, the man who theorized that Vedic Soma was *Amanita muscaria*. Many others have been published elsewhere (Salzman et al., 1996), including a recent report that copious perspiration can sometimes result from eating this mushroom, making it seem to objective observers as if the imbiber has "swallowed fire" (Heinrich, 1995: 26-27). Similarities with aspects of Cú Chulaind's warp spasms and wasting sickness – as well as with his "salmon leap," the hero's fabled ability to jump across great distances or high into the air – would seem to be obvious.

The emphasis here is on *aspects* of Cú Chulaind's abnormal behavior corresponding to various symptoms of *Amanita muscaria* intoxication. The combined effect of these aspects in Cú Chulaind is exaggerated, making his deeds "more often seem superhuman than heroic," as one scholar noted (Gantz, 1981: 25). There is nothing in the record to suggest that *A. muscaria* induces either lethal battle furies or year-long lassitude. Wasson, for example, was dismissive of the theory that the famous "berserk-raging" of the Vikings during battle was caused by ingesting this mushroom (*see* Ödman, Ch. 9). Such "murderous ferocity," he noted, "is conspicuously absent from our eye-witness accounts of fly-agaric eating in Siberia" (Wasson, 1968: 157). However, it seems to me feasible that Cú Chulaind's warp spasms and wasting sickness represent exaggerated versions of the mushroom's true effects as perceived by observers and subjectively reported by its users, including ataxia (loss of muscle control), frenzied dancing in shamanic rites, amazing feats of strength, dramatic optical distortions, and disruption of sleep patterns lasting for several days with unusually vivid dreams.

Another interesting property of *Amanita muscaria* is that the urine of a person who ingests it becomes psychoactive. Apparently, the chemical constituent that causes its psychoactivity metabolizes only rather slowly. Thus, the Siberians have a long history, noted with astonishment by many Europeans, of drinking their own or another person's urine to prolong the mushroom's usefulness. A witness to this practice described it as follows in 1809:

> *Among the Koryaks...it is quite common for a sober man to lie in wait for a man intoxicated with mushrooms and, when the latter urinates, to catch the urine secretly in a container and in this way to [sic] obtain a stimulating drink even though he has no mushrooms. Because of this curious effect, the Koryaks have the advantage of being able to prolong their ecstasy for several days with a small number of fly-agarics. Suppose, for example, that two mushrooms were needed on the first day for an ordinary*

intoxication; then the urine alone is enough to maintain the intoxication on the following day. On the third day the urine still has narcotic properties, and therefore one drinks some of this and at the same time swallows some fly-agaric, even if only half a mushroom; this enables him not only to maintain his intoxication but also to tap off a strong liquor on the fourth day. (Wasson, 1968: 249-250)

In other words, the psychoactive urine becomes less so with each successive urination unless supplemented by ingesting more *Amanita muscaria*. This diminishing effect, confirmed by scientific research (Ott, 1993: 328), recalls the procedure of dunking Cú Chulaind in three vats of water in order to progressively curtail his warp spasm and restore him to a state of normal consciousness. Another interesting parallel is found in the *Rig Veda*, where hymn 9.97.55 mentions three filters through which Soma must be passed to render it clarified (Wasson, 1968: 51-58).

Also, worth mentioning here is a striking description of Cú Chulaind's hair. According to one of the myths, it was "brown at the base" and "blood-red in the middle," with "a crown of golden yellow" (Kinsella, 1969: 156). Dried caps of *Amanita muscaria* often exhibit the same three colors (Figure 1).

Shamanic Origins

The foregoing adds credence to theories that the ancient Irish Celts may have used *Amanita muscaria* as an intoxicant or in Druidic rituals. If so, this would seem to refute Wasson's (1968: 181) claim that the Celts were one of the peoples "infected with a virulent mycophobia, coming down from prehistory" and his report of finding "no mushrooms in the records that we possess of the shadowy Druids" (p. 176). Perhaps what Wasson interpreted as Celtic mycophobia was actually a strong tabu against the casual mention of a psychoactive mushroom used exclusively in secret magic rituals. I do find it curious that Wasson does not comment on the mysterious "magic egg" (*anguinum*) which, according to Pliny the Elder, was "esteemed by the Druids and believed by them to 'ensure success in law-courts and a favourable reception by princes.'" Pliny described it as "round, and about the size of a smallish apple," with a "cartilaginous shell...pocked like the arms of a polypus [octopus]." The Druids told him that it was composed of secretions and spittle from snakes. Scholars have guessed that it may have been either a sea-urchin minus its spines or the agglomerated egg-case of a whelk (Piggott, 1975: 188). To me it sounds possibly fungal in origin – perhaps a mushroom somewhat altered in appearance by Druidic art.

In any event, it is certain that the Irish bards, themselves most likely Druids, do not openly allude to any psychoactive mushrooms in the ancient myths. We have heard the opinion of Laurie and White (1997) that these myths refer to *Amanita muscaria* only symbolically in order to "communicate teachings about the sacrament's use to initiates, while maintaining a protective veil of secrecy around its identification" (p. 54). That may well have been the case. But they concede that Wilson's theory is another possibility; that Irish Celts may have "pre-

Fig. 1: Dried caps of *Amanita muscaria* (Photo by Thomas J. Riedlinger).

served *soma* motifs in their myths without actually continuing the use of *soma* – just as Christians still cherish many ancient pagan religious symbols, such as Yule logs and decorated trees at Christmas, and fertility bunnies and eggs at Easter, without understanding their original pagan content" (p. 54). Wilson's viewpoint also seems to be supported by the fact that *A. muscaria* does not readily grow in Ireland. Laurie and White think it may have been more common there in ancient times, before the almost total deforestation of that country in the last thousand years. But if so, as they themselves admit, "there is no irrefutable archeological evidence...such as the discovery of an archaic medicine bag filled with psychoactive mushrooms" to prove that Irish Celts used *A. muscaria* or any other "psychotropic substances capable of inducing ecstatic, visionary experiences" (p. 53).

I would add that the effects of *Amanita muscaria* are so unpredictable that the Celts had strong incentive to abandon its ritual use before settling in Ireland. The problem for them was not only that *A. muscaria*'s active ingredients vary "considerably" from mushroom to mushroom (Heinrich, 1995: 15). "Soil conditions and geographic and seasonal factors also affect its hallucinogenic properties" (McKenna, 1992: 108). Imagine the frustration of attempting to gauge the appropriate dose of local mushrooms for different people in varying regions and seasons and climates as the Celts migrated westward for thousands of miles from their point of origin! That alone would suffice to explain why the Celts abandoned using *A. muscaria* in favor of surrogate plants or mythic symbols such as magic food. However, my respect for the intelligence, resourcefulness and pluck of traditional shamans allows that it is possible that once they had settled in Ireland the Celts found some way to mitigate the mushroom's potentially dangerous effects. In that case what Cú Chulaind represented for the bards of ancient Ireland was not *A. muscaria*'s use in their contemporary culture but the difficulties that their Celtic ancestors experienced when using it before they came to Ireland. These difficulties may have been conflated with the memory of battles fought during the Celtic migration. For example, *The Cattle Raid of Cuailnge* (*Tàin Bó Cuailnge*), is believed to derive from a narrative version transcribed as early as the 7th century C.E. But the earliest surviving copy dates from the 11th century and appears to be "a conflation of two 9th-century texts with some extra material added by the compiler himself" (hÓgáin, 1991: 414). According to Dáithí Ó hÓgáin (1991: 414), who regards *The Cattle Raid of Cuailnge* as the "central, and structurally the basic, story in the [Ulster] cycle":

> *This narrative is taken to encapsulate many aspects of the culture of the ancient Ulaidh [var. Ulaid], portraying a warrior-aristocracy organised on the lines of a heroic society and providing an authentic picture of an Iron Age Celtic culture. The military-political situation described in the narrative was explained by a series of "pre-tales" which were put together at a quite early date in support of the Tàin. These "pre-tales" also preserve fragments of myth and ritual from ancient tradition, and thus the general corpus evidences several details which can be compared with what Greek and Latin writers on the Continent attribute to the Celts known to them.*

This reinforces the theory that Irish Celts retained in their traditions ancient memories of things that their ancestors did or saw done either during or before their migration from the Indo-European heartland. At about the same time that they started this westward migration around 2500 to 2300 B.C.E., their Aryan neighbors apparently left the same heartland and started moving southward into India. Reflecting on this process of migration and culturization, the distinguished archeologist Stuart Piggott (1965: 106-107) has observed:

> *More and more the archeological evidence begins to reflect the existence, over most of Europe, of a warrior aristocracy of a type familiar to us from the heroic and epic literature ranging from the Iliad to the Sagas; from the Rig Veda to the Tàin Bó Cuailnge. If we look for an early context for the structure of society already ancient in early Europe, the tripartite social grading in its various forms; the government by king, elders, and assembly; the importance of the warrior class--it fits readily within the framework of what we can infer from the early second millennium B.C. onwards.*

Piggott (1965: 106-107) says that this earlier framework, comprising a "curious amalgam of traditions and techniques, of peoples and ideas," provided the general context from which Celtic culture emerged en route to Ireland. Wasson, whose research exposed him to Piggott's ideas, sent Piggott a copy of his book *Soma* and received in return this endorsement of Wasson's fly-agaric theory in a previously unpublished letter dated 11 October 1970:

> *I would say straight away that I am persuaded that you must be right. One can never arrive at mathematical certainty in such things, but I find the cumulative effect of your always level-headed and judicious arguments quite convincing. I had already thought that, whatever Soma may have been, its use in Vedic religion as an agent to induce ecstacy [sic] meant that we were, in that religion, partly in a world that was not characteristically Indo-European, but which was somehow linked with shamanism in its widest sense. I recently wrote a little book on the Druids, and in it had to consider whether there was a shamanistic element in Celtic religion: with Eliade, I had thought there was none, but his book set me thinking in wider terms. The probability of Finno-Ugrian elements in*

> *Sanskrit would of course give just the links needed between the two worlds.* (Piggot, 1970)

What Piggott (1975: 88) actually says in his book *The Druids* is that the "Celtic tradition in Ireland conserved untouched archaisms in language, ideas and even prosody which have their counterparts in Sanskrit or Hittite, and we must be seeing fragments of a common heritage that goes back to the second millennium B.C." His analysis of these archaisms led him to conclude that "the Irish vernacular sources, especially the hero-tales, are the product of a primitive, illiterate, heroic society with a warrior-aristocracy" that composed them in accordance with its values (p. 100). He also believes that certain Druidic practices of the Irish Celts suggest "a very archaic substrate of belief" reflecting, it may be, an even earlier and more "primitive" influence: Central Asian shamanism, possibly originating near the Altai Mountains that border Siberia (p. 188). Rutherford (1987: 126, 128), too, calls attention to this region as a possible shamanic nexus for the Druids and Vedic Aryans (*see also* Carl Ruck's comments on the Altai Mountains region as a possible source of *Amanita muscaria* in ancient times [Wasson, Kramrisch, Ott and Ruck, 1986: 253]). In addition to observing that "the Brahmanic ritual drug soma... is plainly the descendant of those used by the shamans to assist in inducing the trances" (Rutherford, 1987: 104), he states in a footnote:

> *Evidence as to whether the Druids used ritual drugs to aid trance is wholly circumstantial, coming in part from Pliny's description of the mistletoe-gathering ceremony, though the mistletoe-berry is not a hallucinogen, and partly from a folk-tradition that they extracted an opium-like substance from the poppy. It is also possible that the so-called "magic mushroom" was used for Druidic purposes.* (Rutherford, 1987: 108)

The term "magic mushroom" is usually used in reference to those of the *Conocybe, Psilocybe* or *Stropharia* genera, which have very different effects from *Amanita muscaria* (Wasson, 1968: 162; *see also* Ch. 28). However, Rutherford's guess is more compelling in regard to *A. muscaria*. For even if the Druids did not use the fly-agaric in their practices, the fact that it grows only in mycorrhizal relationship with conifer and birch trees makes it similar in one important respect to mistletoe: the latter is a parasitic plant that grows only on trees. According to Pliny, the Druids considered "nothing more sacred than mistletoe and the tree that it grows on, so long as it is an oak." Beneath such trees, he said, they held religious feasts and sacrificed "with prayers to the god to render [their] offering propitious" (Eluère, 1993: 119). It thus seems feasible to me that the Druids used mistletoe in these rituals as a symbolic surrogate for *A. muscaria*.

Conclusion: The Soma Connection

A similar process occurred, I believe, in the Vedic religion with Soma. However, this change would have happened more slowly because of different circumstances. Migrating westward from their Indo-European heartland, the Celts encountered a wider variety of indigenous cultures than the Aryans did heading southward into India. Each successive conquest of these cultures presumably altered the Celtic religion through a process of syncretic transformation, just as the religion of the Aryans was changed by local Indian religions. Since it took the Celts far longer to reach Ireland (apparently the first of them arrived there in the sixth or fifth century B.C.E. [Gantz, 1981: 6]), it is easy to see why the Irish myths differ so markedly from Vedic hymns to Soma while yet bearing what Mac Cana (1985: 14) calls "the unambiguous marks of a common origin."

One of these marks, I believe, is the Indic god Rudra who figures in some of the earliest Vedic hymns. Clark Heinrich has convincingly explored the possibility that Rudra was initially identified with Soma (Heinrich, 1995: 37-43). I would add that Rudra's attributes remind me of Cú Chulaind. This is clearest in *Rig Veda* 2.33, the "Hymn to Rudra," which acknowledges his power to bestow special favors on those who worship him properly and to injure or at least withhold his favors from those who do not (O'Flaherty, 1981). The first of the favors solicited is "the vision of the sun" (2.33.1). Next come prayers for protection, progeny and health or healing. As with Cú Chulaind, Rudra's powers are potentially useful in battle. He is asked to ensure that "our warriors on horseback remain unscathed" (2.33.1) and to fend off "all assaults of injury" (2.33.3). But also, like Cú Chulaind, Rudra sometimes shows a tendency to injure those he normally protects. Helplessly acknowledging that Rudra's "shattering power" and "vast strength" (2.33.10) make him "like unto a beast" that is "fearful and strong, all ready for the kill," his petitioners pray to be spared the effects of his "fearful wrath" (2.33.5) and "great ill-will" (2.33.14).

Their exhortation, "lay another low – not us!" (2.33.11), sounds identical to what we would expect to hear Cú Chulaind's fellow warriors tell him.[2]

The Vedic priests apparently tried hard to accommodate Rudra's capricious behavior. I believe this accounts for the complex stipulations of their Soma ceremony. If the mushroom itself had an unpredictable potency that threatened unpleasant experiences, there was only one recourse: all other variables had to be strictly controlled. This is consistent with an observation made by Andrew Weil (1978: 152-153):

> *One valuable influence of ritual is to minimize the disruptive potential of [psychoactive] drugs and maximize their useful effects. All drugs have this ambivalent potential: they can trigger positive, helpful reactions or unpleasant and unhelpful ones... Ritual in the use of drugs works to curb the development of panic reactions by standardizing expectation in a positive direction. It helps define reasons for taking psychoactive substances in the first place, and gives participants a framework of order through which to interpret their experience.*

In the Vedic Soma ritual, techniques were developed to standardize both the preparation of the sacrament and the constitutional receptivity of those who ingested it. The Soma plant was pressed to release its juice, a tawny liquid. Then this juice was mixed with milk or curds. Perhaps numerous *Amanita muscaria* mushrooms that had been harvested at different times in different locations were combined at this stage of the ritual, in order to average out their diverse potencies. The Soma drinkers prepared themselves to imbibe the concoction by fasting and chanting the relevant Vedic hymns for several days. This would have standardized the body biochemistry of different individuals as much as possible, while also adjusting their minds to provide an appropriate "framework of order" for enlightenment. Yet despite these precautions, they sometimes, unpredictably, encountered Rudra's devastating wrath and eventually tired of it. Then they, like their cousins the Celts did at least temporarily, gave up using *A. muscaria* in favor of non-psychoactive alternatives that were dependably transmissible from adepts to initiates—most notably the esoteric system of philosophy laid out in the Upanishads. At that point Rudra-Soma joined Cú Chulaind as a literary vestige of the Indo-European heartland.

Notes

1. Paper presented 29 October 1999 at the Mycomedia Millennium Conference, Breitenbush Hot Springs Retreat and Conference Center, Detroit, Oregon. © 1999 by Thomas J. Riedlinger.
2. All quotes from the *Rig Veda* taken from O'Flaherty (1981).

Bride of Brightness and Mother of All Wisdom:
An Ethnomycological Reassessment of Brigid, Celtic Fertility Goddess and Patron Saint of Ireland

Peter McCoy

> *Adjuva Brigitta!*
> (O Brigid Help Me!)
>
> *Brigit, that is, the female poet, daughter of the Dagda. This is Brigit the female seer, or woman of insight, i.e., the goddess whom poets used to worship, for her cult was very great and very splendid.*
> Cormac mac Cuilennáin
> (10th century scholar)

One of the most beloved and enduring goddesses of ancient Celtic culture is Brigid, the red-headed patron of prophesy, fertility, smithcraft, and healing. With her control over the fires of creation and creativity, Brigid has had a more significant influence on Western cultural development than most pan-Celtic deities through the guidance and inspiration she's provided to midwives, poets, and spiritual leaders for thousands of years. Like other Celtic goddesses, Brigid's mythos is filled with intricate tales of miraculous powers that guided the minds of men and transmuted the elements of nature. Yet the importance and density of symbolism attached to her is unlike that found with most of her immortal kin – a reverence that not only emphasizes Brigid's high status in the Celtic pantheon, but that also draws out questions around her rise to prominence.

Unfortunately, as few writings about Brigid were left by the Celts, her full history remains unclear and in recent decades has only been roughly pieced together by historians. Most of these chroniclers focus on dispelling contradictions found in the writings of the Celts' conquerors, such as the Romans. Surprisingly, few modern scholars have looked to the multi-layered symbols found in Brigid's stories for insights into her past, instead opting to dismiss these recurrent themes as only the details of fables.

When given a thorough assessment, however, the symbols that surround Brigid seem to provide clues into her mysterious origins. For among the most notable aspects of the Goddess' legacy – at least, for the mycologically minded – is the recurrence of motifs that have been strongly correlated by ethnomycologists with the most culturally significant mushroom of all time: *Amanita muscaria.*

Throughout Brigid's tales and associated modern customs we find the colors red and white, altered states of mind, and more cryptically *Amanita*-esque motifs (such as crows and single eyes) repeatedly mentioned, often with several symbols being presented in conjunction and in sequences that offer little value as allegories or teaching tools. As the works of Ruck (2007; 2011), Heinrich (2002), Wilson (1999), and others have argued, such visual metaphors can be readily interpreted to suggest that cultures from around the world and throughout history knew of *Amanita muscaria*'s mind-altering effects and used the perseverance of art and story to preserve that knowledge.

Thus, by treating the appearance of these symbols in the tales of Brigid not as mere curiosities, but as lingering truths from forgotten Celtic traditions, we might find answers to how one of the high goddesses of the Celts gained her greatness. And in doing so, further support the existence of an occult human-fungal relationship that once spanned Old World Eurasia.

PART I: Brigid's Reach

To appreciate these connections in full, the contexts and belief systems within which Brigid developed must be understood. The assorted cultures collectively referred to as "Celtic" originated in either central Europe or along the western shores of the Black Sea as far back as 7,000 years ago. During the millennia that followed, the Celts spread across the continent and established cities in central Turkey, southern Poland, Italy, France, Spain, Portugal, the British Isles, and Ireland. As they travelled, their language, customs, and belief systems evolved, with major shifts likely being influenced by the climate and species encountered in their new settlements, as well as by cultural exchanges with outsider groups.

The Celtic culture was caste based, with laborers in the lower classes, and warriors, intellectuals, and the nobility holding greater power. The spiritual and philosophical leaders of the Celts, the Druids, held the top caste due to their extensive knowledge of history, natural cycles, and divination practices (Weber, 2015). Despite being literate, the highly secretive Druids never wrote their teachings down, leaving our current knowledge of these "magico-religious" leaders to be based on a small number of writings by ancient Roman and Christian scholars.

As the political and spiritual biases of these writers likely skewed their depictions, the accuracy of their claims (especially in regard to their negative depictions of the Druids) is still debated. And yet, their texts describe the Druids as highly intellectual, with members specializing as prophets, philosophers, or poets to share their wisdom with the lower castes. Druidic initiates trained for as long as twenty years, during which time they memorized the stories and customs of the order's complex, animistic belief system and honed their arts of logic, magic, and rhetoric (Caesar, 50 BCE).

The Druid's legendary status and mastery of nature's mysteries has seen a revival over the last century in various occult schools and Neo-Pagan practices, with one of the order's most iconic representations found in the Magician card of Arthur Edward Waite and Pamela Colman Smith's tarot deck (Figure 1). Cloaked in red and white and lifting a white wand in one hand while pointing toward the ground with the other, the card's Magus stands beneath a floating lemniscate – an *Amanita*-button-esque figure-eight that signifies the eternal cycle – as he draws together the four elements of water (cups), air (swords), fire (wands), and earth

Fig. 1: The Magician card of Pamela Smith's tarot deck encapsulates the Druids' nature-based spiritual practice through layers of symbols that are strongly associated with Brigid and *Amanita muscaria*.

(pentacles) on his table, all while red roses and white, psychoactive *Datura* flowers grow at his feet.

Druids and the Way of the Oak

Of all the plants of the Old World, the Druids held the oak tree among the most important. With its grand height, expansive branches, and resistance to insects, fire, and disease, the mighty oak was associated with the great gods of many cultures, including Thor, Zeus, and Jupiter from Scandinavian, Old Norse, Roman, and Greek mythologies. For the Druids, oak groves were considered sacred spaces, within which they trained initiates, held secret meetings, and performed divination practices.

It also seems that the Druids paid special attention to other species associated with oak trees, as one of their most sacred plants was a mistletoe shrub (*Viscum album*) found growing on an oak. European mistletoe (Figure 2) is a slow-growing, perennial evergreen plant that commonly lives on the branches of apple, poplar, hawthorn, and rowan trees as a hemi-parasite. It is rarely found on oaks, however, and when it was encountered in these trees, the Druids would ceremonially harvest it six days after a new moon, with the harvesting priest using a golden sickle to cut down the plant while being dressed all in white (Whalley, 1982).

Why the Druids cherished mistletoe is not certain, though it may be due to the plant's notable appearance. Mistletoe has the unusual ability to host both flower and fruit simultaneously – a result of its berries requiring a full year to mature following pollination. Such a trait is rare among flowering plants and was likely regarded as a curiosity, if not a magical trait by the Druids. Additionally, the plant's flowers are triple headed and produce white, round, moon-like fruits in sets of three, a number that is sacred in many cultures and

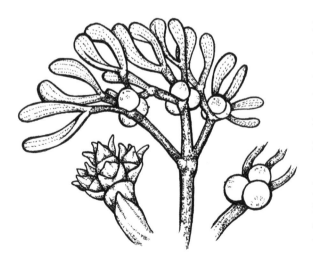

Fig. 2: A mistletoe (*Viscum album*) monograph denoting the three-ness of the plant's flower and fruit structures (Illustration by Peter McCoy).

which recurs throughout Celtic motifs and stories of Brigid.

Curiously, the Druids seem to have venerated the plant for its influence over fertility, for though mistletoe is considered toxic today, they were reportedly able to prepare a potion from it that made barren animals fertile. If true, this implies that rather than avoid mistletoe due to it having unpleasant characteristics, the Druids worked to uncover a beneficial use for this intriguing plant (i.e. to increase fertility), and to incorporate it into their customs.

The *Amanita* in Druidic Rites

If such an emphasis was placed on mistletoe due to its anomalous growth habits, symbolic appearance, notable healing effects, and association with oak, it is likely that the Druids also studied other organisms with similar traits. In *Amanita muscaria*, we not only find all of these features, but also the ability to significantly alter consciousness – a combination that would have been hard for the Druids to ignore as they surveyed the inhabitants of their sacred groves (Figure 3).

To the Druids, the mushroom's unique fire-like radiance must have stood out against the backdrop of the forest, just as mistletoe's unusual fruiting habit was likely appreciated. Similarly, the mushroom's seed-less underground origin must have held a degree of mystery, and would have been especially regarded when patches fruited near an oak each autumn – whether as a direct mycorrhizal partner of that tree, or as a partner of another nearby tree species.

The mushroom also hosts other visual traits that the Druids may have noted as symbolically significant. The growth of the mushroom can be said to come in three stages, where after starting as a white egg (which looks a bit like a large mistletoe berry), the mushroom breaks out of its veil and extends, only to soon after uplift before collapsing in age. As many researchers have noted, the red-and-whiteness and general form of the fruit body can also be symbolically interpreted in several ways (Heinrich, 2002). The two halves of the mushroom have been likened to the union of various opposites, including the sun (as the red cap) and moon (as the white, bulbous base), as fire and water (or ice), and as a copulating couple, with the male principle represented by the phallic stipe and the female principle seen in the red "breast" that is "penetrated" by the stipe. The all-seeing eye and evil eye have also been suggested to be a reference to the mushroom's glowing cap, just as several one-eyed gods (e.g. Odin) and creatures of mythology (e.g. the cyclops) are suggested to be analogs of the *Amanita* (Ruck, 2007). Acts of beheading in various cultures may have developed as allusions to removing and consuming the mushroom's potent cap (Feeney, 2013). And with its underground origin, the *Amanita* has been likened to snakes exiting their den, just as the mushroom's growth cycle and emergence from its (snakeskin-like) universal veil have repeatedly been attributed to notions of renewal and rebirth.

Many myths connect the mushroom with ravens or crows, birds that are regarded for their high intelligence. Some stories claim the raven is a messenger for the dead, and that its wisdom is gleaned from knowledge of the afterlife. Other cultures call the *Amanita* "raven's bread" (Ruck, 2011), suggesting the mushroom was historically perceived by some as the source of the bird's knowledge. To the Druids, who believed in reincarnation, the bird was considered a wise elder, one which they trained to aid in their acts of prophesy. Today, "to have a raven's knowledge" remains an Irish proverb that means to have a seer's supernatural powers. As the bird is connected to the *Amanita* in Celtic lore by way of symbolic associations in the tales of Brigid and her kin, we must wonder if the Druids also believed that ravens consumed the mushroom to gain their wisdom.

As we will explore in Part II of this chapter, all of the above symbols recur throughout the myths of Brigid. When assessed in isolation, such connections can be fairly dismissed as coincidental. However, when considered holistically, the density of these symbols in each tale of the red-haired Goddess is too great to ignore and must be considered as references to *Amanita*-based knowledge held by ancient Druids, and potentially other Celtic castes.

Upon a *myco-literate* review, it seems quite likely that at some point in the history of the Druids, members of the order worked with *Amanita muscaria* to determine how to mitigate its nauseating potential and incorporate the mushroom's effects into their practices, just as they had with mistletoe. As existing accounts of the Druids make no direct mention of any *Amanita*-related practice, we will never know for certain how the Celts utilized the mushroom's effects on the body, or if they might have used the *Amanita* to alter consciousness. And yet, we can fairly suppose that either the Roman and Christian chroniclers chose not to record such activities, or, more likely, that the Druids kept their knowledge of the mushroom – and especially its mind-altering properties – a secret from anyone outside of their order.

Fig. 3: Ancient Irish Druids likely encountered oak trees that hosted mistletoe in their branches and *Amanita muscaria* fruitbodies near their roots. The unique traits of all three of these species was likely not overlooked by the Druids, but instead incorporated into their animistic belief system and spiritual practices (Illustration by Peter McCoy).

Such mycological elitism was not uncommon historically, as many advanced ancient cultures – including the Egyptians, Chinese, and Greeks – treated mushrooms as a "food of the gods" that was reserved for priestly and noble classes (Berlant, 2005). Some researchers have suggested that this class division may have been due to a secret understanding of the psychoactive effects of some mushrooms by ancient elites, which they sought to occlude from the lower tiers of their culture.

In the otherworldly myths of the ancient Celts, we find what seems to be a highly creative method for obscuring an *esoteric* understanding of *Amanita muscaria* in fanciful *exoteric* tales that can be easily disregarded as childish by the uninitiated and mycologically unaware. As the Druids transmitted their knowledge orally, these tales may have historically served as cryptic time capsules for their wisdom, with coded language that perhaps became more elaborate toward the end of the Druids' reign, so as to bypass the cultural filters of their conquerors. Such attempts to seemingly transmit lessons on the *Amanita* are especially apparent in the stories of Brigid, which are replete with mushroom-like imagery and subtle suggestions of altered states of mind.

The Wanderings of a Goddess

In the Celtic myths that survive today we find a tapestry of gods, faeries, and otherworldly creatures engaging in battles that convey direct and indirect lessons on the power of nature. Among the many deities described, the Goddess Brigid stands out as one of the most beloved, as well as one of the most historic. While it may be that she is a unique outgrowth of the Celtic migrations, her similarity to other Earth Mother goddesses found across India, North Africa, and Asia has been suggested to be a result of cultural exchanges between early Celts and other groups (Weber, 2015). If this was the case and the prototype for Brigid was inherited, we must wonder if her *Amanita*-associated symbols were also of non-Celtic origin – and thus intrinsic, or even fundamental, to certain Old World representations of the fertility goddess and female principle – or if they were attributes added by the Druids and later shared with their peers in nearby cultures. The former progression seems most likely, and is supported by research on the possible common ancestry between Celtic and Vedic cultures and their highest castes of Druids and Brahmans (Ellis, 2003).

The name Brigid also adds to the confusion of her origins as it is thought to derive from the Indo-European word *bhrghnti*, further suggesting she arose early in, or outside of, Celtic culture. Similarly, the Celtic term *Brig* or *Brid* translates to "exalted," "high one," or "high lady," and may have also been used by the Celts as a title of respect for sacred objects, places, and concepts, as well as for women that held positions of power. It may also be that the Celts avoided directly stating the names of their deities out of fear, and instead used surrogate titles, such as Brig, when referring to any goddess (Wright, 2009). Such naming

Fig. 4: This illustration is based on an existing British statue of Brigantia that is replete with Minervan symbols (e.g. a grasped orb, spear, and wings) that can also be interpreted as references to *Amanita muscaria* (Illustration by Peter McCoy).

complications likely underlie the reason why Brigid came to encompass such a wide range of qualities, as well as why her name has taken on a variety of adaptations across Celtic cultures, including Bride, Ffraid, and the following:

- French – Brigette, Britt, Britta
- Gaelic – Brighid ("bree-id")
- Gallic (Gaul) – Brigandu, Brigindo
- Irish – Brigit
- Manx (Isle of Man) – Bahee, Brede, Breeshey, Veeshey
- Northern English – Brigantia
- Scottish – Brede, Breo-Saigh, Bride, Bridean, Bridi, Brüd
- Swedish – Birgitta, Bridget
- Welsh – Ffraid ("frry-ed"), Ffred, Fride

The Warrior Goddess and the Romans

The height of Celtic culture started around 600 BCE and began to decline during the Roman conquests initiated under Julius Caesar (100 BCE–44 BCE) (Weber, 2015). Unlike the Christian missionaries that came after them, the Romans did not attempt to eradicate the spiritual practices of the Celts, but rather to lessen the leadership status of the Druids and merge Roman gods with Celtic deities. These efforts aimed to undermine Celtic leadership, while attempting to find a spiritual common ground that reduced the need for violent confrontations between the groups.

One example of this merging is in the apparent Roman adoption of Britain's version of Brigid: the warrior Brigantia, from whom the country and many of its towns and natural features take their name. It is thought that the Romans found similarities between Brigantia and their own warrior goddess, Minerva, as surviving statues of Brigantia depict her wearing a helmet, hosting wings, and holding a spear and globe, all of which are Minervan features (Figure 4).

Curiously, these traits are also somewhat suggestive of *Amanita muscaria*, as a helmet is analogous to a mushroom cap, a globe is akin to the egg stage of the mushroom's life cycle, a spear is much like an elongated mushroom, and wings appear similar to the gills of an agaric mushroom when cut in half. Some Brigantia statues also depict her standing in a pillared and arched portal that looks both similar to a mushroom's silhouette, and, more suggestively, to an *Amanita* mushroom encased within a universal veil.

Other statues of Brigantia show her with a snake-headed Gorgon on her chest. As snakes are often connected with mushrooms, researchers have suggested that the Gorgons were yet another symbol of the *Amanita* (Ruck, 2011). Similarly, at least one metal bust of Minerva dating from the mid-18th century has been found with an agaric mushroom embossed on its chest (Irvin & Rutajit, 2005). Though this sculpture is of more recent origin, it may allude to a historic Minerva-mushroom connection that has been poorly recounted – a connection that may have helped increase Celtic acceptance of the Brigantia–Minerva merger. While many scholars believe Brigid existed in Ireland for centuries before the Roman invasions began in Celtic lands, some suggest her stories came to Ireland around 71–74 CE by way of British Druids fleeing Roman persecution (Wright, 2009).

The Christianization of Brigid

Though Roman invasions covered most of the Celtic territories, the Irish Celts were never conquered by them, enabling their culture and stories to evolve mostly unimpeded until the fifth century CE, when Christian missionaries finally arrived. Soon after, the Christians converted many Celts to their religion, often by "Christianizing" Celtic beliefs, by building churches atop sacred sites, or by creating saints based on Celtic gods and goddesses.

Being such a central deity to the Celts, Brigid seems to have been one of the few goddesses that the Christians chose to incorporate into their teachings, as soon after their arrival in Ireland miraculous tales of a Saint Brigid healing the sick began to circulate. Most of these stories host details and *Amanita*-like qualities that are quite similar to those found in myths of the Goddess, and yet modern devotees of the Saint believe she was a real person born around 439–452 CE – or around the time that the Christians came to Ireland.

Saint Brigid is said to have performed many miracles, and as such an abbey was founded in her honor in County Kildare, on what was potentially the historical grounds of a temple for the Goddess Brigid. Here, an eternal flame was lit for the Saint, which was almost consistently tended for a thousand years until the Reformation when Henry VIII (1509–1541) deemed it unholy.

The assertions of Saint Brigid having existed are based on two accounts of her that were written over 100 years after her death – evidence so sparse that the Vatican officially stopped recognizing her as a Saint in 1969. Regardless, in 1992, modern followers of the Saint reignited her flame near its original site, which remains watched over by a group of women today.

Saint Brigid is now considered a patron Saint of Ireland, with a status so great and endearing that she is second only to Saint Patrick in terms of cultural significance. Despite the inexplicable chronology, she is even said to have been the midwife to Mary during the birth of Jesus, and to have anointed the head of the newly born Christ Child with three drops of water.

As Christianity later spread to the Caribbean, Saint Brigid's importance was incorporated into Voudou practices, where she became the culture's only white skinned and red-haired *loa* (or *lwa*, spirit), known as Maman Brigitte, the Goddess of cemeteries and death (Weber, 2015). Among her many traits, Brigitte is said to drink a special type of rum called *piman*, which is fiery-hot due to being infused with spicy peppers.

As with the Goddess Brigid, Saint Brigid is referred to in various ways, such as with these diminutive pet names:

- Irish – Breda, Breege, Bridhe
- Irish and Scottish – Bree, Bridie
- Irish and English – Biddie, Biddy

Or by titles of respect that include:

- Bride of Brightness
- Bride of Joy
- Brighid of the Immortal Host
- Flame of Two Eternities
- Mother of All Wisdom
- Mother of Songs and Music
- St. Brigid of the Flame
- The Flame in the Heart of All Women
- The Lady of the Sea

Such a retention of Brigid's connection to fire in these titles speaks to the centrality of this trait in the Goddess and in the Celts' memories of her during Christianization. As a likely stand-in for the *Amanita*, this aspect of Brigid's mushroom embodiment is the most visually enduring, as the fruit body's red cap beckons from the duff like a flame while, upon consumption, it embodies traits of the life-giving mother in its ability to heal the minds and spirits of those who embrace its many gifts.

Imbolc, The Day of Brigid

She who put the beam in sun and moon,
She who put the food in earth and herd,
She who put fish in stream and sea,
Hasten the butter up to me.
Pray Brigid, see my children yonder,
Waiting for the buttered buns,
White and yellow.
　　—Traditional Irish Prayer for Imbolc

Three months after November 1, when the holiday Samhain marked the end of the harvest season, the ancient Celts celebrated Imbolc on February 1 to honor the beginning of Spring. After spending much of winter with their entire family confined to a small and often single-room home, the Celts would commemorate the first signs of lengthening days by opening their home's doors and welcoming in both the warmth of the returning sun and the goddess most connected to that return: Brigid.

Coming out of the coldest months of the year, when survival was sustained on increasingly limited rations, Imbolc was celebrated as a time of the first fresh foods. Among the harvests was sheep's milk, which began to flow around Imbolc due to sheep being autumn-mating animals ("short-day breeders") that only gestate for five months. This cherished nourishment is the reason that the full moon of February is referred to as the Milk or Nursing Moon. It is also why the holiday obtained its name, as the Old Irish word *imbolg* (em-bol/g or immol'g) means "in the belly" and the reportedly related Saxon word *oimelc* ("oy-melk") means "ewe's milk" (Chadwick, 1970: 181).

The Imbolc season is also when mistletoe hosts its flowers and drops its fruits, and when ancient Druids likely performed harvests of this sacred plant. Though *Amanita muscaria* fruits in the fall, the mushroom could have been dried and stored over winter, potentially to be consumed during Imbolc ceremonies.

This day of renewal and rebirth was also the time of celebrating Brigid as the primary fertility and Earth Mother goddess of the Irish Celts. With her many powers, it was she who guided life in the land back from its long wintery death. Today, February 1–2 is observed under the names Imbolc, Brigid's Day, or Candlemass, depending on the practitioner's spiritual beliefs, and is primarily a day meant for honoring life, fertility, renewal, and the sun's fiery light, as well as for cleansing and letting fresh air into the home. Traditional Imbolc celebrations include welcoming Brigid into one's house and using various rituals and talismans to increase fertility in the home through the blessing of the Goddess. Aided by the fire of Brigid, the day is a time of birth and revival that can be likened to the butterfly's emergence from its cocoon, or to an *Amanita's* eruption from the confines of a universal veil.

PART II: Brigid Unveiled

The images of Brigid handed down to us today developed within the Druids' animistic worldview: a perspective wherein each element of the natural world is seen as imbued with a unique spirit. Though we do not fully know how the Druids expressed this belief, we can suppose that the unique qualities of certain plants and mushrooms were perceived as arising from the personality of each species' spirit. Perhaps this is why mistletoe was harvested in such a ceremonial fashion: not only did the Druids seek to respect the plant, but to also appease the soul that it hosted.

Considering the unique traits of *Amanita muscaria*, it is compelling to suppose how the Druids might have described the spirit of such a mushroom. Would its appearance have been seen as an incarnation of the fires of creation, or its effects on the user as magical powers that only a goddess could possess? How might the Druids have described the spirit of the *Amanita*, or addressed it when harvesting, preparing, or consuming its flesh? In the symbols and stories of Brigid we seem to find answers to these questions. For if these tales are considered as lessons on how to engage with the mushroom, it seems that the *Amanita* commanded great respect from the Druids and that they held it sacred to similar degrees as the mistletoe.

Indeed, the importance of Brigid in Druidic practice is so strong that many researchers consider her to have been their patron goddess. In many ways, her prowess made her the greatest of all of the Earth Mother goddesses in Celtic culture – the ultimate provider as well as an embodiment of the primal and uncontrollable forces of nature, much like the unpredictable appearance of a radiant mushroom from the hidden webs of wild mycelium.

Daughter of the Dagda, Oak, and Boyne

Brigid's legend begins with one of her most commonly told origin stories, wherein she is the daughter of the Dagda (Daghdha) and (most likely) the Morrigan, deities that both host their own *Amanita*-like qualities, and of whom Brigid is said to be a combination.

The Dagda was known as "the Good God," meaning he was highly skilled at all things, and was associated with abundance, fertility, generosity, magic, and wisdom. Also known as Ruad Rofessa ("Red Man of [Occult or Druidic] Knowledge") and Aedh Alainn ("Fiery Lustrous One") – names that easily call up images of the red *Amanita* – he is one of the chief gods of the Tuatha dé Danann, a race of supernatural beings who dwell in the magical land of the Otherworld (or *Tír na nóg*), which is accessed by going under water or underground.

The Dagda was known for his great, bottomless (i.e. life-giving) cauldron and for his magical harp that created three types of music. The Strain of Lament (*Goltrai*) was a sound that caused listeners to cry uncontrollably, the Strain of Laughter (*Geantrai*) brought about great happiness and intense laughter in the audience, and the Strain of Slumber (*Suantri*) could lull listeners into a deep, magical sleep. Notably, all three of these effects allude to consciousness altering, and are similar to common features of *Amanita muscaria* intoxication.

The Dagda was also known for his massive club, one end of which was used to kill – thus making it bloody and red – while the other end was able to bring the dead back to life. The notion of reincarnation was fundamental to Druidic teachings, and is also a common theme in *Amanita muscaria* symbolism found in other cultures. While many aspects of the natural world could have led the Druids to believe in reincarnation, we can suppose that a type of spiritual awakening process brought about by the *Amanita*, coupled with observations of the double-ended and club-shaped mushroom's life cycle, reinforced those beliefs or helped influence this symbol of the Dagda's power over death and rebirth.

The Morrigan (or *Mór-Rióghain*, which potentially means, "Phantom Queen," "Queen of the Dead," or "Great Queen") is a triune goddess who is presented as a one-eyed crone, a raven caller, and shapeshifter who could turn into a crow known as the *badb* – all symbols that have been repeatedly connected to *Amanita muscaria* by other cultures. The Morrigan is also said to have owned a magical white cow that had red ears, a combination of features that is not well explained by scholars on Celtic mythology.

In the story of the Tain Bó Regamma, the Morrigan is described as red headed and as wearing a red dress while she rides a one-legged red horse into battle (one-legged-ness being akin to a mushroom that "stands" on one "leg"; *see* Wasson, 1971b). Her warrior attributes have been transmitted by some writers to Brigid, often interpreting this trait as being more akin to a mediator than a fighter. That is, the Morrigan acted as a deity able to talk between both sides of an argument, or navigate between worlds, much like the effects of the *Amanita* that can cause one to feel like they are in another world, or in a battle between states of mind.

Another origin story states that Brigid is the daughter of the goddess Boann, a one eyed, one-legged, and one-armed deity that is said to have drowned in a flood that was caused by her approaching a magical well, after which she became Ireland's river Boyne. This well was in the (underground) Otherworld, where it was off limits due to being filled with the Salmon of Knowledge: subterranean fish that obtained great insight by eating hazelnuts of wisdom that fell from nine magical trees above their well. These hazelnuts have been suggested to be references to *Amanita muscaria* (*see* Laurie & White, Ch. 11), as the mushroom can be likened to a nut beneath its common associate the birch tree, which is in the same family as hazelnut trees (the Betulaceae), and shares a few loosely similar visual qualities with it.

Other origin stories tell of Brigid being born from an acorn beneath a great oak – a connection that may also allude to *Amanita muscaria*, which has a cap texture that looks similar to the tops of acorns. In many tales, Brigid is also described as a triple-aspect goddess, which may potentially be due to the confusing use of Brig as a generic title for many older deities. Today, her three sides are often described as the sisters *Ban leighis*, the Woman of Healing (or Leechcraft), *Ban goibnechtae*, the Woman of Smithwork, and *Ban fhile*, the Woman of Poetry.

Saint Brigid's Fiery Birth

The origin of Saint Brigid is as magical as that of the Goddess Brigid in its connection to Druidry, fire, and walking between worlds. In the most commonly told account, the Saint is said to have been born while her mother, a Druid's slave, stood in the front doorway of her master's home at sunrise. Much like a mushroom emerges from the darkness of the underworld into the light of a new day – while also often appearing on the border between an open meadow and a dark forest – Saint Brigid was born straddling the boundary between magical and mundane worlds. Such a symbolic birth can be interpreted as an exchange of the older form of Brigid for her Christianized appearance, as well as an allusion to the *Amanita* experience, which can bring about an interplay between perceptions of reality.

When she was born, it is said a great pillar of fire shot from the Saint's head and consumed the Druid's home, but did not burn it. This image not only demonstrates the Saint's connection to the heavens and to the fiery qualities of the Goddess but is also suggestive of the red cap of *Amanita muscaria*, which is born from an enveloping veil that is white like the hottest fires.

In the celebrations of Imbolc, the importance of the doorway is reflected in the custom of welcoming Brigid into the home through one's front door, either by active statement, or through ceremonially welcoming in a person playing the role of Brigid after they knock three times. Another Imbolc custom is stepping through Brigid's girdle: a ring, or portal, of woven rushes. In doing so, practitioners are symbolically reborn into the fresh light of spring – albeit in an act that looks much like an *Amanita* mushroom growing out of its universal veil.

Brigid of Smithcraft and Fire

As a fertility goddess, Brigid is linked to the etheric and immutable life force that guides creation – a mysterious power that is embedded into her iconography through depictions of a radiant fire that consumes her head of red hair. Brigid is, if nothing else, an embodiment of the fire of life and its timeless ability to create, destroy, and rebuild all that we find in the natural world. And through her many celebrations, Brigid's followers venerate this power over fertility and seek to bring its innate abundance into their lives and landscapes.

Alongside this role, Brigid is considered the steward of smithcraft, a skill that, in the eyes of the untrained Celt, would have been a miraculous act that used the fires of creation to transform raw stone into braided torc collars and ceremonial cauldrons. In the *Amanita*'s growth we find a strong metaphor for this process wherein the fungus' mycelium orders the chaos of the soil web into a fiery, phoenix-like fruit body with the potential to transmute an unrefined soul into something more substantial through its effects on the spirit.

As metal is heated in a forge, its tip becomes bright red. And when the smith plunges the tip in water to cool it, the metal becomes covered in flakes that are reminiscent of snake scales and the universal veil remnants (warts) of an *Amanita muscaria* cap. The colors of red, white, and black are common to smithing and are also found in *Amanita* mushrooms arising from dark soil. Likewise, they are the colors most associated with Imbolc, as well as those that are used by modern followers of the Goddess Brigid to celebrate her (as in the use of colored candles on a ritual altar).

Brigid's connection to fire was reinforced through various associations between her and the sun, most notably in the solar symbol made during Imbolc known as Brigid's Cross (Figure 5). This talisman is crafted by weaving together dried reeds or rushes into a four-armed cross, each of equal length. (As reeds were also commonly used as flooring in ancient Irish homes, the choice of this material would have historically reinforced a connection for the Celts between the Goddess and fertility of one's land, or, for the initiated, between her spirit and its power to transform earth into a sun-like mushroom.) The Cross is one of Brigid's most iconic symbols and is thought to bring about her protection and to increase fertility when hung over beds and the entrances of homes and barns. These Crosses are kept in place for one year before being burned in the hearth fire on Imbolc Eve and replaced with new Crosses the following day.

In one story of the Saint, we are told of Brigid miraculously hanging her wet cloak on a sunbeam to dry. On its surface, this tale stands out, as it is unlike her other miracles, which mostly relate to healing. However, when considered as a disguised reference to the *Amanita*, the story can be seen in one light as a reminder to sun dry the mushroom, potentially when its universal veil hangs from the fruit body's red-tipped shaft. Likewise, the tale also seems to be a reverential metaphor for the miraculous appearance of the mushroom after a rain, or, potentially, after a lightning strike, which can be described as a thick

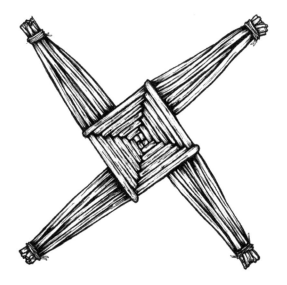

Fig. 5: Brigid's Cross is made by weaving reeds on the night of Imbolc Eve as a means to increase fertility around the home and to invite Brigid's protection over her followers (Illustration by Peter McCoy).

sunbeam that bridges the heavens and earth.

In many ancient cultures, the appearance of certain mushrooms (including *Amanita muscaria*) was attributed to the fiery power of lightning strikes, more so than the rain that often accompanied them (Wasson, et al., 1986). At the same time, oak trees, with their great height, are often the first to be struck by lightning in aged forests – an affinity that contributed to the oak being associated with lightning gods, such as Thor. This relationship with lightning would have likely been noted by the Druids in their work with these trees. It would be quite interesting to know, then, if the Druids held their own connection between lightning and mushrooms, as other cultures of their time had done, or if they had attributed special significance to lightning-related mushrooms that occurred near their sacred oaks.

Brigid of Healing and Water

One of the primary traits of both Goddess and Saint Brigid is their miraculous healing ability. The Goddess oversaw the healing arts in Celtic culture and thereby was connected with sacred wells that were said to have restorative properties. The British Isles hosts over 7,000 wells, with approximately 3,000 in Ireland, 1,000 in Scotland, 1,200 in Wales, and at least 2,000 in England (Wright, 2009). It is thought that many of these are directly connected to Brigid, though the exact number is unknown.

Mushrooms are also strongly connected with this element as their fruit bodies are often comprised of 80–90% water, and thus appear in greater numbers after periods of rain. As the *Amanita* matures, its cap uplifts, creating a cup shape that can collect rainwater like a miniature well. This water can even become tinged red in time (much like turning water into wine) and may be able to take on some of the medicinal or psychoactive compounds found beneath the mushroom's cuticle. As with many mushrooms, *Amanita muscaria* has also been traditionally used across Eurasia as a medicine, both for internal and external maladies (Rogers, 2012).

One of Saint Brigid's miracles tells of her creating a well by dropping a burning coal on the ground. This odd connection between fire and water not only recalls similar associations found in the Goddess, but also draws up images of the red-and-white mushroom, which, when young, looks somewhat like a white-hot coal topped by red flames. Two sacred wells dedicated to Saint Brigid are located near her church in Kildare and her followers still visit and decorate these sites each year during the time of Imbolc.

Brigid is also associated with water through the art of brewcraft, which the Goddess is said to oversee. Like smithcraft, the creation of ale requires the skilled use of water and fire to transform base materials into something new and more powerful. To the ancient Celts, consumption of mind-altering ale was connected to acts of prophesy, just as work with *Amanita muscaria* could have been. Stories tell of Brigid turning milk into wine, a potential reference to upturned caps of the *Amanita* and the magical waters they contain. And through her connection to wells, Brigid is tied to the underground waters in the Celtic Otherworld that host the speckled Salmon of Knowledge and their magical hazelnuts of wisdom.

One of the recounted miracles of Saint Brigid states that her blood once cured the muteness of two girls. If we are to interpret this rather odd act as a veiled *Amanita* allegory, this tale may offer a lesson on the red mushroom's ability to enable one to speak their truth more clearly after a psychoactive experience.

In another tale, we are told of an encounter that Saint Brigid had with two lepers at her well in Kildare. After meeting them, the Saint used her gifts to heal one of the lepers, under the requirement that he wash the other leper's wounds thereafter. When the healed leper refused, Brigid brought his sickness back upon him and then wrapped her white cloak around the other, completely healing him.

While this story can be interpreted in many ways, an *Amanita*-minded analysis suggests that the tale is a reminder of the importance of heeding the lessons gleaned from an altered state of consciousness. For though such an experience can offer great potential for growth and healing, if those insights are not used to help oneself or others, then the original illness is likely to return. Conversely, when individuals allow themselves to be enveloped in the message of the mushroom – signified by Brigid's white, universal veil-like cloak – they are more likely to retain their healing insights for a longer period.

Brigid of Poetry and Air

Along with her connections to the elements of water and fire, Brigid is known for her gifts associated with air, as offered by her invention of various types of sounds. The Goddess is said to have been the first to *keen*, or offer a lament or song of bereavement in the wake of a death. Her first keen is said to have occurred after her red-haired son Ruadán was killed in a battle – an act reminiscent of the

Dagda's ability to relay sorrow through sound. The true origin of keening is not known, though the Christians later outlawed it as a pagan act. This is surprising as mourning is also a part of the Christian bereavement process. We might wonder, then, if keening traditionally accompanied other Celtic practices that undermined the church, such as attempting to communicate with the deceased or shifting one's state of consciousness.

Brigid is also said to have invented whistling, so that people could communicate in the night, as well as writing, so that words could be heard with the eyes, even in faraway places (McGarry, 2005). Such an ability to increase communication is much like the results that can come from *Amanita* experiences, where perceiving the world in new ways can lead to the creation of new forms of art, stories, and symbols.

One of Brigid's primary attributes was as the Goddess of poetry, which in Celtic times was an important art for communicating knowledge. Poetry was also closely connected with prophesy and altered states of consciousness as repeatedly chanting poems was used by the Celts to bring about a trance-like state. The moment of poetic inspiration or frenzy was even said to be caused by a "fire in the head," a phrase that some researchers suggest is an allusion to the consumption of *Amanita muscaria,* which can cause warmth and sweating in the body (*see* Laurie & White, Ch. 11). The mushroom's fire-red "head," or cap, also contains the bulk of its psychoactive compounds, a fact that potentially led the Druids to their belief that the human head is the seat of the soul, despite most other cultures of the time believing the heart to be the spirit's primary residence.

Female druids (*ban Druaid*) were especially regarded for their skill at prophesy, so much so that their insights were used to guide the decision making of the kings and nobility. Recorded methods for divining included looking deeply into water (*fàisnich uisge*) or flames (*fàisnich teine*), though it is likely that the Druids also used other practices to obtain novel ideas, such as by altering their state of consciousness.

Divination is also a major aspect of Imbolc, with one common practice being the act of reading the ashes of a hearth fire. If marks appear in the ashes on the morning after Imbolc, it is said that they were caused by Brigid's footsteps and signify that the Goddess had spread the ashes across the family's land during the night, so as to increase fertility for future crops. In Scotland, weather divination was an important aspect of Imbolc as it was said that snakes (a common symbol for mushrooms), would emerge from their dens to foretell the qualities of the coming spring. This belief was later reinvented as Groundhog Day (February 2) in the United States, where a groundhog is said to foretell the weather by the presence of its sun-cast shadow.

In the 1993 film *Groundhog Day* (Ramis, 1993), we find hidden references to the holiday's original purpose of celebrating growth, renewal, and transformation. In the film, a weatherman (i.e. a person who forecasts the future) reincarnates into the same day hundreds of times until, after killing himself several times, he matures intellectually, emotionally, and spiritually, eventually turning from a selfish ingrate into a wise and caring person. The allegorical nature of the film is even referenced in one scene where the protagonist, Phil (a name shared with the divining groundhog), eats an abundance of food made from milk, butter, and bread, and calls himself an immortal – a god.

Brigid of Fertility and Earth

The final element of earth is also strongly tied to Brigid, as she was a fertility goddess connected to the creation of life and its annual renewal each spring. Her relationship with procreation was so strong that her other common name, Bride, came to denote a fertile woman who is soon to be married and, thereafter, with child.

In the stories of Imbolc, Brigid is said to draw upon her fiery qualities to reignite life in the land after winter by tapping a white wand made of birch or willow on the ground. This act was said to counterbalance the work of the Cailleach (or "veiled one"), the fearful winter hag and Goddess of the Underworld who caused the coming of winter by touching her own white, club-shaped wand to the ground during Samhain.

The Cailleach is often considered the crone aspect of Brigid, as well as her opposite. She is said to have red teeth, white hair, and to wear a veil, which itself signified occult knowledge in Celtic mythology. Like the Morrigan, the Cailleach is sometimes represented by the raven, and thus tied with death, darkness, and secret wisdom.

With her striking *Amanita*-like qualities and connection to Brigid, the Cailleach seemingly represents the mushroom as well. Considering the autumnal timing of her work then, it is possible that the ancient Druids saw the fruitings of *Amanita* mushrooms at this time of year as an appearance not of Brigid's spirit but of the powerful Cailleach, who, through her fearsome description, embodied the mushroom's darker qualities. If this was the case, then perhaps the fabled triumph of Brigid over the Cailleach in the spring was not solely a representation of the return of longer days, but also a metaphor for the battle between the darker and brighter aspects of one's mind and spirit, as might be experienced under the influence of *Amanita muscaria*.

The annual dance between the Cailleach in the fall and Brigid in the spring may be the forgotten reason behind an odd game played during Imbolc, wherein two people stand back to back, lock arms, and dance about. Apart from simply being something fun to do on a cold day, this act may have roots in a symbolic representation of the dance of opposites at this time of year. With their bodies arched away from each other, the dancers can be said to reflect the balance of warmth and cold, maiden and crone, life and death, and autumn and spring – all while appearing somewhat like the arcs of the sun and moon, or the top and bottom orbs of a young, red-and-white *Amanita* button.

With her ability to bring about fertility in the spring, the Goddess and Saint Brigid host many tales wherein foods that look similar to the *Amanita* play a central role. In one story, the Saint was gifted a bounty of apples by a woman who refused to offer any of her crop to lepers. Despising such disregard for the sick, Brigid cursed the woman's orchard to be barren from then on. As the red-outside, white-inside apples of the story look much like the caps of *Amanita muscaria*, we can consider this story to allude to one's need to respect the gift of an *Amanita* bounty and to share the benefits it provides with those in need of healing and purification.

In other tales, Brigid is connected with milk. When Saint Brigid was a child, she could only consume the milk of a magical cow that had a white body and red ears (similar to the cow owned by the Morrigan). In her later life, the Saint was also said to own a magical cow that never ran dry. The Goddess Brigid was likewise seen as the guardian of livestock as she possessed four regal animals, one being the King Ram, *Mag Cirb*, which, as the counter to the sheep of Imbolc and their life-giving milk, reinforces Brigid's connection to fertility, birth, and nursing. Likewise, the plant most closely connected to Brigid is the dandelion (*beàrnan Brìgdhe*), which produces a healing, milk-like latex when cut and hosts a flower that, in its sun-like radiance, is somewhat similar to the cap of a fiery *Amanita*.

In the tale that describes how Saint Brigid acquired the land for her temple, we find imagery that can readily be seen as influenced by fungi. In one version of the tale, the greedy King of Leinster told the Saint that she could have as much land as her white cloak would cover. Thinking he had gotten the best of her, the King was shocked when he watched Brigid's attendants each take a corner of her cloak and walk in four directions until the cloth had magically grown across much of the Curragh, an open plain in Kildare. Here, Brigid's Temple of Kildare (or *Cill Dara*, the "Church of the Oak Tree") was finally situated, reportedly next to a great oak. Along with reinforcing the connection between the Saint and the oaks of the Druids, this tale sounds strikingly like a metaphor for the equidistant growth of a white mycelial network, one that perhaps originated from an important mycorrhizal tree partner in the Curragh.

In the traditions of Imbolc, we find this story remembered through the practice of hanging Brigid's magical cloth, or mantle (*brat Bhride*), outside the front door of one's home on Imbolc Eve. This strip of cloth would sit overnight to collect healing waters in the form of dew, which were thought to be blessed by Brigid as she passed in the night. This important cloth would later be used during healing activities or births, as the Goddess was also closely tied to the practice of midwifery.

Though Brigid's connection to fertility is often minimally highlighted in modern assessments of her (despite it being one of her primary attributes!), this trait was likely of great importance to the ancient Celts, who depended on the vitality of their partners, land, and livestock for survival. We can imagine, then, that Imbolc was historically not merely a time for celebrating the return of spring, but a period for intentionally calling in the power of life's creation through an intensive use of rituals and icons.

In his assessment of the Dead Sea Scrolls, John Allegro suggests that *Amanita muscaria* was worshiped as a highly important icon of fertility cults in the Middle East around the time of the Celtic culture's peak (Allegro, 1973). Similarly, Clark Heinrich argues in his analysis of the Vedic Pine Forest Story that the mushroom was honored as a fertility symbol due to its combination of phallic and vulvic features (Heinrich, 2002). As we find similar connections seemingly made by other traditional cultures, it seems quite possible that the mushroom was considered to play a role in the creation and sustainment of life by many peoples across the Old World during the Celtic era. If the Celts inherited these connections from their ancestors or those they traded with – or came to them in their own time – the fertility rituals of Imbolc would have likely incorporated this potent mushroom in some form – a motif that was seemingly obscured by symbolism in the centuries that followed the decline of the Druids.

One of the contemporary customs of Imbolc seems to be a relic of such fertility worship, wherein a *biddy*, or doll (*Brídeóg,* "Little Bride") that represents Brigid, is made from reeds and placed in a small reed bed alongside a phallic-shaped white wand, which sometimes has a pinecone attached to one of its ends. Though this modern form of a likely ancient fertility ritual is rather family-friendly, we should wonder how versions of this ceremony looked during Celtic times, and whether the fertility-associated *Amanita* or mistletoe (which many kiss under today during winter celebrations) played a more prominent role in these rites.

If this was the case, then it is most likely that *Amanita* mushrooms harvested and dried during Samhain were stored until Imbolc, at which time their caps were removed and consumed during rituals of fertility and perhaps divination upon the coming year. Further, as this was the season of celebrating new births, the caps could have been consumed in conjunction with sheep's milk, just as the *Amanita* has long been eaten in parts of Siberia with caribou milk. The following excerpt from a traditional Imbolc hymn seems to hint at such a "beheading" practice:

> *Early on Bride's morn the serpent shall come from the hole,*
> *I will not molest the serpent nor will the serpent molest me…*
> *On the Feast Day of Bride,*
> *The head will come off the Cailleach*

Brigid Endures

Brigid's endurance through the trials of time is arguably her greatest gift. In the hope she offers her modern followers, the Goddess stands as an eternal flame who carries on life's many mysteries and signals in the night with a whistle, keen, and poem. In her various tales we are drawn to imagine a time when the forests were dense and inspired, when the coming of each spring was venerated in the highest order, and when the magic of mushrooms was never ill regarded. In her modern celebrations we kindle that power of return in customs around the hearth, all of which point back to a great Earth Mother's bounty, as well as her innumerable prehistoric incarnations.

In the Celtic version of the great goddess handed down to us by Druidic bards and Christian reformers, we find connections to the great mushroom of fertility, *Amanita muscaria*, thinly veiled in icons of death, birth, wisdom, and magic. In these layers of symbolism, hidden traits of the ancient Celts seem to shine forth from occulted wells to suggest that these great peoples knew more about fungi than many in the West today.

Where and when such knowledge entered the development of the various Indo-European cultures is uncertain, but the commonalities of the *Amanita*'s presentations across the Old World suggests it to be a relationship deeply seated in the life ways of the distant past. Having been one of the last indigenous European cultures to be dissolved by the dictates of the church, the customs and beliefs of the ancient Irish Celts offers a rare glimpse into this all but forgotten era of aligning with nature's patterns and demands. By considering the potential that the red mushroom was a central element in the practices of their spiritual leaders, we expand the mushroom's reach from Beringia and India to the western-most islands of the European continent. And in doing so, find means to reincarnate the mushroom's heritage through a reassessment of our contemporary February customs.

This legacy can perhaps best be summarized in a story of Saint Brigid where we are told that upon being asked by a suitor to marry, the Saint plucked out a single eye, so as to ward off the man who attempted to make her his own. Though often interpreted as a metaphor for Brigid's autonomous power and vast wisdom (conveyed here with her *Amanita* button-esque all-seeing eye), this tale may also serve as a reminder that no person or religious organization can ever control the fiery *Amanita*, who's spirit has guided the Celtic culture, and all peoples of the forest, for so long.

Chapter 14

Mail-Order Mushrooms:
An Interview with Mark Niemoller

[The following interview between Kevin Feeney (KF) and Mark Niemoller (MN), former proprietor of JLF: *Poisonous Non-Consumables* (1986-2005), was conducted electronically during the summer of 2018.]

KF: I first became aware of JLF: *Poisonous Non-Consumables* in the spring of 1995 when I came across a classified ad for *Amanita muscaria* mushrooms in the back of *High Times Magazine*. While there were other classifieds advertising Morning Glory seeds and *Psilocybe* spore prints for mail-order, to my knowledge JLF was the only business at that time to offer dried *Amanita muscaria* mushrooms for sale. Can you provide a little background about JLF: How it got started? And how it came to offer *Amanita muscaria* and *pantherina* mushrooms for sale?

MN: I believe the first classified ad in *High Times Magazine* for *Amanita muscaria* mushrooms came out in 1986 or 1987. That's when JLF started. After researching this mushroom for ten years prior, I had realized the opportunity. Since there was great demand but no supply for this product, I figured I could provide it and fund further research. So, I drug a friend along who agreed to cash the checks under his name, and the company became his initials. Off to southern Mississippi we went on our first mushroom hunting expedition in search of the nearest red variety to put in the product line, already having a small inventory of the dried yellow variety from back home in Indiana. My predictions were right and the first JLF business trip paid off. We hit the November season just right and picked and dried about 27 five-gallon bucketfuls. Drying was done in the over-cab sleeper of a beat up 1975 class C motor home, using a dehumidifier and fans and homemade screens.

The next mushroom hunting mission was to acquire the *real* red variety in

Colorado, at which we also lucked out on the first attempt. The Greenhorn Mountains in August of 1988 were full of them. The actual hunting and transporting was done with our two dual-purpose motorcycles and giant backpacks. It was very hard and long work, but it was adventurous. We probably rode our cycles over 100 miles a day each.

After careful consideration, we decided the safest way to distribute the dried mushroom product was as 'poisonous' and 'not for consumption', hence the rest of the name *Poisonous Non-Consumables*. Although, a couple years later my partner got tired of the mundane office work and sold out to me; but the JLF name stayed. I haven't seen him since.

KF: Wow, so JLF started in the mid to late-1980s. That's a little earlier than I had suspected. While I want to talk more about the history of JLF and your experiences with the company, I am curious to know more about what first sparked your interest in this mushroom. Did this interest arise out of the general counter-culture interest in mushrooms during the 1970s, or was there something else? Do you remember your first experiences with this mushroom? Or the first time you came across a fly agaric in person?

MN: Yes, memories of my first fly agaric experiences are pretty distinct. Here are a couple. I'd always been a nature boy, interested in things like fossils and reptiles and insects. The Golden Field Guides were my favorite books. In the mid-1970s, just out of high school, I became interested in wild edible plants. One day out foraging for plants, I "rediscovered" toadstools, which were all over the place on a hillside in a Brown County [Indiana] forest. I had always been told to stay away from them, but as I looked at their perfection and symmetry, I wondered if any of them could be edible. I went straight to the county library for some field guides. As I was looking through the fungi section, I came across Gordon Wasson's (1968) *Soma*. I was amazed at the subject matter, which not only seemed out of place in this section but also in this small conservative county library. So, I searched the card catalog and found three other out-of-place books: John M. Allegro's (1970) *Sacred Mushroom and the Cross*, Mary Anne Atwood's (1960) *Hermetic Philosophy and Alchemy*, and a book called *Alchemists and Gold* by Jacques Sadoul (1972). I read these books over and over, thinking I was losing my mind, seeing the same mushroom at the heart [of] Hinduism *and* Christianity *and* Alchemy. But it was a good kind of insanity.

I was also fascinated and excited. I had heard about *Psilocybe* mushrooms from friends, but this was different. I sensed something very special about this mushroom. I had to learn more and go search for them. But finding them proves less than quick or easy for a lone newbie in this area, which does not harbor them naturally. I didn't find any until the fall of 1979 on a trip to southern Michigan with my brother (whom I had visited in California the previous year, where I found none – because it was summer). It actually took me ten years to find them locally, at least in quantity and with certainty (I think I picked my first one here in 1978 but did not know it since it was yellow, and I was unaware of the color varieties at the time). During this ten-year period, I joined the North American Mycological Association, went on forays, became a semi-expert on wild mushrooms, checked out psilocybin, went to the Telluride Mushroom Festival a couple times, etc. – all the while hunting my butt off for the Red Fly Agaric.

I also had experiences of another kind, which are quite memorable. Before I had found local *Amanita muscaria*, a couple of times I was supplied with a few dried specimens of red *muscaria* from other people for consumption. I also consumed some of those yellow ones from Michigan in 1979. Without getting into "just another trip report," suffice it to say that I had sufficient subjective experiences to assure myself that I was on the right track. That brings us up to 1985 when I sold my first bulk shipment of self-collected/dried *muscaria* to a foreign chemical company, which was the technical beginning of "serious business."

I still wonder who was to thank for getting those four books into our library and changing my life.

KF: It is amazing the impact that a couple of good books can have on an individual. I was not so fortunate to have such volumes available in my local library, though I do remember one of the early books that kindled my interests in mushrooms and ethnobotanicals was Jeanne Rose's (1972) *Herbs & Things*. I remember the book was filled with strange anecdotes, recipes and psychedelic drawings; however, Rose's book was quaint in comparison to the next book that made a major impression on me, Jonathan Ott's (1993) *Pharmacotheon*. And by the time I received my first JLF catalog in 1995 (Figures 1 & 2), after mailing my two crumpled dollar

Fig. 1: Cover of JLF Catalog #10, distributed circa 1995.

bills, *Pharmacotheon* was an indispensable reference for viewing and interpreting the expansive list of ethnobotanicals included in the JLF catalog.

Perhaps you can speak a bit to the journey of JLF from a small two-man operation in 1986, focused on *Amanita muscaria*, to becoming one of the first comprehensive ethnobotanical suppliers in the U.S. by the early to mid-1990s. What happened? Was the demand for *Amanita muscaria* sufficient to support the expansion of the JLF catalog? And how did you maintain sufficient stockpiles of *Amanita muscaria* to make JLF a viable business (assuming the fortunes of two foragers would not have been sufficient)?

MN: Yes, Ott's *Pharmacotheon* was the last in a classic trio, starting with the little booklet *Legal Highs* (Gottlieb, 1973), then Schultes' (1980) *Plants of the Gods*. Those were the bibles of the religion. After obtaining an inventory of fly agaric, the next task was to obtain as many things from those books as possible – knowing that the books created the demand.

As far as obtaining enough fly agaric, I basically did that all by myself; at least in the early years. I could usually put away around 50 dried pounds per year, spending long hours during fruiting seasons picking and slicing and drying in the old Realite camper while on the road and in the old modified school bus at home. It was hard work, but I loved it more than any other part of the job. Picking that much *muscaria* is an experience in itself. Over the years, I alone probably picked somewhere around a quarter million fly agaric mushrooms. The species can be very prolific, and a little knowledge about their growth patterns proved very profitable.

In the later years, I started buying *muscaria*, but only because I could not pick enough of the red variety myself; which did not grow closer than Colorado. And the climate in Colorado is very hit-or-miss. I made too many mushroom collecting trips there that were complete failures. So yes, the fly agaric mushrooms definitely helped the company grow, but my overhead was always very low. I operated it from my home and barn and did most everything from a shoestring, never even taking out a loan. Starting with nothing and never having over three employees, in 15 years the company had acquired a million customers and even more dollars. And even though a lot of that success came from other items offered in the catalog, *Amanita muscaria* always remained the featured product and one of the best sellers.

KF: Fifty dried pounds a year is quite a lot! Probably enough to send a couple thousand people into orbit – if they knew what they were doing. Between the different varieties of *muscaria* and *pantherina* that JLF carried, how many pounds would you estimate that JLF sold during its years of operation?

MN: That would be hard to estimate. Hundreds of pounds? A ton? More?

KF: The internet has done a lot to democratize information about psychoactive substances in the last 20 years, but in the late 1980s and early 1990s most individuals interested in the psychoactive properties of the fly agaric probably only had experience with *Psilocybes* as a reference point. I imagine this created confusion for some customers, particularly when *Psilocybes* are active in the range of 2 and 4 dried grams and *muscarias* typically require somewhere between 5 and 20 dried grams.

Another point of confusion, I suspect, was with the different varieties of fly agaric (red, orange, yellow, peach, brown, white). I remember being quite surprised to see these different varieties and being unsure what to make of it. I had never heard of *Amanita muscaria* var. *formosa* or var. *flavivolvata*, and the local library was no help in providing illumination or clearing up this confusion.

While JLF products were marketed as "poisonous non-consumables" and the catalog provided a lengthy, but entertaining, disclaimer about the use of JLF products (*see* Appendix 1) I assume there must have been a number of customers that called seeking additional "information" about the fly agaric, or perhaps to complain that what they purchased did not "work," so to speak.

Although both *A. muscaria* and *A. pantherina* were and are legal (with the recent exception of Louisiana, where they are prohibited) the federal government has typically taken the position that anything that is used to purposefully alter consciousness (outside of coffee, tobacco, sugar and alcohol) can be treated as suspect, and potentially lead to legal entanglements. In this murky legal area (thus the tag: *Poisonous Non-Consumables*) how did JLF handle questionable customer inquiries and complaints? Were there many?

MN: Yes, there were many such questions. That's what led to the long disclaimer. But even after, there were still such questions. And that's what led to the required recorded release statement, swearing that the customer understood and agreed to the "not for consumption" condition for purchase.[1] We had a whole file cabinet full of cassette tapes of releases. Essentially, we did not give out usage information; we just sold the raw material. And if a customer persisted and/or admitted intent to consume/abuse, we refused to sell to them and put them in a blacklist database. Quite a bit of time and effort (expense) was put into enforcing this policy.

   Astonishingly, at the trial, the prosecutor argued that such a long and ridiculous disclaimer meant the opposite; and since I got convicted on eight counts, which were contingent of the intent to sell for consumption, the jury must have believed the backward logic of the prosecutor. Well, at least the disclaimer prevented any civil lawsuits; and in today's business world, selling such risky items, I figure that's no small accomplishment.

KF: It sounds like you were pretty meticulous in maintaining and promoting your product line as "non-consumable" goods. Regarding the disclaimer, one could argue that it was tongue-in-cheek, however, one could also interpret it as a satirical commentary on the excessive litigiousness of American society as well as social trends towards the rejection of personal responsibility. My perception was that it was a little of both, but regardless of one's interpretation it would be hard to argue that it wasn't an unequivocal and thorough disclaimer.

MN: I wrote the disclaimer with no tongue-in-cheek intent. It was purely an attempt to be thorough, regarding the standard point of any *terms of service* disclaimer; for the purpose of safety for the customer and liability protection for the seller. It is interesting because the current situation seems to emulate my attempt: Just look at any owner's manual for an electronic appliance, and you will likely see 12 pages of warning and cautions and disclaimers *before* you get to any use instructions – 12 pages of the most ridiculous warnings imaginable ("do not use this hair dryer in the bath tub").

KF: The trial you referred to, that's the one that eventually led JLF to close its doors, right? I have gathered bits and pieces of what happened, some of which I think was shared on JLF's website, but never heard the whole story. My understanding, however, is that the charges that were brought were related to other items in JLF's catalog and did not specifically relate to *Amanita muscaria* or *pantherina*. Would you be willing to briefly summarize what happened? Or share more about the decision (if indeed it was a decision) to close down JLF?

MN: Right; none of the charges included *Amanita muscaria/pantherina* products. As I had expanded the catalog offerings based on customer requests and legally purchased some new items from legitimate companies, it turned out that some of them were also actual prescription drugs and/or chemically related to prohibited drugs; neither of which was an easy thing to ascertain back then. While proof of guilt was supposed to be based on a knowledge element, I did *not* know what the prosecution claimed I did (that those things were 'illegal'). Anyway, there's no way to briefly summarize such an insane legal proceeding (that took 7 years) without getting off track. Yes, the legal issues caused the end of the company. I kept it open for a few years after, but just for principle. And since everything was paid for, why not?

My interest in *Amanita muscaria* was the foundation of the company, survived its demise, and continues to this day.

KF: So JLF opened its doors around 1986 and was active for about 15 years, until 2001 or 2002, is that correct? It strikes me that the market realities, economic opportunities and level of competition must have been dramatically different at the time JLF closed compared to the time it began. My recollection is that there were not a lot of individuals or companies involved in the sale of ethnobotanicals until a certain critical mass of Americans had arrived on the internet; sometime in the mid to late-90s. By the late 1990s there was quite a bit of online activity, with special interest listservs and discussion boards where people shared information and traded fly agaric, *Salvia*, and other items, and businesses specializing in ethnobotanicals were popping up like, well… like mushrooms!

How did JLF experience this changing business environment? Did you feel the increase in competition? Did it affect sales? Did you have to change any business strategies? Or were changes during this time more minimal?

MN: JLF actually stayed open until around 2005, even though business dropped off considerably after the raid in 2001. As far as competition and the internet in the late 1990s, it was the reverse of what your question asks. Things really took off in this era with the internet and our website, and we never really felt increasing competition. Maybe it's because there was plenty of business to go around, but it felt like we were on top. The only changes we felt were positive; such as, vast increases in sales and revenue. The website was able to reach far more people, but I'm sure our biggest advertiser was internet word-of-mouth. Of course, our quick rise in popularity and success not only reached more customers, it also reached unpleasant people in big government.

KF: Would you describe the raids as a focused endeavor or was it more of a fishing expedition? More specifically, did the government come in looking for certain products they considered suspect for seizure or did they end up confiscating large sections of JLF's product-line that were ultimately unrelated to the goals of the raid? If they seized any unrelated products were these ever returned, or was this simply a loss the company was forced to bare?

MN: I'm sure the Feds had specific items listed on the warrant, but they were taking anything they thought looked suspicious. And if you're a 'Drug Fed', many things look suspicious; like any brown jar with a chemical name. Yes, they seized a lot of stuff that day; probably a quarter million in assets and inventory value. Although my lawyers made them give a smidgeon of it back after the case was over, it probably cost me more in attorney fees than it was worth. They also seized the bank accounts, of which they gave about half of it back – four years later. That kind of raid would have immediately destroyed any normal small business, which I'm sure was their hope.

KF: I wonder if the government was disappointed with the outcome of the case. I am not aware of any similar raids against other ethnobotanical companies despite business practices that lacked the discretion practiced by JLF. Many of the ethnobotanical companies that have proliferated online in the last 20 years have featured psychedelic imagery on their webpages, provided background on cultural and historical uses of the items they sell, and have more suggestive business names, incorporating terms such as "shaman," etc… The JLF website, in comparison, was simple and discrete.

There was also a lot of DEA interest in *Salvia divinorum* at one point, but since popular interest in this herb has died down it seems that federal interest in regulation has also waned. Do you think the government has moved on from the ethnobotanical market, or do you suspect that this is something that's still on their radar?

MN: Yes, at that time, we [at] JLF had a stricter no-consumption policy than any of our competitors; yet we were targeted. I think it was mainly due to popularity and publicity. But not too long after that, many competitors were targeted. They even had a catchy name for the 'operation', but I can't remember it now. And some of those business owners actually went to prison.

It definitely upset the feds that I got no prison time. To a large degree it's what they consider a 'loss', especially with such an uncooperative defendant. My only two instances of 'cooperation' was an 'agreement' to strike 12 products from the catalog in exchange for being released [from] jail upon my arraignment, and the plea agreement. During the course of the 7-year ordeal, attorneys first told me I was looking at 20 years in prison, then 7, then 3.5, then 1.5. But come sentencing day and I was able to read my statement, the Judge obviously saw the light and gave me no prison time; at which the prosecution was very upset and was overheard complaining how much they had spent on this case.

As far as the current legal situation with ethnobotanicals, I'm no expert, but it looks to be some of both. Even though many such businesses have continued to exist and thrive, *Salvia divinorum* is now a scheduled substance in many states as are many of the tryptamines that we used to sell.

KF: That is a good point; there have been some legal changes with some tryptamines and other compounds at the federal level, and with specific psychoactive plants in certain states. It seems that many of these changes were brought about, at least in part, as a response to the rapidly-growing ethnobotanical market. I am not familiar with the other cases you mentioned, but this is certainly an area worthy of further exploration, particularly in terms of understanding the history and current legal standing of the ethnobotanical market in the U.S. and how it has been approached by federal and state officials. But, for our purposes, I would like to head a slightly different direction.

Over the years, during my own research on the fly agaric, I have found that the name Mark Niemoller will occasionally and unexpectedly (though perhaps not unsurprisingly) show up, usually in the context of mycology. One example is some footage I came across of an *Amanita muscaria* discussion panel from the 1992 Telluride Mushroom Festival that featured you as a panelist along with several other notable speakers, including Drs Andrew Weil and Emanuel Salzman (Global Village Video, 1993). You mentioned earlier that you had been an early attendant at this event, but I wonder if you can speak to how you came to be a panelist for this talk, and perhaps a bit more about your role in the mycological community and your reputation as an expert on the effects of this particular mushroom?

MN: I have no idea how I got on that panel, other than Emanuel Salzman (the organizer of the Telluride Mushroom Festival/Seminar) had recently become enthusiastic about *Amanita muscaria*. It's a horrible video; and by the way, that is *not* a good recipe/formula for preparing *Amanita muscaria* for consumption.

I had been a regular attendee at the festival and 'Manny' [Emanuel] knew I was also into the mushroom and had a business that featured it, and we had had conversations about it all. Beyond that, I was no more an 'expert' than many other festival goers. But Telluride was kind of informal like that back then. Anyway, the *real* experts were/are typically negative on the entheogenic potential of *Amanita muscaria*. Salzman was no exception prior to that. But for some reason in the early 1990s he became enthusiastic. He even organized a tour for a small group to Kamchatka Siberia soon thereafter, to study indigenous shamanic use of the mushroom. Maybe you've seen some of those videos too, with the supposed shaman dancing around for them in her red and white polka dotted dress.[2]

I think Manny had heard that *muscaria* was good for the elderly and he was getting up there. Anyway, a couple years later he reverted back to his preceding low opinion of it. I know that, because I was there for many years after and heard his lectures. I imagine he probably tried consuming some and had a lousy experience – like happens to most. Yes, the interest of many experts has waxed and waned over the years, but mine remains.

KF: Interesting. I wonder what led to his initial enthusiasm. The other significant reference I found for you was in an article co-authored by Salzman (1996, *see also*

Ch. 6) recounting the trip to Siberia you mentioned, which would have occurred a couple years after the Telluride panel. In the article the authors discuss consulting you for dosage information as well as for a personal bioassay on some Siberian specimens they sent you. Did the Siberian specimens seem different from their North American counterparts in any significant way?

MN: At the time, if I recall correctly, I reported the Siberian sample as superior to American strains, but in retrospect I would not do that. I mean it was only a single test of a small sample, of which nothing of much significance should be derived. One cannot put much significance on a slight subjective feeling. Of course, the background for this is that one of the more popular 'rumors' surrounding *Amanita muscaria* is that Siberian strains are superior [in] quality, regarding entheogenic effects. In my opinion, it's a myth. Many Siberian strains have been distributed commercially since those early days, and I don't see the trip reports generally confirming the rumor. In my opinion, the main factor is not the mushroom 'strain'; it's the mushroom *preparation*.

KF: Yes, it appears that the degree of interest people have in this mushroom as a psychoactive seems to be highly correlated with the degree of "success" they experience. Preparation was a key part of Gordon Wasson's theory identifying *Amanita muscaria* as "Soma," but he also appeared to get side-tracked trying to prove that urine-recycling, as practiced by some tribes in Siberia, was a key step in preparing Soma rather than exploring some more obviously described steps, such as blending the Soma beverage with milk.

You mentioned that the recipe you described in your talk at Telluride is "*not a good recipe,*" but when I first saw the talk I was intrigued by your incorporation of yoghurt and other dairy products in your recipe. JLF would later go on to offer a "Lactic Ferment Extract," which also seems to be based on the idea that dairy played an important role in the preparation of Vedic Soma. Could you speak more about your thought process in developing these different recipes? Have you found any approaches or preparations you would qualify as "successful"?

MN: The *Formula* is and was always my primary goal, but in the spirit of classic alchemy and proprietary research, I will say the least about what's possibly the most important. And even then, I may say too much. *Once the pearls leave the hand, control of where they end up is lost*. Suffice it to say that a scientific basis for the 'lactic acid' formula has recently been established [by] Dr. Trent Austin (2014; *see also* Ch. 4). At that date, a *pearl* was cast.

KF: Without getting into proprietary specifics can you speak to your own success following Vedic inspired preparation techniques? And based on your own successes or failures what are your current thoughts on Gordon Wasson's theory? Was he on the right track with *Amanita muscaria*?

MN: For anyone trying to learn the exact preparation of the mushroom, the Vedic Soma texts are a chaotic mess, not to mention huge losses in translation. Among the voluminous material, about the only decipherable parts of the process were milk, fire, pounding stones and/or filtration. I certainly believe Wasson was on the right track and proved beyond reasonable doubt that *Amanita muscaria* was the base of Soma; but as you noted, he failed to focus on the process that *made* it into Soma. He only touched on it, even though a good portion of the Soma Mandala is obviously devoted to some kind of process. I imagine his failure was because it was more common back then to dismiss stuff like that as 'religious/meaningless ritual'. But of course, the absolute proof that Wasson was right about the base is only revealed through a 'successful experience'. I believe there were many formulas and Soma was only one. Many information sources provide clues, some more cryptic than others, and each appearing to be a different recipe; but they all seem to have common denominators.

KF: Earlier you stated that you believe the differences between Siberian and North American varieties of *Amanita muscaria* to be minimal or negligible. Recent research shows that these two varieties are genetically distinct (Geml et al., 2006; Geml et al., 2008), belonging to different *clades*, but I wonder how relevant these genetic distinctions are when looking at the level of psychoactivity among the different varieties of *Amanita muscaria*.

JLF offered a number of the different North American varieties for sale, including red, orange-red, yellow and brown; did you come to have personal experience with these different varieties? If so, did you discover any discernible differences in the levels of psychoactivity, nausea or muscarinic effects? And did you ever develop a preference for a particular variety?

MN: I haven't seen or perceived a chemotaxonomic difference among the strains of *muscaria* that is significant enough to warrant putting extra effort toward more research. Compared to the processing, a small difference in the content quality or quantity of the carpophore is not a factor. That also goes for var. *persicina*, which has recently been reclassified as a separate species (*Amanita persicina*: Tulloss *et al.*, 2015).

KF: I imagine your response will be similar for the different varieties of *Amanita pantherina* that JLF carried, however, most of these are now considered separate species rather than varieties. Did you ever discover any significant differences in activity within the Pantheroid group?

MN: I only have subjective experience with what used to be called *Amanita pantherina* var. *pantherinoides* from the Rocky Mountains, and it is limited.

KF: During our conversation you have observed that many peoples' interest in this mushroom seems to wax and wane but have described your own interest as *persistent*. What is it about the fly agaric that has sustained your interest all these years?

MN: Lots of aspects; modern and ancient, obvious and cryptic:
- The appearance.
- The experience.
- The science.
- The history.
- The mythology.
- The legends.
- The traditions.
- The literature.
- The art.

I've always felt something very special about this mushroom, and apparently so did quite a lot of people a long time ago. At least that's how I translate the clues they left. It was very important for a long span of civilization. Why? Incredible qualities were ascribed to it. So much so, it became blind tradition. For so much mystery to surround one object, how can it not be interesting? But that *is* the problem; it's just too incredible. In fact, for me it beggars description, and for most it beggars belief.

KF: Now that JLF is behind you, how has your relationship with this mushroom changed? Where have your investigations taken you?

MN: Less focus on picking and drying mass quantities; more focus on researching the *Formula*. Things keep unfolding, but progress is slow and meticulous.

KF: The fly agaric continues to inspire the interest and curiosity of people around the world, from photographers and artists, to mycophiles, herbalists and those with an interest in entheogenic plants and fungi. What advice, if any, do you have for novices interested in learning about and exploring the various aspects of this multi-faceted and multi-dimensional mushroom?

MN: The *Amanita muscaria* either represents the greatest mystery of all time, or the greatest wild goose chase. If you feel you have been called, follow the dream and enjoy the quest, but not too much. You might discover the secret of life, or you might waste your life on a fallacy.

Recent research into the history of this mushroom seems to provide a scientific explanation for religion. If the religions can be believed, surely the science can. But what if the science reveals a *more* unbelievable story? May your ability to believe and accept the possibilities, and the hard truths, be stretched to the limit. Good luck seeking the source of luck! No one person can tell you you're on the right track, but listen to the ancients. The clues they left became our traditions.

KF: Mark, thank you for taking the time to share some of your experiences as one of the founders and proprietors of, *perhaps*, the first business to offer these extraordinary mushrooms to the masses, and for sharing your personal knowledge and expertise on all things *muscarioid* and *pantheroid*.

Notes

1. JLF Ordering Agreement Statement & Required Release: "I, the customer, am over 18 years old and have read all the text in this catalog and understand that all products sold here are poisonous and are not to be taken internally or used in any manner prohibited by JLF. I purchase all merchandise from JLF at my own risk and will not hold JLF liable for any personal harm or misfortune incurred on myself from any accidental or intentional misuse or abuse of said merchandise."
2. Tom Stimson and Lena Chabin, who joined Emanuel Salzman on the Kamchatka expeditions in the early-1990s filmed several of their encounters with Even shaman Tatiana Urkachan. A lecture given by Tatiana in her red-and-white garb (reminiscent of a fly agaric mushroom) can be found at: https://www.youtube.com/watch?v=3oFlXHUUmXg A performance by Tatiana of the story of Umemqat and the Fly can also be found at: https://www.youtube.com/watch?v=PBK7s-4DevNE&t=15s

Fig. 2: Cover graphic from JLF Catalog #3, Summer 1993.

Appendix 1

(Below are excerpts from catalog #10 for JLF: *Poisonous Non-Consumables*, distributed circa 1995.)

WELCOME TO THE WONDERFUL WORLD OF POISONOUS NON-CONSUMABLES

Thank you for your interest in the JLF product line. You are to be commended for your taste in such non-consumable products of real substance. JLF is not your average small mail order company. JLF is a very different entity. JLF is unique because the natural products which are offered are so diverse and the format in which they are sold is so unusual. JLF is proud to offer many items which you probably have never seen for sale before or even heard of. Rare, common, beautiful, ugly, divine, evil, wonderful, dreadful, practical, useless, priceless, worthless, appealing, disgusting, important, insignificant, famous, notorious, profound, mundane, mysterious and obvious are some of the adjectives which may be used to describe merchandise sold in this catalog.

Nevertheless, JLF will refrain from using any of the above descriptions for promotional advertising purposes to increase sales. Gimmicy sales tactics of any kind are not employed by JLF. In fact, no beneficial claims of any kind are made about any of the items sold here. The products simply exist, are identified, listed, and sell themselves. Any processing is kept to a necessary minimum. Actually, JLF is more in the habit of disclaiming things about the products than claiming things about them. Part of JLF's policy is to be relatively informal and non-specific about authorized/intended uses of the products while being very strict and specific about unauthorized uses (abuses). The basic disclaimer is that all products are sold as poisonous non-consumables. The unauthorized uses would obviously be internal consumption in any manner. Below are some commonly asked questions supplied with the official JLF policy answers to hopefully help (and not further confound) the somewhat confused.

CUSTOMER: If your products are poisonous, what do people do with them?

JLF: They buy them.

CUSTOMER: But what are they for?

JLF: They are for sale.

CUSTOMER: No really, what do they buy them for?

JLF: For themselves.

CUSTOMER: Can you give me any legitimate uses for these things?

JLF: 'Legitimate,' being synonymous with 'authorized,' you will need to refer to the "Authorized Uses…" paragraph in the catalog. Foreseeing the imminently forthcoming debate, I will open and close by asking you what the official legitimate uses are for rocks, sticks, or fire (yours truly enjoys lengthy debates if you are paying for the call).

CUSTOMER: Could you tell me about some of your products I've never heard of?

JLF: Not unless you are more specific about your question. I can tell you about the colors, shapes, sizes, smells, etc. but not much else. JLF is in the business of dispensing raw materials, not information.

CUSTOMER: I've heard and read that this or that product of yours is hallucinogenic, tremorgenic, carcinogenic, tetratogenic, narcotic, emetic, tonic, hemostatic, cytolytic, hemolytic, hemorrhagic, analgesic, hematoxic, neurotoxic, hepatoxic, myotoxic, sedative, laxative, inflammative, convulsant, disinfectant, muscle relaxant, antibacterial and/or insecticidal. Could you elaborate on that?

JLF: I am not a botanist, mycologist, chemist, toxicologist, pharmacist, or a formally educated professional of any kind, and therefore claim no ability to accurately comment on the details about the toxins contained in any of the products, whether it be types, amounts, names, actions or symptoms resulting from ingestion. Besides, these types of questions are irrelevant – our products are not sold for consumption.

CUSTOMER: Which products do you recommend?

JLF: JLF does not make any recommendations. The choices are completely yours. You make the decisions.

We see no need, nor are we required, to educate the uninformed about the products sold here by reprinting everything that has ever been said or published about them. That is not only not required, but not even remotely possible. One advantage of this method of operation is that no products are irresponsibly misrepresented, thereby promoting the safety of both the purchaser and the seller. The products need not be, nor will they be, embellished. The selling point of the product will remain its name only. The extent of the information given out by JLF will be the scientific taxonomic (the classification of organisms in categories based on common characteristics) descriptions. Other information included with the products consists of Latin names, common names, weights, a guarantee, disclaimers, warnings and cautionary notes. The letters "JLF" stand for nothing but what they are; the name of the company. They stand for no secret phrase, slogan or meaning. As so, the products are sold for what they are; a named existing product of nature with no alternative secret purpose, meaning or usage. All products are considered equally poisonous, ruling out consumption as an authorized use. The main reason these products are sold is because they can be; people buy them. In other words, there exists a demand and a business to satisfy it. Reason enough? If you need more than that, see the "Authorized Uses…" paragraph. JLF is capitalism in a purer form and proud of it.

The only guarantee JLF makes is one of authenticity. The identities of the products are guaranteed to be correct.

THE LONGEST DISCLAIMER YOU'VE EVER SEEN

Any seed prohibited as a noxious weed, or other, by law, is supplied to the consignee for non-sowing purposes only. By ordering such seed, the customer agrees to the use of such seed for non-sowing purposes only. All merchandise sold in this catalog is poisonous and not intended for human internal, or other, consumption. Keep all products out of the reach of children and those not responsible for their own actions. All items are classified as non-consumable. Do not eat. Do not use in any manner prohibited by JLF. JLF assumes no liability for the mis-use or abuse of its products. For example (for those litigious types who exist solely to keep the human aspect of Murphy's Law alive): Do not take orally as a food, beverage, nutritional supplement, medicine, recreational drug or as an agent for suicide. Do not inject in any way or form. Do not directly inhale any smoke or vapors from any of JLF's products. DO not stick, put, insert or throw into your or another person's mouth, nose, ear, eye, anus or any other orifice or port of entry that may exist on your or another person's body. Do not snort. Do not wet any dried organic with water, allow it to rot, then smear the black maggot-infested stinky slop on your or another person's body or insert it into any of the orifices previously mentioned. Do not deploy any of JLF's products as dangerous high-speed projectiles aimed at people or property. Do not use as weapons of war. Do not light with fire and burn yourself or another or any private or public property. Do not leave lying on the floor to trip over or slip on to incur personal injury. Obviously, there is not enough room here to list all the possible "do nots." If you still can not keep from harming yourself or others or property with JLF's products then you should probably go back to bed and stay there for the rest of your life. No, then you would get bed sores and bring suit against the bed or sheet manufacturer, or against JLF for telling you to stay in bed.

AUTHORIZED USES OF JLF's PRODUCTS

Some products may be used as educational tools to aid in taxonomic studies in mycology or botany, such as identifying wild specimens in the field, or examining microscopic and macroscopic characteristics. Some may be used as voucher specimens for reference in hospitals, poison centers, herbariums, educational institutions or other research centers. Some may be viewed as collectibles to collectors of curios, novelties or rarities of any sort. Some may be used as art supplies in manufacturing homemade crafts for decorative purposes. Some may be framed, hung on walls of homes, and viewed for their natural artistic value. Some may be carried for good luck and/or to ward off evil spirits. Some may be used/worshipped as sacred objects in certain magical-religious-pagan-spiritual-new age-occult circles. Some may be used as incense or pot-pourri ingredients. Some may be touched, smelled or worn as a fashion statement. All products may be used as conversation pieces or social ice-breakers. All products may be used in ways similar to uses of the once famous Pet Rock. *Amanita muscaria* and *pantherina* products may be used as ingredients in the formulation of the once famous housefly poison. Some seeds may be used for cultivation purposes. The bibliography may be used as reading material or woodstove fuel. Some extracts may be used as charcoal-burning resin-incense. Toad skins may be used to manufacture miniature toad-hide doll clothes and/or accessories such as shoes, belts, hats or purses. The *Amanita Muscaria* Lactic Ferment Extract may be used by devout students of Hinduism as an offering to their god, Indra, in their traditional religious ritual, the Soma Sacrifice; being a reasonable facsimile of the original Soma, having been manufactured according to ancient Rg vedic scriptural descriptions. For that matter, any of JLF's products, including this catalog, may be offered to, or sacrificed in the name of, any God or gods. Offer them to yours! The above suggested uses are offered to those readers who feel they need them for clarification.

ABOUT *AMANITA MUSCARIA*

Amanita muscaria (the Fly Agaric Mushroom) is JLF's most popular item; the star attraction of this catalog. It is the genuine red and white spotted toadstool of worldwide notoriety. Yet many people today do not even realize it to be a true living plant, only knowing it from its caricature in children's fairy-tale books, on greeting cards, kitchen towels, key chains, curtains, wallpaper, as candles, ceramics and other various assorted nic-nacs and what-nots. It is a most beautiful creation of nature and can only be found growing there; it can not be fruited in the lab. It has to be collected from the wild. When it comes to hunting, this mushroom can be most elusive. It is generally considered to be most mysterious. It is, at least, absolutely unique. The late R. Gordon Wasson identified it as the Soma of the Rg veda; the first Hindus worshiped it as a god. The late John M. Allegro proved that the original Christianity/Judaism also worshiped it by deciphering that the Bible is full of mushroom puns and word-plays; many proper names in the Bible are actually ancient folk names for this mushroom. The *Amanita muscaria* has probably had more common (folk) names given to it than any other life form throughout history. Many of these names have been clouded in secrecy and confusion for theorized reasons. The Fly Agaric seems to have played some very significant roles in many ancient cultures in some incredibly fascinating ways. In brief, it appears to have served as a symbol for fertility, good fortune, health, wealth, and happiness (physical & spiritual). Mycologists have so far [circ. 1995] named six different taxonomic variations. These "sub-species" names are *muscaria* (red cap), *flavivolvata* (red-orange cap), *regalis* (red-brown cap), *formosa* (yellow-orange cap), *persicina* (melon cap), and *alba* (whitish cap). Each variation usually has its own geographical growth range. The mushrooms have been mechanically warm air dried at under 110 degrees F using an original technique developed by JLF for large amounts. This technique, which preserves the mushrooms in the most perfect condition possible, requires slicing for efficiency. The stems have also been detached from the caps but are packaged together at the original ratio of about 1 to 2 (by weight).

JLF always wants new mushrooms, plants and substances to add to the product list. Please call or write if you have any leads on anything interesting you may have seen growing, for sale or available in any way, whether it's already listed in this catalog or not. Always open for discussing details on selling or trading. You may want to become a supplier. If you have any requests for items that are not listed in this catalog, please send them or call them in. All suggestions are welcome, as is any good, informative rap.

Chapter 15

Glückspilz:
The Lucky Mushroom
Kevin Feeney

In Germany, *Amanita muscaria* is commonly known as either *Fliegenpilz* (fly mushroom) or *Glückspilz* (luck mushroom). The first of these names is easy to explain as it is a direct translation of fly agaric or fly mushroom, and highlights an association between flies and *Amanita muscaria* that has been recognized since before 1256 AD, when the folk-practice of using *Amanita muscaria* as an insecticide in Europe was observed and recorded by Albertus Magnus in his *De vegetabilibus*. The association between the fly agaric and "luck" or "good fortune," an association that extends well beyond Germany, is much murkier and appears to be of more recent origin.

The term Glückspilz appears to have its origins in the 18th century where it was used to refer to "up-starts" or "status seekers," those who sought to change their lots overnight much like a mushroom seems to appear overnight, as if from nothing. Despite having a somewhat pejorative origin the term evolved over time and by the second half of the 19th century the term applied to those blessed with "good luck" and "good fortune" (Kluge, 1989; Pfeifer, 1995). While the association with mushrooms and "luck" springing forth overnight is clear, it is less clear why this term should have become synonymous with *Amanita muscaria*. It is possible that this association arose simply because *Amanita muscaria* is one of the most widely recognized mushrooms and is visually striking and memorable. However, it has also become popular to speculate that the association between the fly agaric and luck may stem from its hallucinogenic properties and from ancient and forgotten shamanic uses of the mushroom, which could suggest a deeper historical connection.

While there is evidence suggesting that the fly agaric may have been used for its psychoactive properties in Europe, as explored in other chapters of this book, being psychoactive is not a *sine qua non* for symbols of luck. Four-leaf clovers and horseshoes are not psychoactive but derive their special status from elsewhere. Neither are any other "psychedelic" or psychoactive substances known to be considered as good luck charms. The only bit of evidence that may suggest further inquiry into the relationship between the fly agaric's psychoactive properties and its "lucky" nature is the relationship between the fly agaric and the chimney sweep as companion symbols of luck (Figure 1).

According to Christian Rätsch and Claudia Müller-Ebeling (2006), the chimney represents a doorway or portal to the otherworld, and part of the chimney sweeps' importance comes from his role in clearing and maintaining this passageway. Rätsch and Müller-Ebeling note that "[c]himney sweeps climb through the chimney just like ancestor spirits, witches, sorcerers, shamans… and Santa Claus" (p. 167). While the chimney sweep and his status as a symbol of luck are largely forgotten in the modern world, his "luckiness" and his connection to the "other world" were both immortalized in the song *Chim Chim Cher-ee* from the film *Mary Poppins* (Stevenson, 1964). The main verse from the song follows:

*Chim chiminey,
chim chiminey,
chim chim cher-ee
A sweep is as lucky
as lucky can be
Chim chiminey,
chim chiminey,
chim chim cher-oo
Good luck will rub
off when I shakes
'ands with you*

Fig. 1: An early-20th century postcard with a New Year's greeting. Here the chimney sweep, who appears to be a child, is featured alongside a fly agaric mushroom and a horseshoe; all symbols of luck. The clock in the background has struck midnight indicating the New Year has arrived. Artist unknown.

In the film, Jane, the young girl, remarks how the inside of the chimney is dark and gloomy. But Bert, the chimney sweep, provides a different perspective when he responds, "That there is what you might call a doorway to a place of enchantment." This statement is followed by the following illustrative verse:

> *Up where the smoke is all billered and curled*
> *'Tween pavement and stars is the chimney sweep world*
> *When the's 'ardly no day*
> *Nor 'ardly no night*
> *There's things 'alf in shadow*
> *And 'alf way in light*
> *On the roof tops of London*
> *Cool, what a sight*

When taken together, the dialogue from the film and the lyrics from the song appear to contextualize the chimney sweep as someone who travels between

Fig. 2: Another early-20th century postcard with a Holiday greeting. The angel (or fairy?) carries a basket of fly agarics and four-leaf clovers and appears to be leaving them as gifts of luck for the New Year. Horseshoes are also featured. Artist unknown.

Fig. 3: A German postcard wishing the recipient a "Happy New Year." The card features dwarves dancing with "lucky" pigs around a fly agaric while surrounded by four-leaf clovers and horseshoes. Artist unknown.

Fig. 4: A German postcard wishing the recipient a "Happy New Year." The card features several important symbols of luck, including the fly agaric, four-leaf clovers, and piglets. Artist unknown.

worlds. Perhaps it is the sweep's access to the other world that makes him the source of *luck*. If we accept that the chimney sweep's *lucky* status derives from his otherworld journeys it becomes more probable that the fly agaric, which also provides passage to the otherworld, might be *lucky* for a similar reason. If this is the case, then it would appear that the association of luck with the fly agaric precedes the appearance of the term Glückspilz, and the exact origins of this association would likely be difficult to trace.

Regardless of the precise origins of the fly agaric / luck association, we can say, with some certainty, that the fly agaric was a well-known symbol of "luck" and "good fortune" throughout Central Europe (and beyond) by the end of the 19th century, and became a popular component on holiday postcards where it was frequently featured along with other symbols of luck and good fortune, including four-leaf clovers, horseshoes, piglets (*Glücksschweinchen*), chimney sweeps, and bags of money (Figures 2, 3, 4). Usually, these symbols were combined to wish the recipient good fortune in the New Year and were also typically accompanied by dwarfs or young children. In investigating these postcards, I found holiday greetings in a number of different languages, including: Dutch, English, Estonian, Finish, French, German, Hungarian, Italian, Latvian, Norwegian, and Slovakian; suggesting that recognition of the fly agaric as a symbol of luck was widespread. While the presence of the fly agaric may not be as common in holiday greetings as it once was, it continues to be a common decorative motif of the holiday season in Germany and other parts of Central Europe (Hoffman & Hoffman, 2001; Marley, 2010; Rätsch & Müller-Ebeling, 2006).

During my investigations into the fly agaric as a symbol of luck I came across another interesting find, a mushroom shaped insignia badge (Figure 5) that was worn by members of Germany's 205th Infantry Division during WWII. The

205th was nicknamed the *Pilzdivision* (mushroom division) and the fly agaric was apparently the symbol or mark that identified the division. It is not clear how the 205th came to be represented by the fly agaric, or if the mark carried any particular significance, but given its reputation as a symbol of luck and good fortune it seems probable that the mark was intended to bring the division "good fortune" in battle.

Others might speculate that the badge was intended to inspire or invoke the "battle-fury" of the *berserkers*, legendary warriors who were said to be as strong as bears and impervious to injury by "fire or iron" while under the influence of *Amanita muscaria* (*see* Ödman, Ch. 9). Adolf Hitler and other prominent Nazis were reportedly obsessed with the supernatural and the occult (Kurlander, 2017), and they may well have been familiar with Ödman's theory connecting the fly agaric to the berserker rages described in the Nordic Sagas. The connection between the berserkers and Odin, one of the Nordic-Germanic Gods, may have sealed the deal for the Nationalist Nazi party in terms of making the fly agaric a symbol of or inspiration for ruthlessness on the battlefield. In fact, Carl Jung (1936) once argued that Wotan (Odin), the medieval Germanic God of war and poetry, had become a deeply engrained part of the German psyche and that the re-emergence of Wotan-like "fury" from the German unconscious could be used to explain the extremism

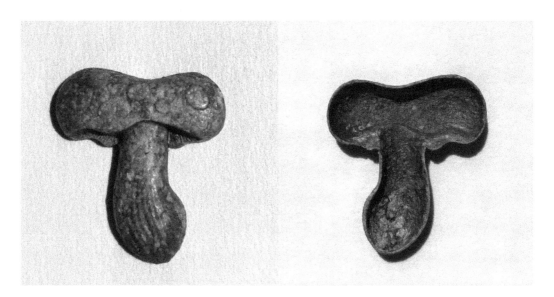

Fig. 5: The badges, which are made of zinc, measure just under 3 cm across and 3 cm tall and feature the characteristic warts of the fly agaric on the cap. The featured badge was discovered in the Courland Peninsula where German forces were surrounded by the Russians in 1944 and forced to surrender at the end of the War in 1945.

of the burgeoning Nazi movement. Jung, however, never revisited these assertions in his later writings, and his assertions have been dismissed as rooted in deep misunderstandings of both the mythology of Wotan and the philosophies of Nazism (Metzner, 1994: 111).

While a possible berserker connection is intriguing, such a connection is speculative at best and seems quite unlikely. Were the mushroom badge to symbolize the berserker "battle-fury" one would think that the prevalence of the symbol would have been much greater. It also seems unlikely that Hitler, or any of the other prominent and powerful Nazis with occult interests would have bothered to assign this mark to a single infantry division. It seems more likely that the symbol was adopted internally within the division, or at a very low and local level in the party hierarchy, making it more probable that the symbol was a product of common folk beliefs in the fly agaric as a lucky talisman. Regardless of the origins of this mark, whether a symbol of "luck," "battle-fury," or something else entirely, its mere existence suggests a high degree of cultural saliency behind the image of the fly agaric.

We may never know when or how the fly agaric came to be a symbol of luck and good fortune in Europe, but the reputation this mushroom has gained as a symbol of luck remains an intriguing aspect of the cultural history and traditions surrounding this singular fungus. Perhaps future investigations will shed more light on this topic.

Chapter 16

The Lucky Mushroom:
A New Fairy Tale Story

Original Verses by Marie Meissner
Illustrations by Karl Schicktanz
Adapted by Sandra Grecki

Preface

Der Glückspilz: Ein Neues Märchen, or "The Lucky Mushroom: A New Fairy Tale Story," was originally published in 1910 in Duisburg, Germany by J. A. Steinkamp. The story is one of childhood adventure, an adventure that begins with a couple of familiar protagonists, Hänsel and Gretel, who find themselves lost in an enchanted wood during their search for a "lucky" mushroom. The children are guided by the *Man in the Moon* to Rapunzel, who lives in a fly agaric house, or, the "Lucky Mushroom" house. While the reader will be familiar with stories of the long-haired Rapunzel, made famous by the Brothers Grimm, the Rapunzel featured in this story is the *original* German Rapunzel, a bearded dwarf who lives in the woods and leads Hänsel and Gretel on their adventure home.

The original German prose is written in rhyming verse, which makes a direct translation difficult. Instead, what follows is an adaptation, with some liberties taken for purposes of readability. Another difficulty with the translation has to do with the title, *Der Glückspilz. Glückspilz* is a common name for the fly agaric in Germany, and can be translated as "Lucky Mushroom," alternately, however, it can be translated as "Lucky One." As a result, it appears that the title may refer both to the fly agaric, where Rapunzel makes his home, and to Rapunzel himself. It is tempting to read a convergence of meaning into the title, where Rapunzel might be considered a personification, of sorts, of the fly agaric mushroom, though a double meaning is more likely intended. Regardless of whether the title is translated as

"The Lucky Mushroom" or "The Lucky One," *luck* and the search for it are central to Hänsel and Gretel's peculiar adventure. While Hänsel and Gretel leave the *lucky mushroom* in the woods they return home with a basketful of four-leaf clovers and a marzipan piglet (*Glücksschweinchen*), a treat frequently gifted on Christmas and New Years in Germany as a token of good luck.

-KMF

A New Fairy Tale Story

In an old enchanted forest, there is a large mushroom called the Lucky One. Everyone longs to find this mushroom, for the luck that it confers, but those who don't know their way around these woods would walk around day and night and still never find it.

It has been said that only on the longest day of the year will lucky children be able to find the Lucky Mushroom. That's how Hänsel and Gretel, the lively twins, ended up in the middle of these woods. Usually, they know their way around the woods very well and always find their way home, but today is different.

Today it seems as if something is wrong for, they can't seem to find their way out of the forest. Every bush and tree look as if enchanted and the forest appears strange and new. Little Hänsel tries to keep himself together and act like a man, but Gretel begins to cry.

The dark of night begins to descend around them. The bunnies jump into the bushes, the singing birds quiet and huddle into their nests, and the owl just hoot, hoots, as if to say, "Good Night!" Even the tiny wasp flies through the dusk and the deer cuddle up with one another for their slumber.

The full moon shines its bright light through the leafy canopy and finds the children all alone. But before the children realize what's happening, the Moon is already standing before them. "I am the Man in the Moon, you see. What are you two doing in the middle of the woods so late?"

Startled, Hänsel exclaims "Mr. Man in the Moon, we've gotten lost!"
"Please be so kind and quickly bring us back home!" Gretel begs.
"Dear children, that is way too far to travel in the darkness, and I don't have time for that tonight," the Man in the Moon replies.

"But I shall light up the path for you. The Lucky Mushroom is not far from here. That is where Rapunzel lives, and he enjoys the company of good children. You will surely find your luck there." Then the moon climbed back up into the night sky.

The Moon lights the way for the children and carefully watches over them. Fireflies sparkle near and far, as they dance through the night. Before they know it, they find the Lucky Mushroom and Rapunzel is already looking out the door, welcoming them.

Oh, the house is so cozy on the inside! Rapunzel walks over to the cabinet and brings the children milk and bread, and two very clean and shiny cups. As soon as the children are finished, Rapunzel returns to his work and everyone helps as well as they can.

Animals, large and small, come to Rapunzel in their despair, for he is their doctor. When there is suddenly a growl from underneath the table, Rapunzel just laughs. "This is Brown, the bear. He has not eaten anything for three days. His sweet tooth has caused him to eat too often of the bee's honey and now he has a tummy ache. Now he has to swallow bitter medicine instead, but tomorrow morning he will be well again, and I will set him free."

"The white buck is also in bad shape. He is very distraught, because he crushed his antlers during a fight with a brown buck and now, they must be set in a cast," Rapunzel explains. "Oh no!" Hänsel yells "I know how bad it hurts, when you run into each other like that."

The children learn that the bunny is also in a lot of pain, because a dog bit his leg. And while it was dancing a squirrel stepped on a splinter. The treefrog sang all night by the lake and now his throat hurts terribly. "All of these animals have come for my assistance," Rapunzel explains.

"Enough for today! Let's all get some sleep. You may share a bed inside my little room for the night," Rapunzel tells the children. He carefully puts out all the lights and then it becomes dark in the Lucky Mushroom House.

The next morning, as soon as the rooster crows, everyone is awake again. After their breakfast, Rapunzel takes the children down the river in his little boat. They come across the mill, but the miller there does not seem very happy when he sees them.

During their ride, the water dripping from the oar glimmers with silver sparkles and lovely waterlilies and forget-me-nots grow everywhere they look. The mill's wheel turns with a clattering noise, as wagtails come for their morning bath.

Then all three of them climb onto the shore, because they see the king of the enchanted world. He just came from his castle and wants to go for a stroll. When he sees Rapunzel approach, he stops to say "hello."

The king waves to the children and he seems very friendly. Gretel kneels courteously and then hands him a bouquet of flowers. Hänsel is also very well-mannered, as he stands there politely with his hat in his hands.

They all admire how beautiful the king's golden crown sparkles, which is adorned with precious stones! The king is wearing a beautiful crimson coat with ermine fur at its seams! And his shoes are just as exquisite! Then there is his castle in all its glory, guarded by owls all around.

The king says "Because you are such good children, I would like to share with you something entertaining and funny. I am on my way to see my herd right now, and you may accompany me." Hänsel and Gretel are very happy to follow him.

Off they go, and soon they see a gate. It springs open from the inside, and from within lucky little piglets spill out, squealing, running and tumbling about. They are very cute to look at, particularly with the way they twist their curly little tails.

Lots of people, big and small, old and young, are drawn to this field, where the lucky clover grows. The king says "Go and catch yourselves one of those pink piglets. You may take it back home with you. It brings good luck to good people.

Also, collect some of the lucky clover. Dig it out with its root attached. And back at home, plant it all around your parent's house. You should also give this little plant away as a gift to others. Yours will grow better if you do that."

Hänsel and Gretel cheer loudly and begin to run around everywhere. However, these lucky little piglets run very fast, and they are very hard to catch. Finally, they catch up with one, and it is particularly dainty and small.

When it's time to go home, they lead the little piglet along on a leash, but it gets tired very quickly. So, Hänsel picks it up and takes it into his arms to carry it through the woods.

The children don't find the walk to be very far, because everybody comes along with them. Rapunzel leads the procession arm in arm with Hänsel and Gretel. The king rides proudly on his horse and a parade of little woodsmen follow, each making music, so that the procession would march in time with the beat.

One musician blows a blade of grass. Lily of the valleys sound so good, almost as if all those fine flowers were silver bells today. A large drum comes along too, though the little woodsman struggles with his heavy load.

There is drumming, tooting, singing and whistling and everyone does as best as they can. In the lime tree, there is even a flock of birds chiming in with their song, and a cuckoo, a finch and a nightingale all greet them with cheer too.

The procession soon reaches the end of the enchanted forest, where the Lucky Mushroom can be found. And Hänsel and Gretel finally recognize the path that will bring them home. "Farewell, farewell!" the children call. "We thank you very much!" Rapunzel waves and calls "Goodbye!"

The twins walk home hand in hand, only now realizing that their parents must be worried sick. And so it was, because during the night, until the early morning, they had been scared and worried and hardly slept a wink. They had thought something bad had happened to their children, but now that they are back their parents rejoice.

"Mama, Papa! We are back!" the children yell. "We were lucky to find Rapunzel's Lucky Mushroom House!" Gretel still carries the lucky clover and Hänsel is holding the lucky piglet. But when they look at it more closely, it does not move its head nor tail. Mysteriously it is now made of the finest marzipan! Amazed, they all marvel at the miracle.

Since Hänsel and Gretel returned from their adventure, it has become a custom to give away lucky little piglets made from marzipan at Christmas time. Because the four-leaf clover came from the same enchanted forest where the Lucky Mushroom House stands, the lucky clover is highly coveted too.

Hence the reason, why we return to the fairyland every once in a while, to listen quietly to the magic forest, all us children big and small. For that is where good luck might blossom for us. Who will find their luck? And who will bring it back from the enchanted forest?

The End

Part III:
Archaeological Evidence

Mushroom Effigies in Archaeology:
A Methodological Approach
Giorgio Samorini

In 1989 I became the librarian for the World Rock Art Archive (WRAA), the UNESCO body dedicated to rock art. Under the direction of Emmanuel Anati, then director of WRAA, I participated in the drawing up of the first list of fifty prehistoric rock art sites declared a UNESCO World Heritage Site. This work gave me the privileged opportunity to observe the prehistoric rock art from all over the world, and to deepen my research in the field of archaeo-ethnomycology. In this chapter I propose some methodological considerations regarding the archaeo-ethnomycology of psychoactive mushrooms, with the aim of delimiting the area of scientific research from what I have previously defined as "phanta-ethnomycology" (Samorini, 2001b: 175-9): a literary vein that has produced and continues to produce pretentious theses based on superficial or preconceived observations. This article provides an updated review of two archaeological sites I studied during this period, which are located in the heart of the Sahara Desert and in South India.

Methodological Aspects

Mushroom-like images are frequently observed in rock art and on archaeological artefacts around the world. There are a great variety of forms – the stem can be stocky, long, wavy or filiform, and the hemispherical, bell-shaped cap, umbonate or pointed. There is also considerable contextual variability – with mushroom-like objects drawn alone or grouped, held in the hands, on the head, or coming out of the body of anthropomorphic or theriomorphic beings.

Interpreting these objects requires determining whether their mushroom-like form was intentional, or not, what type of mushroom is depicted, and whether it is edible, poisonous, 'insignificant' (mycological terminology considers any non-edible and non-poisonous mushroom 'insignificant'), or intoxicating. Issues with identification are generally minor, because in most cases the morphological details, or the ritual and religious contexts in which they are represented, lead to their identification as psychoactive fungi.

Depictions of non-psychoactive fungi are rare. This observation led me to establish the following *axiom of archaeo-ethnomycology*: "if in ancient art mushrooms are depicted, then they most probably concern psychoactive and non-edible or poisonous mushrooms". This conforms with the basic motivation for the production of ancient art, which in most cases is dictated by religious, shamanic, initiatory and ritual motifs. In other words, mushrooms depicted in a religious or initiatory context are much more likely to be a psychoactive fungus that allows contact with the afterlife, with the divine or with the spirit world, than an edible or poisonous species with no such potential.

Interpreting mushroom-like objects as true psychoactive fungi and deducing ancient mushroom cults on the sole basis of the analogy with its form, produces an excessively speculative and weak hypothesis. A methodologically rigorous ethnomycological interpretation is almost always supported by some other element: a morphological detail or a scenic context that includes an ethnographic, mythico-religious, ritual, or ecological association with psychoactive fungi.

Another factor usually necessary, although not sufficient, to prove an ethnomycological interpretation, is the *repetitiveness* of the fungal reproduction, as is seen with the well-known Mayan mushroom-stones, dated between the first millennium BC and the first millennium AD. Their identification as objects of a fungal cult would have been weak if it had been based on only one of these finds, but instead, this explication is based upon the four hundred mushroom-stones discovered so far (Mayer, 1977: 2).

Scenic *variability* can also bring useful interpretative clues, as is the case with the same Mayan mushroom-stones, where adoration scenes, women kneeling on a millstone, ecstatic or dreamy individuals upside down, etc., are depicted in the area corresponding to the stem of the mushroom solar gods. The importance of scenic variability in the context of archaeological finds is analogous to the importance of variability in the different versions of a myth or a folk tale, which often contain different and sometimes apparently contradictory elements. It is precisely this plurality that helps us to better understand aspects and themes of the story that otherwise would remain enigmatic. As Wendy Doniger (1977: 19) stated in a study of Hindu mythology, "no myth, taken individually, contains the key: it is given by the totality of variants of the myth".

To confirm an ethnomycological interpretation of archaeological findings, the *morphological details* associated with mushroom-like images should, where possible, be identified. The two hallucinogenic mushrooms *Amanita muscaria* (fly agaric) and *A. pantherina* (panther cap) have certain characteristics that, when depicted in ancient art, leave little doubt about the identification of the species drawn: a ring around the stem, the maculation present on the cap, and the birth of these mushrooms from an ovule. In the case of psilocybin fungi belonging to the genus *Psilocybe*, *Panaeolus*, etc., a frequent feature, although not required, is the presence of a protuberance (a papilla or umbo) on the upper part of the cap. Another characteristic feature is the bluish hue of spots that appear on the stem and more rarely on the cap of psilocybin fungi when they age or following harvesting. This distinctive feature will not have escaped the attentive observers and consumers of these mushrooms, but has hardly been represented in ancient art, or at least in that which has been preserved. In the case of rock paintings, it should be considered that the original color may have been subject to mutation due to oxidative processes that lead, for example, to a change in the bluish hues in greenish tones or other colors, depending on the type of organic or inorganic pigment employed (Soleilhavoup, 1978: 83).

The maculation on fly agaric and panther cap – which is due to the remains of the veil that originally surrounds the ovule – is indicated in two ways in two-dimensional artistic representations: either with "dots" or small round patches that cover the inside area of the hat, as seen in the rock carving of Mount Bego (Figure 4), or with small bumps scattered on the upper edge of the hat. The latter can be seen on some Siberian rock engravings, where so-called "mushroom-men" are depicted. These anthropomorphic figures have a 'head' in the shape of a mushroom hat, which as Marianna Devlet noted (2001: 11), contain a series of small bumps on the upper part of the 'head' that must have been purposefully engraved (Figure 1). An ethnomycological interpretation of these "mushroom-men" is further confirmed by other stylistic elements and ethnographic correspondences, long recognized and re-discussed by other authors in this same volume.

Fig 1: Rock engravings: a) Aldy-Mozaga, Yenisei (Dikov, 2004: f. 65/4); b,c) Kalbak-Tash I, Siberia (Kubarev & Jacobson, 1996: f. 208, 284).

In plastic (three-dimensional) representations there are two ways to represent the "dots" of fly-agaric and panther cap: through protuberances that come out on the upper surface of the hat – as in the case of the Nayarit (Mexico) terracotta dated to about 2000 years ago, studied by Peter Furst (1974: 60; Figure 2), where an individual is depicted sitting under a "mushroom-tree"; or otherwise by making small grooves or holes on the upper surface of what corresponds to the mushroom cap. The latter can be found on a small stone object excavated in the Pátzcuaro Basin, Michoacán, Mexico, and which the late Gaston Guzmán gave me the opportunity to hold and carefully examine. This object is approximately 5 cm tall and has an ovoid shape with the two hemispheres separated by a narrowing. On the surface of the upper hemisphere numerous small concavities are engraved (Figure 3). The interpretation of this object as a newborn ovule of *Amanita muscaria* is quite indisputable (Guzmán, 1997: 14).

It should be noted that most of the time these morphological details are not depicted – or when they are depicted, only one of these details is drawn. This is the case with the Nayarit sculpture previously mentioned, where the protuberances on the hat, though not on the ring around the stem, are shown – which indicates that their presence is not necessarily required for fungal identification. Often the prehistoric artist represented a mushroom in a standardized form, sufficient for its immediate interpretation as a "generic" mushroom, similar to the standardized way in which he depicted a fruit, a branch or any other vegetable element, without reproducing the finer details for a more precise determination. The reasons for this lack of detail are various: from practical ones – such as a difficulty in reproducing them due to lack of space or the excessively small size of the feature to be reproduced – to the lack of interest in reproducing them as they were considered unnecessary for the purpose of the visual message the artist intended to communicate. It is plausible that the artist considered it entirely obvious that if he reproduced a mushroom in an artistic context, it was meant to represent a specific intoxicating mushroom, and that the difficulty in interpreting the image is only a problem for modern scholars (Samorini, 2001a).

An interesting case of the contemporary presence of the ring and the "dots" on a fungal object comes from the Mount Bego rock engravings, in southern France. Along the main valley of this mountain, the Valley of Wonders, there are thousands of rock engravings dating back to the Bronze Age, starting from 2500 BC. The main feature of these rock engravings, which are located at an altitude of 2000-2500 m, is the significant, one could say obsessive, presence of horned zoomorphic images, in particular the bovine bucrane, which consists of the frontal part of the skull and the horns. A further characteristic is that all of these horned figures are engraved in such a way that the tips of the horns point towards the top of the mountain. This is not accidental. It has been observed that Mount Bego is one of the mountains of the Maritime Alps most frequently affected by lightning, and it has been repeatedly suggested that this mountain was chosen by prehistoric populations as a "sanctuary" precisely because of this meteorological feature. In ancient times it was generally believed that the sacredness of certain places derived from this feature. It is also appropriate to observe the zigzag shape of many horns engraved on Mount Bego, which recall the shape of lightning (Bicknell, 1972; Marro, 1945-46).

Fig 2: A Nayarit (Mexico) terracotta (Furst, 1974: 60).

In the highest part of the Valley of Wonders, there is an engraved stone, considered one of the most significant finds in the region's rock art, popularly called the "Rock of the Altar". This is engraved with a scene called the "Chief of the Tribe", dated to approximately 1800 BC (Lumley, Béguin-Ducornet, Échassoux, Giusto-Magnardi & Romain, 1990: 45). On the surface of the rocky "altar" some daggers are engraved, a scaly design, a small "prayer" and a larger anthropomorphic figure which has been given the name "Chief of the Tribe" (Figure 4a). This last figure has been executed in two phases: in the first, three bucranes have been engraved which, going up from the bottom upwards, are reduced in size; in the second, the three horned figures have been joined together in such a way as to give an anthropomorphic form to the whole.

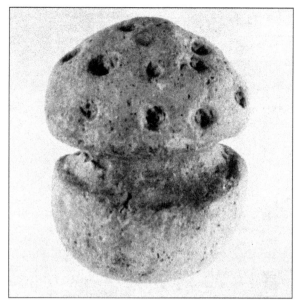

Fig 3: Stone artefact, Pátzcuaro, Michoacán, Mexico (Guzman, 1997: 14).

Among the various interpretations proposed for this anthropomorphic figure – including a tribal chief, an officiant or a sacrificial victim, because of the supposed dagger pointed at his head – the most interesting is that given by Patrick Duvivier, who sees the depiction of a shaman, in addition to interpreting the other significant object that I am about to present, as a mushroom of the fly-agaric species: "Exactly like with the Siberian or the Ojibway shamans for whom the power of the sacred *A. muscaria* was also closely linked with lightning, our 'Chief of the Tribe' is enlightened (symbolized by the lightning bolt) with the power of the mushroom" (Duvivier, 1998: 34). The fungiform object (Figure 4b), considered firstly as such by Duvivier himself, is at the center of the scene, and has been mostly interpreted as an abstract design, a dagger or a bucrane. Also in this case, as for that of the "Chief of the Tribe", the object appears to have been executed in two phases, where in the first a bucrane had been engraved, subsequently completed in the upper part by joining the two horns with a line and internally engraving small cupels.

It is worth considering that in several other engravings from Mount Bego, objects have been engraved starting from previous representations of a bucrane, a fact that underlines the symbolic importance of bovine horns (Lumley et al. 1990). In the drawings reproducing this scene in archaeological books and magazines, the number of small cupels – which would correspond to the "dots" of the mushroom cap – were frequently recorded incorrectly and, as I could verify by looking directly at the preserved original at the Musée des Merveilles of Tenda (France), their exact number is eleven. The presence of these "dots", together with the original head of the bucrane, which with the completion of the object during the second phase of elaboration would correspond to the ring on the stem, would seem to lead to an immediate interpretation as *A. muscaria* or *A. pantherina* (Samorini, 1998).

In addition to images of real mushrooms, in ancient art we observe the so-called "mushroom-men". These are anthropomorphic figures that have a head shaped like a mushroom hat and a stem-shaped neck, which gives them a decidedly fungal-shape. This iconographic scheme is spread all over the world – from Siberia to Central America, to North Africa– and, in cases where it has been possible to ascertain an ethnomycological interpretation, it appears to represent the idea of anthropomorphization or anthropomorphized deification of psychoactive fungi.

Nikolai Dikov pointed out that these representations and the associated knowledge of psychoactive fungi are a common thread running through central and northern Asia, and that this could be evidence of a Paleo-Eurasian substratum of

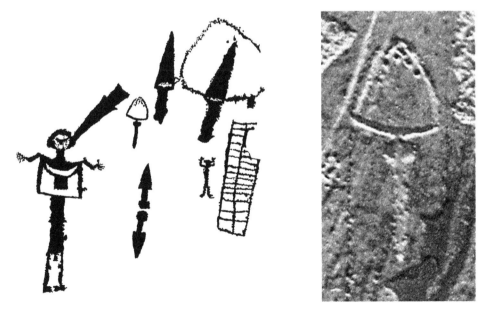

Fig 4: a) design from the rock engraving "The Chief of the Tribe", Mount Bego, France (Lumley et al., 1990: f. 63, p. 59); b) particular of the engraving (photo by author).

ethnic and cultural migrations. The same author, together with other Russian scholars, added that this "fungal" theme would also include the images of "mushroom-men" of the Stone Age found in the Iberian peninsula, as well as the origins of the Vedic Soma and, indeed the knowledge and anthropomorphisms of psychoactive fungi in the pre-Columbian cultures of Mexico and Central America (Dikov, 2004: 124-7). However, this generalization is likely to be a stretch. Attention must be paid not to force diffusionist hypotheses concerning the knowledge of psychoactive plants, as this knowledge may have been acquired in different places independently and at different times, which would explain the iconographic analogies as a phenomenon of *cultural convergence*. The showiness of the fly-agaric, with the red cap sprinkled with white spots, renders it more than plausible that its intoxicating properties were independently discovered in different chronological and ethnic environments (Samorini, 2012).

Returning to the "mushroom-men", there is a variant widespread in Siberian rock art, in which a mushroom is depicted above the head of the anthropomorph, rather than the head itself being mushroom shaped. Among the rock engravings of the Pegtymel river, the fungus is almost always designed hanging above the head (Dikov, 1971), and in cases where it is not suspended but attached to the head, it is clear that this is due to problems of creating the rock engraving, due to the miniature dimensions of these anthropomorphic figures, which are often no higher than 5 cm.

The detail of the suspension of the fungal object might have meant that the mushroom was an attribute of an anthropomorphic figure understood to be divine. But it is also opportune to consider, that one of the ways to depict an individual under the influence of a certain intoxicant source in ancient art was to draw the inebriating source suspended on the head of the anthropomorphic figure, and at other times, beside it. I refer to the case of the blue water-lily flowers (*Nymphaea nouchali* var. *caerulea*), that in Egyptian Pharaonic art are often depicted on the heads of women, clearly detached from the head, such as to exclude the representation of a flower actually resting on the head. This suspended flower most likely indicated that those women were under the effect of a female aphrodisiac – the petals of the blue water-lily – and not by chance the context of these representations is often of a sensual, sexual, or even obscene nature (Samorini, 2012-13). It is possible to consider the empty space between the head and the vegetable or fungal source as a *semantic void*, as this "void" communicates something: that the mind is under the effect of that intoxicating source.

Further evidence for an ethnomycological interpretation is provided by *ethnographic correspondences*. Fungal-like images can be accompanied by objects or symbols on the body or around the body. An exemplary case is that of the "mushroom-men" engraved on the Siberian rocks of Ortaa-Sargol, studied by Devlet (1982) and dated to the second half of the second millennium BC (Figure 5). Several of these "mushroom-men" have a protuberance protruding from the area of the belt, ending in a roundish shape. This object has variously been interpreted as a zoomorphic attribute - such as a tail, a war bat, - either real or ritual (many "mushroom-men" hold a bow or spear), or a shamanic drum (*see* discussion in Devlet, 2001a: 34). More convincingly, Devlet has interpreted this protuberance as a bag made from leather or an animal bladder, an object used by Asian nomadic ethnic groups as containers of liquids; but it could even more specifically be a "medicine-bag", held by the shamans and containing intoxicating substances used for the "journey" into the other world. The intoxicant *par excellence* of northern Asia is the *mukhomor*, the fly-agaric, which would be precisely the fungus that is represented by the "mushroom-men". Dikov (1971: 118) suggested that this leather bag contained an infusion of fly-agaric. In this case the fungal image of the "mushroom-man" head and the "medicine-bag" mutually reinforce each other in the ethnomycological interpretation.

Among the ethnographic correspondences there are also those of a mythico-religious nature. Remaining in Siberia, the rock engravings of the Pegtymel river present a scenic scheme in which two anthropomorphic figures are depicted, one with a large mushroom suspended above the head and the other without. The one with the mushroom is often or always female, and is shown taking the hand of the figure without the mushroom on the head (Figure 6). This corresponds to a common theme among the Siberian populations that use the fly-agaric, of the mushroom-induced vision of anthropomorphic spirits - the spirits of the amanita

Fig. 5: Rock engraving from Ortaa-Sargol, Yenisei, Siberia (Devlet, 1982: 118).

called "mannequins" (Saar, 1991a: 162) - and the notion that these "women-amanita" take the fly-agaric eater by the hand and carry him through the afterlife (Bogoraz, 1904-09: 282). Other ethnographic correspondences include the recognized use, ancient and/or modern, of psychoactive mushrooms in the region where the archaeological find was found (as is the case of the Mayan mushroom-stones described above).

Fig. 6: Rock engraving from Pegtymel, Siberia (Dikov, 1971: 93).

Further evidence to validate an ethnomycological interpretation can be drawn from the ecology in which the artefact is found: it is always advisable to verify the territorial presence of the psychoactive fungi identified in the archaeological findings when drawing conclusions. In the case of the *kuda-kallus* of Kerala, India – huge mushroom-stones that I will describe further – the fact that both *Amanita muscaria* and *A. pantherina* grow a few tens of kilometers away from the *kuda-kallu* presence area is an ecological correspondence that reinforces the hypothesis that the *kuda-kallus* were intended to represent these fungi.

Another important ecological factor is the fact that numerous species of psilocybin fungi grow in association with the dung of large quadrupeds – bovids, cervids, pachyderms, etc. This association is widespread throughout the world and is represented in one of the prehistoric paintings of the Sahara. In the Tin-Abouteka site, in the Algerian Tassili, an anthropomorphic figure is depicted forward bending, with a pair of fungiform elements that seem to come out from the lower back of the anthropomorphic subject (Figure 7). Assuming that those fungiform objects were really meant to depict intoxicating fungi – as corroborated by other details that I will present below – this bizarre association of a scatological nature could find justification in the possible dung-habitat of the mushrooms painted in the Saharan scenes.

Another important source of information is the *scenic context* in which the fungal representation is inserted. A context of adoration of the mushroom is strong evidence of the fact that it is a psychoactive mushroom, following the axiom of the archaeo-ethnomycology that I have previously described. Exemplifying this is a terracotta sculpture from the Colima culture of western Mexico, where a "mushroom-tree" is depicted surrounded by four individuals holding each other in a circle, like in the child's game "ring-around-the-rosie" (Furst, 1974: 62; Figure 8). In this case, supporting the identification of the tree as a psychoactive fungus are the following elements: the umbo on the cap, a characteristic common to many *Psilocybe* species (morphological detail); the presence of similar Colima sculptures in which fungal objects are shown held in the hand or in scenes of a cultic nature (repetition of the representation) (De La Garza, 2012: 63); the geographical area to which these findings belong, Mexico, where the ancient and modern use of psychoactive mushrooms is recognized (ethnographic correspondence); and lastly, the fact that the four individuals are embraced can be considered an added value, both for a certain analogy of the intoxicating effects of the mushroom with vertigo in the game of ring-around-the-rosie, and as a possible gesture of collective adoration (scenic context).

In several cases of fungal depictions, it is often sufficient to have one of the details I have listed – the "dots" on the hat or an unequivocal ethnographic correspondence – to exclude alternative interpretations: graphic elements that I

Fig. 7: Rock painting from Tin Abouteka, Tassili, Algeria (photo by author).

indicate as "killer-details", in the sense of details that remove ("kill") any doubt. One of the most striking of these "killer-details" will be discussed in the description of the prehistoric Saharan paintings.

As for the psychoactive fungi of the genus *Amanita* – which fall into the biochemical class of isoxazole fungi, as producers of isoxazole alkaloids (ibotenic acid and muscimol) – in archaeo-ethnomycological interpretations there is a general tendency to attribute the fungiform identifications to *Amanita muscaria*, undoubtedly the best-known and most striking species, to the detriment of the *A. pantherina* congener; the latter being generally smaller in size and less showy than fly agaric. In reality this last species is frequently present in the same areas occupied by fly agaric. Furthermore, the chemical analyses developed on samples of both species collected in different geographical areas, frequently show a greater concentration of the active ingredients in panther cap compared to fly agaric. For example, analyses described by Feeney and Stijve (2010; *see* Ch. 23) on samples collected in central Europe (Germany, France, Switzerland) showed a muscimol concentration of 0.01-0.22% on the dry weight in fly agaric, and of 0.025 -0.31% in the panther cap.

Fig. 8: Colima terracotta, Mexico (Furst, 1974: 62).

Furthermore, the concentrations of muscarine – an alkaloid with toxic properties – are considerably lower in the panther cap than in the fly agaric. It is therefore possible that panther cap is generally more psychoactive and less toxic than fly agaric, and thus more reliable for use as an intoxicant. When studying the ancient use of psychoactive fungi it would seem appropriate to consider the panther cap, more than has been done so far.

Below I present two cases of ethnomycological interpretations of archaeological finds that I have personally studied during my research – in the Sahara Desert and in South India – and that have been accompanied by my research missions on site. In this regard, the importance of directly observing the archaeological findings cannot be underemphasized. Depending solely on the observation of photos or graphic reliefs is limited for various reasons, one of which being that the findings have often been imprecisely graphically reproduced, as I have personally noted for several rock engravings.

The "Round Heads" of prehistoric Sahara

For over 100,000 years the huge basin of the Sahara has been subject to climatic cycles that have transformed it from an area rich in water sources (rivers, lakes) and consequent luxuriant flora, to a partial or total desert. During the most recent wet phases, dated between 9000 and 3000 years BC (White & Mattingly, 2006), the Sahara has been inhabited by human populations that have produced a rich set of rock engravings and paintings distributed along the banks of the ancient rivers and on the mountain buttresses, which have been preserved in an exceptional way because of the subsequent drying up of the territory. This artistic production has been studied for over a century, predominantly by French and Italian researchers.

Among the prehistoric Saharan paintings, those that stand out for their richness of color and scenic originality are the paintings of the so called "Round Heads" phase, concentrated in the plateau of Tassili (Algeria), with minor presences in other mountainous regions of the central Sahara such as the Tadrart Acacus (Libya), the Ennedi (Chad) and the Jebel Uweinat (Egypt) (Muzzolini, 1986).

The chronology of this stylistic horizon has been hotly debated over decades between the supporters of a "high chronology" (before 6000 BC; Mori, 1968) and those of a "low chronology" (4000-1000 BC; Muzzolini, 1991). However, recent analyses of micro-samples of the paintings and deposits found at their feet, have determined a date not earlier than 8000-7000 BC (Mercier, Le Quellec, Hachid & Agsous, 2012) and, for the pictorial phase of the "Round Heads", the period of 6000-4500 BC (Le Quellec, 2013).

On the Tassili plateau, at approximately 2000 meters high, images of gigantic mythological beings in human and animal form, together with a myriad of smaller creatures with horns and feathers, mostly appearing in dancing positions, cover the rock shelters, which in certain places intertwine in a play of rocks so as to constitute real "citadels", with streets, squares and terraces. The religious and initiatory context of the art of the "Round Heads" has been firmly established by scholars. Fabrizio Mori (1975: 346) highlighted the close relationship that had to exist between the artist of the paintings and the figure common to all prehistoric societies, whose main characteristic is the role of mediator between the earth and the sky: the priest-sorcerer or the shaman; while for Henri Lhote (1968: 280) "it seems clear that these painted shelters were secret sanctuaries". For Umberto Sansoni (1994: 25) in these paintings "the imaginary and perhaps the dreamlike and the ecstatic enters forcefully".

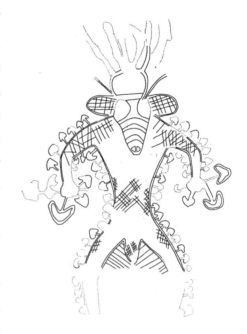

Fig. 9: Rock painting anthropomorph from Tin-n-Tarim, Tassili, Algeria (illustrated by author).

In different scenes we observe mushroom-like objects. These are rarely isolated and in most cases are held in the hands of, or emerging from the body of, anthropomorphic figures, or are inserted into the masks frequently worn by them. The masks appear to hold a particular value, since in many cases they are painted in isolation, or almost covering the entire anthropomorphic, or divine figure. Throughout the scenes there are a wealth of figurative constants that allow us to glimpse a definite conceptual, mythological and religious structure. To exemplify this are the two characters of the southern Tassili – located in In-Aouanrhat or In-Aouanghad, and Ti-n-Tarin (apparently erroneously cited as the location of Matalem-Amazar by Lajoux (1964); *see also* Fouilleux & Mouchet, 2010: 132) (Figures 9, 10) – both of which are about 0.8 m tall, bearing the typical mask of this pictorial phase, with a similar bearing (legs bent and arms bent downwards), and with the body entirely reticulated. Another common feature is the presence of fungiform objects that depart from the forearms and thighs, while others are held in the hands. In the character found at Ti-n-Tarin these objects entirely cover the outer contour of the body.

Other anthropomorphic characters are characterized by a mushroom-shaped head, some pointed (in a couple of cases with a bluish shade, although it is not clear whether it is the original color or its color change from subsequent oxidation), while others bear a leaf or a vegetable branch in their hands (Figures 11a, b). In a painting at the Tin-Abouteka site we observe a curious and unique representation, that appears at least twice, of a fish with two large mushroom-like objects painted opposite each other on either side of the caudal fin (Figure 12). In a scene from the Techekalaouen site, an anthropomorph in a slightly bowed position appears to be holding a large mushroom in one hand, in the act of offering it to a second human figure, while in the other hand it holds what appears to be a toad or other small animal being (Figure 13). In the case of a toad, this would surprisingly confirm the ubiquitous symbolic association between the toad and the fungus already recognized in Eurasia (Wasson & Wasson, 1957, I: 65-91). However, the "killer-detail" that led me to a definitive ethnomycological interpretation of the "Round Heads" rock paintings can be found in a complex scene painted in a shelter at Tin Tazarift, where five horned masked individuals are painted, in a line or procession and in a hieratic/dancing arrangement, surrounded by long and lively festoons of phosphenic designs. Each dancer holds in his right hand a fungal object, from which two parallel dashed lines branch out, reaching the central area of the head, where

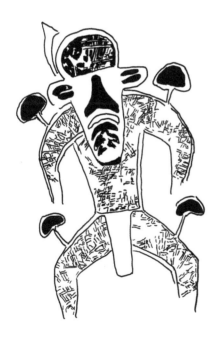

Fig. 10: Rock painting anthropomorph from In-Aouanrhat, Tassili, Algeria (illustrated by author).

the two horns emerge (Figure 14).

The double dashed line could mean an indirect association or an immaterial fluid passing between the object held in the hand and the human mind, which is well suited to the interpretation of these objects as psychoactive fungi. In a previous study of mine (Samorini 1995b) I pointed out that the grapheme "linear sequence of points" is an ideogram representing something immaterial in archaic artistic horizons, which seems to have kept the meaning constant from the Paleolithic to the present day, and thus could be considered as a "fossil-guide" in the semiotic approach to rock art.

A clear example of this is a Tamgali rock engraving, found in Kazakhstan, and dated to the Bronze Age. This engraving depicts an individual with zoomorphic features (the tail and animal features in the face), interpretable as a shaman who is turning into his animal guide, which is drawn next to him (Figure 15). In this case the shamanic transformation - an event of a psychic nature - was represented by the concentric series of points that depart from the head of the individual. This design well expresses the modified state of consciousness experienced by the individual.

Another example of this can be found at the Pahi site, Kondoa, Tanzania, in the painting of a flute player and the specific way in which the sound of the

 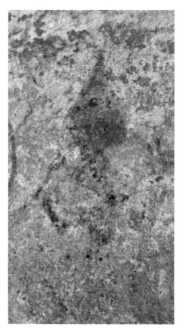

Fig. 11: a,b: Rock paintings from Tin-Teferiest, Tassili, Algeria (photos by author).

flute is depicted. The sound of the flute, which is understood to be immaterial and unseeable, is represented with a series of dots (Figure 16). The immaterial value of the series of points can also be seen in the most ancient Paleolithic art, the geometric drawings of the Castillo Cave, in Spain (Figure 17), where on several occasions linear sets of dots are painted alongside full-contour quadrangular figures. These juxtapositions have been seen as representations of opposing feminine and masculine values (Anati, 1989: 173), but it is

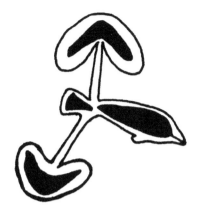

Fig. 12: Motif from a rock painting, Tin-Abouteka, Tassili, Algeria (illustrated by author).

also possible to interpret them as combinations of material elements – the full-contoured figures – and immaterial ones, the latter being the sets of points.

The dotted, dashed, or otherwise interrupted lines represent something immaterial, unseeable, but penetrable (like the dashed lines drawn on the roadways, indicators of the possibility of overtaking). In modern day comic books, when the author wants to make it clear that a character is saying something aloud, then what the character says is surrounded by a closed line (the words in the comics are considered visible). But when the character is in thought, an action that is considered invisible, the thought is represented within a dotted line. In these modern cases, the same grapheme – the linear series of points – has retained the meaning that originated in the Paleolithic: to represent something immaterial, invisible, a thought or a mental process.

Returning to the scene of Tin Tazarift, it would appear that the dashed lines uniting the mushroom with the head of the dancer are meant to represent the effect that the mushroom has on the human mind, making this depiction one of the most surprising and exemplary cases of "killer-details" of global archaeo-ethnomycology.

This scene is imbued with a profound ritual meaning and represents a cult event that has occurred and been periodically renewed over time. Perhaps we are faced with a realistic representation of one of the most salient moments in the religious and emotional life of the populations of the "Round Heads". The repetition of physical characteristics and attitudes of the five dancing figures reveals a coordinated collective understanding of scenic representation for collective contexts. The dance depicted here has all the air of being a ritual dance and, perhaps, from a certain moment within the ritual, of being ecstatic.

In a painting in the Jabbaren site, which is also located in the Tassili, five individuals are depicted one behind the other, kneeling down and with their arms stretched forward, placed in front of three upright figures, two of which are clearly anthropomorphic (Figure 18). It would appear to be a scene of adoration in which the three figures standing represent gods or mythological characters. Two anthropomorphic figures are equipped with large horns, while the third, behind these, has the upper part of the body in the form of a large mushroom. If the interpretation of this as an act of worship were correct, this scene could be revealing of a "divine trinity". The trine production of the images is frequently depicted in the art of the "Round Heads", and an important case concerns the three masks painted on the Sefar site. As already anticipated, the isolated representation of a mask would seem to symbolize a divine presence, and the three masks of Sefar, placed side by side in an isolated shelter, may indicate the presence

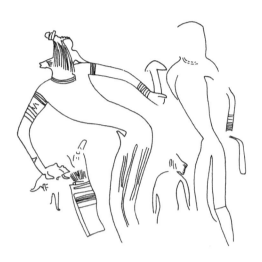

Fig. 13: Scene from a rock painting, Techekalaouen, Tassili, Algeria (illustrated by author).

Fig. 14: a) Scene from a rock painting, Tin Tazarift, Tassili, Algeria (illustrated by author).

Fig. 14: b) detail from Fig. 14(a) (illustrated by author).

of a divine trinity. It is significant that one of the three figures "adored" in the Jabbaren scene has the upper part of the body in the form of a mushroom. This could be related to the anthropomorphic figures, perhaps divinities, of the previously described In-Aouanrhat and Ti-n-Tarin paintings.

Fungal images are observed in the artistic production of the "Round Heads" of other regions of the Sahara. In a painting from the Uan Muhuggiag site of the Tadrart Acacus, in the Libyan Fezzan, a procession of individuals is painted in which the lower part of their bodies are hidden by a longitudinal band, initially interpreted by Mori (1975) as a probable boat (Figure 19). From this band (or boat) the figures protrude halfway up, as if they were sitting, and the serpentine line inside the area corresponding to the hull could indicate the surface of the water. Subsequently Mori (1990) interpreted this scene as a reproduction of the same rocky shelter, in which the painting is housed, seen from the outside, where the wavy line would indicate the surface of the water of the river that flowed alongside the times of painting production. The presence among the characters of an individual painted in an upside-down position and with spread-out legs would seem to represent a deceased, and the whole scene would be characteristic of a funeral. Above the procession there are four fungal images, of red color, which Mori interprets as arrowheads or oars. However, the ethnomycological interpretation, strengthened by the previous interpretations

Fig. 15: Scene from a rock engraving, Tamgali, Kazakhstan (Anati, 1989: 220).

of the Tassili "Round Heads", would fit neatly with these fungal images being reproduced in a funeral context.

The fungal objects of the art of the "Round Heads" have been interpreted by scholars as arrowheads, oars, undefined vegetables, flowers or as undefined enigmatic symbols (Samorini, 1989). The set of details presented here suggests that we are in the presence of a very ancient cult of hallucinogenic mushrooms and related mythological representations. The mushroom species represented are not easy to determine, belonging to a flora that disappeared or withdrew from the now desertified Saharan basin. From the paintings it seems deducible that there were at least two species: one small in size, in some cases endowed with a papilla at the upper end of the cap, which is characteristic of many hallucinogenic *Psilocybes;* and another of greater dimensions, such as *Boletus* or *Amanita.* The colors used are white and different shades of ochre, and in rare cases the pointed shape is painted blue (although the latter could be the possible result of oxidation of the original color).

Fig. 16: Design from a rock painting, Pahi, Kondoa, Tanzania (Anati, 1989: 206).

The *kuda-kallus* of Kerala

In southern India there are archaeological finds of a megalithic culture whose origins are still uncertain; these are distributed mainly in the territories of Karnataka, Kerala and Tamil Nadu and belong to the Iron Age of the Indian peninsula.

In Kerala, the megalithic production has been dated to the beginning of the first millennium BC (Satyamurthy, 1992). The large stone monuments would have been erected in a late phase of the megalithic culture, from 550 BC. to 100 AD, and the artefacts found in association with these vestiges belong to the "Black and Red Ware" culture (McIntosh, 1985). A monument made of laterite stone, characteristic of this culture and specific to the region of Kerala, is the *kuda-kallu* ("umbrella-stone"), which resembles a large mushroom (Figure 20). It makes a certain impression to walk among the dozens of *kuda-kallus* present in the megalithic sites, and their gigantism transports the observer to the land of Alice, as described in the famous novel by Lewis Carroll.

Despite being the main cultural symbol of Kerala, the *kuda-kallus* continue to be mistreated by local archaeology. It is sufficient to note what was reported in a recent review of these megalithic finds: "Though excavations of Kudakall were attempted by various institutions like the Kerala State Department of Archaeology, the Archaeological Survey of India and Mahatma Gandhi University, no detailed reports of the excavations are available. Even reports of the recently excavated sites are not published. Hence much of the discussions on Umbrella Stones are still dependent on the colonial writings" (Peter, 2015: 291).

Fig. 17: Design from a rock painting, Castillo Cave, Spain (Anati, 1989: 172).

The authors would seem to agree on the hypothesis that the builders of South Indian megalithic works belonged to populations speaking Dravidian languages. Even today the states of Kerala and Tamil Nadu are inhabited by Dravidian ethnic groups preserving, in a rare case of historical continuity, megalithic traditions and customs. An example of this is the custom of erecting dolmens in honor of those who died in non-natural ways, as is the case of the Malarayaran of Kerala (Chinnian, 1983). In other regions of India menhirs and similar stone structures are erected alongside common burials. Among the Gonds of the Odhisa region, large stones that recall only the upper part of the *kuda-kallu* are sometimes erected. These "umbrella stones" are considered by the Gonds as the places where the deities sit (Mendaly, 2017: 940). Although the findings associated with megalithic structures are indicative of a strong continuity with previous Neolithic cultures, megalithic architecture highlights Asian and Western influences (McIntosh, 1979).

In Kerala, the most important concentrations of *kuda-kallus* are located in the regions of Trichur and Palghat, north of Cochin, within the coastal region. The area is gently hilly, and its laterite rock is easily carved. The main sites where the *kuda-kallus* are found are: Chataparamba, Cherumangad (or Cheramangad), Porkulan, Ariyannur (or Aryannoor), Ummichipoyil, Kalkulam, and Anakkara. In the same

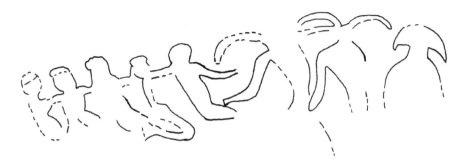

Fig. 18: Design from a rock painting, Jabbaren, Tassili, Algeria (illustrated by author).

archaeological sites other megalithic structures are present: dolmens, menhirs, *topikals* ("hat-stones"), stone circles, caves carved in the rock, and *hood-stones*. Some *kuda-kallus* stylistic classifications have been developed. The most recent, promoted by Jenee Peter (2015), considers the two subgroups, one with the upper stone in the form of a dome (hemispherical) and the other with a flat upper stone.

Although for several authors the *kuda-kallus* appear as a tuft of giant mushrooms (Babington, 1823: 324; Longhurst, 1979: 11; Menon, 1991: 40; Subramanian, 1995: 679; Sudyka, 2010: 380), nobody proposed the idea that they could actually represent mushrooms. The only exception would seem to be that of Malinal (1981), in which he sees a derivation of their forms from that of fungi, although he does not specifically consider the possibility that they might represent psychoactive fungi.

The *kuda-kallus* measure on average 1.5-2 m in height and 1.5-2 m in width. They consist of four stones cut in the form of a half clove and gathered as a base, supporting a fifth stone, flat on the resting side and convex-uniform in shape from the other (Iyer, 1967). The *kuda-kallus* are not burials, and no furnishings have ever been found (Anujan Achachan, 1952-53; John, 1982; Narasimhaiah, 1995). According to Longhurst (1979), their function was that of a "memorial" to the dead, and they were probably erected to mark the place where the body was cremated. The same author associates them to the later *stupa*, a monument of hemispherical form enclosing the relics of the Buddha or of Buddhist saints, or even just commemorating important events in the life of the Buddha. More recently, the *stupas* have been seen as an evolution of the round mound, which is the simplest and perhaps most original form of the monumental megalithic expression, and which in Kerala and

Fig. 19: Design from a rock painting, Uan Muhuggiag, Tadrart Acacus, Lybia (Mori, 1975: 352).

South India is expressed in the *topikal* or "hat-stone" (Menon, 2016).

The most common association of *kuda-kallus* reported by scholars and local people is with the umbrella, known as an archaic symbol of power and authority, as well as of sacredness, widespread in ancient Egypt as well as among the Assyrians and of other Oriental civilizations of later epoch. In some Buddhist countries the umbrella is an object of veneration. In India it acquired a religious significance. Buddha images never appear in early Buddhist art; he is represented by symbols such as a wheel, a throne, a pair of footprints, and these are placed under one or more honorary umbrellas. Even on the top of the *stupas*, wooden and fabric umbrellas are erected (Longhurst, 1979). There are those who wanted to see in *kuda-kallu* "a stone model of umbrellas of palm leaf used by the local people" (Sathyamurthy, 1992: 3). Local tradition attributes a Buddhist origin to megalithic monuments, which are seen as the homes of hermits in the days when Buddhism and Jainism were popular in Kerala (Iyer, 1967: 25).

The term *kuda-kallu*, which literally means "umbrella-stone", is of Malayalam language origin, currently the most widespread language in Kerala, which differed from the Tamil language only in the 9th century AD. It is an undoubtedly late appellation, after the time of the erection of monuments, and there is no reliable evidence that the name designated to these structures by the populations that erected them included the same meaning of "umbrella". Furthermore, as Longhurst (1979: 16-7) states, in all probability it was only during the Asoka period, several centuries after the erection of the *kuda-kallus*, that the royal umbrella was associated with the *stupa*, of which the *kuda-kallus* are seen as the precursors, both from an architectural and symbolic point of view. Perhaps, the umbrella was associated with the *kuda-*

kallu following the migration to southern India of Jainists and Brahmins, which began during the same period as the Asoka. There is also a substantial difference in shape between the *kuda-kallu* and the classic honorary umbrellas represented in Egyptian, Assyrian and Indian bas-reliefs: the latter are characterized by a supporting rod, thin and equal in all its length, by a generally flat umbrella on both sides (the so-called "wheeled" umbrella), often fringed at the edge, and by a short central pivot exiting from the upper part. The *kuda-kallu* has a much more robust and compact appearance, is free of plumes or other striking decorations (unless they were built with perishable material) and morphologically recalls some large mushrooms of the genus *Amanita* or *Boletus*.

Noting therefore the late association of the *kuda-kallu* with the sacred symbol and sovereignty of the umbrella, and following my research in the region, I proposed the hypothesis that similar constructions were meant to represent mushrooms, whose form is remarkably close to an umbrella (Samorini, 1995a). If the *kuda-kallu* were created in order to represent fungi, then according to the axiom of archaeo-ethnomycology that I previously noted, they most likely represent psychoactive fungi, with visionary properties, which can facilitate visions of the beyond and of

Fig. 20a: Kuda-kallus from Cherumangad, Kerala, India (photo by author).

the underworld, making them a probable candidate for association with the cult of the dead.

The presence of *A. muscaria* and *A. pantherina* is currently recorded in coniferous forests in the Kodaikanal region, in neighbouring Tamil Nadu, about 80 km from the sites of the *kuda-kallus* (Natarajan & Raman, 1983: 176). According to Wasson, their presence was attributed by mycologists to the implantation of exotic conifers in the last century (Wasson, Kramrisch, Ott & Ruck, 1986: 136); but this is a non-referenced statement that I have not found confirmed in the literature (for example it is not considered in Natarajan & Raman [1983: 7-8] where the different types of forests in South India are described). In his pioneering ethnomycological study, Wasson mentioned the presence of "mushroom-stones" in Kerala (and Nepal) in a passage in which he discussed the symbolic association between the mushroom and the umbrella, but he mentioned it by hearsay. If he had instead seen them in person, he would certainly have been very impressed. He noted that in classical Sanskrit the term for mushroom was *chattra*, whose primary meaning is "parasol", and has for its root the verb *chad*, "to cover". The parasol, as an object used to shelter from the scorching sun of southern India, was unknown to the Nordic peoples, and

Fig. 20b: Kuda-kallus from Ariyannur, Kerala, India (photo by author).

"when the Aryans invaded Iran and India, they gave this new instrument an Aryan name, *chattra*, and later they extended the meaning of that name to mushrooms with fleshy hat" (Wasson, 1968: 63-6). Wasson himself observed how the umbrella or parasol could have been associated with the fungus since the origins of its symbolic values, emphasizing the affinity of form between the mushroom and the umbrella and noting that "the fungus has lamellae that suggest the uprights of a parasol" (Wasson, Kramrisch, Ott & Ruck, 1986: 61).

In the course of my observations at the Cheramangad site, where there are dozens of *kuda-kallus*, I have noticed an important detail for the determination of the mushroom species that the *kuda-kallus* could have represented. A structure present in the same archaeological sites is the *hood-stone*, consisting of a single large dome-shaped stone, with the flat part resting on the ground, placed at the end of a burial. The *hood-stone* could be seen as a *kuda-kallu* without a pedestal ("stem"), in which the stone that constitutes the "hat" rests directly on the ground, where it acts as a cover for a cylindrical pit, with the base pear-shaped, in which a cinerary urn is deposited. With a little imaginative effort, we could perceive of the missing foot of the *hood-stone* as the shape of the cylindrical pit placed below the ground level, and in this way the similarity in the form of the two structures appears evident.

In the *hood-stone*, the pit carved in the laterite is large enough to contain a red terracotta urn with a piriform base, matching the bottom of the pit. This type of piriform background resembles the terminal part of the stem of numerous large mushrooms, in particular those that are born from an ovule, such as the species of *Amanita*. Even the appearance of the stones making up the *hood-stones* is similar to that of the upper stones of the *kuda-kallus* (the "hats"), although generally smaller.

On a good part of the *hood-stones* that I was able to observe in the Cheramangad site there are deep holes, which however do not reach the opposite end of the stone (the one in contact with the ground): one or two holes that presumably had the purpose to facilitate the placement or removal of the stone by inserting wooden poles on which to pry. A detail on which my direct observation was focused, and which I have not found described in any of the archaeological publications concerning the Kerala megaliths, concerns some *hood-stones* in which there are numerous recesses excavated over the entire aerial surface of the stone. Although the surface is rather coarse, due to the type of lateritic rock, their presence appears undoubtedly man-made. Measuring 4-5 cm deep and 4-10 cm wide, these grooves seem to have had a decorative function or to highlight a distinctive feature of the object that the *kuda-kallus* intended to represent (Figure 21). This detail, which could be considered as a "killer-detail", directly refers to the technique of representing the "dots" of fly-agaric and panther cap in a three-dimensional artefact by performing recesses on its upper surface. At this point, the hypothesis that the *hood-stone* and the *kuda-kallu* intended to represent just the fly-agaric or panther cap mushrooms becomes more consistent.

Moreover, the possibility that the *kuda-kallus were* intended to portray other species of mushrooms should not be excluded, and that their squat shape was due to structural requirements, to give strength to the artefact. In southern India the presence of some powerful species of psilocybin fungi is recognized, such as *Psilocybe aztecorum* var. *aztecorum*, *P. aztecorum* var. *Bonetii*, *P. cubensis*, *Copelandia cyanescens*, *C. tropica*, and *C. bispora* (Natarajan & Raman, 1983).

In case the *kuda-kallus* intended to represent *A. muscaria*, the question may arise: what relationship exists between this megalithic cult and the cult of the Vedic *soma*? According to the well-known hypothesis of Wasson (1968), in

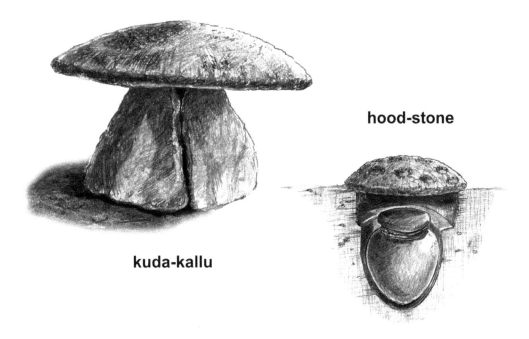

Fig. 21: Depiction of a kuda-kallu and a hood-stone from the Ariyannur site, Kerala, India (illustrated by author).

its original form the *soma* should be identified with a psychoactive source derived from the fly agaric. The knowledge of the psychoactive properties of this fungus would have been spread by the Aryan populations in the context of Indo-European migrations. This priority of the role of Indo-European migrations must not however be considered as a definitive consequence of the Wassonian hypothesis. Whilst it is true that there seems to be a geographic-cultural fulcrum to spread the knowledge of the fly agaric – roughly central-western Asia – it may be misleading to believe that this was the only original area of diffusion of this knowledge, or that this knowledge was promoted solely by the Indo-European populations in their long run migrations.

There seems to be no direct relationship between the *kuda-kallus* and the *soma*, in the sense that these megalithic monuments do not appear to be emblems of a cult that either originated or was influenced by the cult of *soma*. The cult associated with the *kuda-kallus* began to develop in a period prior to the contact of the Aryans with southern India, which occurred around 300-200 BC (Menon, 1991: 43-4). When they reached the Dravidian populations of southern India, they had lost the knowledge of the original *soma* and practiced worship with the use of substitute vegetable sources. Once again, care must be taken not to force the diffusionist hypotheses against cultural convergence phenomena. The discovery of the intoxicating properties of the fly agaric could be much older than the Indo-European times, and could reach as far back as the Stone Age. The recent discovery, along with various plant elements, of fragments of a mushroom fabric – perhaps a kind of *Boletus* – in the dental calculus of the remains of a woman who lived 18,700 years ago that was found in a cave in the Cantabrian Mountains, in Spain (Power, Salazar-García, Straus, González Morales & Henry, 2015), highlights how the Paleolithic man of the Magdalenian period, besides taking advantage of a vegetable diet alongside the animal diet, was able to discern between edible and poisonous mushrooms, making it plausible that he had already addressed attention to one of the most striking mushrooms in the woods, the fly agaric.

Chapter 18

Beyond the Ballgame:
Mushrooms, Trophy Heads, and the Great Maya Collapse
Carl de Borhegyi

While over a century has passed since the first small stone statues carved in the form of mushrooms came to the attention of Mesoamerican archaeologists, they often lacked reliable context. Most had been found or purchased by private collectors in the highlands and Pacific coastal areas of Guatemala and El Salvador. Several archaeologists have endeavored to determine the age, identity and purpose, ceremonial or practical, of these enigmatic objects, with skeptical investigators suggesting that they may have been phallic symbols, small stools, or molds for rubber balls for the Mesoamerican ballgame. It was not until the early 1950s, that Mesoamerican archaeologist Stephan F. de Borhegyi proposed the existence of a hallucinogenic mushroom cult among the ancient Maya of Guatemala, Mexico, and El Salvador, and that the cult was associated with trophy heads, ritual decapitation, human sacrifice, and the pre-Columbian ballgame. In this chapter, the author extends his father's research by illustrating a few threads in the complex fabric of Maya, Aztec, and Toltec art and mythology dating back to great antiquity, which show a relationship between these "mushroom stones" and hallucinogenic mushrooms, principally *Amanita muscaria* (also known as fly agaric), with the ballgame, the god-king Quetzalcóatl, the Venus cult associated with warfare and the Classic Maya Collapse.

Background

In the early 1950s, my father, Mesoamerican archaeologist Stephan F. de Borhegyi (better known as Borhegyi) was the first to carry out a comprehensive study on the enigmatic small stone sculptures called "mushroom stones" that had turned up in collections and in a few archaeological excavations in the central highlands of Guatemala (Wasson & Wasson, 1957: 276). Although these artifacts had been known for a number of years, Borhegyi was the first to suspect that they may have been used in an ancient hallucinogenic mushroom cult. After cataloging over 100 by type and provenience (Figure 1), he dated their earliest appearance to approximately 1000 B.C. According to their archaeological context, mushroom stones were associated from their first appearance with a trophy head cult and the Mesoamerican ballgame.

Borhegyi's research on mushroom stones brought him into contact with ethno-mycologist Robert Gordon Wasson. The two scientists worked in close cooperation and shared a voluminous correspondence of over 500 letters that are today housed in the Tina and R. Gordon Wasson Ethnomycological Collection of the Harvard University Herbaria Botany Libraries (Wasson Archives). Several of these letters and others archived at the Milwaukee Public Museum are published here for the first time to shed further light on their pioneering thoughts. As a result of their collaborative efforts, the two surmised that if the mushroom stones of Guatemala did indeed represent a mushroom cult, then the mushroom itself was an iconographic metaphor, and the mushroom stone effigies would supply the clues necessary to decipher their meaning.

Wasson and his wife Valentina included Borhegyi's chart of types by provenience and age as an addendum entitled "Mushroom Stones of Middle America" to their monumental book *Mushrooms, Russia and History*, published in 1957. Wasson noted that "Dr. Borhegyi's chart suggests to us that hallucinatory mushrooms were the focus of a cult in the highland Maya world that goes back at least to early Preclassic times, to B.C. 1000 or earlier... Beyond that horizon may we project the mushroom agape back through millennia, to the Eurasian home-land whence our Indians' ancestors migrated?" (Wasson & Wasson, 1957: 329). Acceptance of this Eurasian origin theory turned out to be quite controversial for several decades, but my research and recognition by some archeologists is proving the validity of this ground-breaking collaboration between Borhegyi and Wasson. The renowned Mexican mycologist Dr. Gastón Guzmán also confirmed the belief shared by Borhegyi and Wasson that these mushroom stones from Mesoamerica were indeed modeled after the *Amanita muscaria* mushroom (Guzmán, 2013: 489-504).

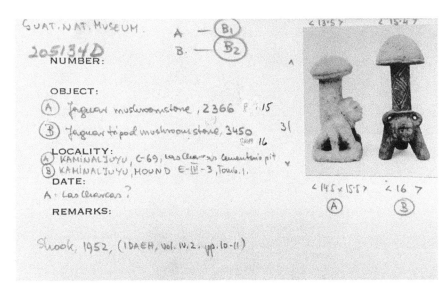

Fig. 1: Borhegyi's catalog card above of Type A and B mushroom stones noting their provenience (Milwaukee Public Museum Archives).

Linguistic and Cultural Roots of the Mushroom Religion and Quetzalcóatl

In furthering my father's research on mushroom stones, I have discovered that hallucinogenic mushrooms are not only frequently identifiable in the prehistoric art of both the Old and New World, but that in Mesoamerica in particular, they played a major role in the development of indigenous religious ideology and mythology, and that both the *Amanita muscaria* mushroom and the *Psilocybin* mushroom were worshiped and venerated as gods in ancient Mesoamerica.

The two most important linguistic and cultural streams to emerge during pre-Columbian times from the developed civilizations of Mexico and Central America are the Nahua and Maya cultures. This area is known today as a singular "Mesoamerica", a term conceptualized by Paul Kirchhoff (1943: 92-107), that recognizes that these cultures shared similar ideologies and mythologies derived from earlier Olmec cultural roots. The author believes the sacred mushroom ritual shared by these cultures was intended to establish direct communication between Earth and Heaven (Sky) in order to unite man with God. Rulers or priests bestowed with this sacred knowledge were believed to be Gods, incarnates of their creator god know among the Nahua as Quetzalcóatl.

Quetzalcóatl's essence in the world as a culture hero was to establish this communication by teaching his followers to eat sacred mushrooms and make blood sacrifices in order to achieve immortality. This was recorded in the Mixtec *Codex*

Fig. 2: Page 24 of the *Codex Vindobonensis* shows the Wind God Ehecatl-Quetzalcóatl (circled) holding an axe in one hand and the decapitated head of the underworld Death God in the other. Here, Quetzalcóatl gives instructions to a young Xochipilli who is depicted with tears in his eyes as he holds a pair of divine mushrooms in his right hand (Photo from The British Museum).

Vindobonensis Mexicanus housed in the National Library of Vienna, Austria, which is one of the few pre-Hispanic native manuscripts that escaped Spanish destruction (Figure 2). It was produced in the Postclassic period for the priesthood and ruling elite. A thousand years of history is covered in the Mixtec Codices, and Quetzalcóatl, known to the Mixtecs as 9-Wind, is cited as the great founder of all the royal dynasties.

Spanish chronicles tell us that the Aztecs and Toltecs attributed their enlightenment to Quetzalcóatl. In the 16th century, Franciscan Friar Bernardino de Sahagún recorded in his *Florentine Codex* (Historia General de las Cosas de Nueva España, 1547-1582) that:

> They (the Indians) were very devout. Only one was their God; they showed all attention to, they called upon, they prayed to one by the name of Quetzalcóatl ... the one that was perfect in the performance of all the customs, exercises and learning (wisdom) observed by the ministers of the idols, was elected highest pontiff; he was elected by the king or chief and all the principals (foremost men), and they called him Quetzalcóatl. (Sahagún, 1950, vol. 10: 160).

Spanish chronicler Friar Diego Duran, wrote that:

> All the ceremonies and rites, building temples and

altars and placing idols in them, fasting, going nude and sleeping on the floor, climbing mountains, to preach the law there, kissing the earth, eating it with one's fingers and blowing trumpets and conch shells and flutes on the great feast days – all these emulated the ways of the holy man, Topiltzin Quetzalcóatl. (Duran, 1971: 59)

The immense popularity of Quetzalcóatl is indicated by the lengthy descriptions accorded to him by almost all of the early chroniclers of New Spain, today Mexico. Quetzalcóatl is alluded to in Nahua myth as the great civilizer and King of the Toltecs, and in Maya legends was known as Kukulcán or Gucumatz, also meaning "Lord Feathered Serpent". All three culture heroes were reputed to be the inventors of the science of measuring time as serpents represented the bondage of time and its cyclical nature. Additionally, the *Annals of Cuauhtitlán* (Nahua manuscripts) record that it was Topiltzin Quetzalcóatl who invented the ballgame, and wherever a temple stood dedicated to Quetzalcóatl, there existed a ballcourt (Nicholson, 1967: 117) (Figure 3).

Topiltzin Quetzalcóatl was regarded as "The Father of the Toltecs", and became ruler of Tollán/Tula, and by his inspired enlightened way he encouraged

Fig. 3: Ballcourt at Monte Alban in Oaxaca, Oaxaca, Mexico. This Zapotec ballcourt is dated back to 500 BCE (Photo by Kevin Feeney).

the liberal arts and sciences, and was revered for the cultural advancement of his people. His life of fasting and penitence, his priestly character, and his benevolence toward his followers, are evident in the material that has been preserved in the 16th century Spanish chronicles and in the hand-painted books of the indigenous people. He was also known as the lawgiver and, according to Spanish historians, he was unwilling to harm any human being, despite the temptation from demons to perform human sacrifice.

Aztec chronicles tell us that the Toltec priest-ruler Topiltzin Quetzalcóatl sailed across the Gulf of Mexico toward Yucatan at approximately A.D. 978. There is also ample evidence in the archaeology of Yucatan for a sea-borne invasion by the Toltecs in the late tenth century (Hedrick, 1971: 262).

It is the author's belief that Topiltzin Quetzalcóatl traveled to Chichén Itzá after he founded the city of Tula, and that he brought with him Toltec culture and a mushroom Venus cult associated with the feathered serpent. The Itzá Maya of Yucatan called this ruler Kukulcán (meaning "holy spirit" or "god" and "serpent and sky"), and it is believed he rebuilt the great city of Chichén Itzá, and later founded the Maya capital city of Mayapán. Toltec influence on the Maya of Yucatan can easily be seen in the architectural design of temples, palace monuments and ballcourts at the impressive ruins of Chichén Itzá.

Archaeologist J. Eric Thompson proposed that the Itzá who came into northern Yucatan were Chontal Mayan speakers (Thompson, 1970: 3-5). Thompson described the Itzá's as the Putún Maya, a group of Mexicanized Chontal Mayan speakers from the Gulf coastal area, who were sea traders who controlled Chichén Itzá shortly after A.D. 900. Most historians believe that the God-king Kukulcán and the Toltec priest-ruler Topiltzin Quetzacóatl, both meaning "Plumed Serpent," were one and the same man.

The Great Toltec diaspora into the Guatemala Highlands was most likely contemporary with Quetzalcóatl's (Kukulcán's) invasion into Yucatan during the Early Postclassic period A.D. 900-1000 and associated with the takeover of the city of Chichén Itzá. Borhegyi believed the Quiché and Cakchiquel Maya were also Nahuatl-influenced Chontal Mayas as both were linguistically related and shared a common Toltec-inspired genealogical origin (Borhegyi letter to Wasson, March 22, 1954). The Chontal-speaking Maya who called themselves Itzá, were devout followers of the god-king Quetzalcóatl and his mushroom-inspired religion. It's tempting to think that these so-called Mexicanized Mayan speakers who claimed Toltec ancestry may have been responsible for the collapse of Classic Maya civilization.

By the close of the Classic period, most of the lowland Maya ceremonial centers were mysteriously abandoned, and this intrusion into the lowlands would coincide with the abandonment of valley sites as ceremonial centers, and the beginning of hilltop defensive sites in the highlands of Guatemala (Borhegyi de, 1965: 37).

At the end of the Late Classic period (A.D. 700-1000) warlike tribes from the Gulf Coast of Mexico reached the highlands of Guatemala via the Usumacinta River and the Pacific coast of Mexico, and established themselves on the coastal Piedmont of Guatemala (Borhegyi de, 1965: 31). The ballgame may have served as a substitute for direct military confrontation by these warlike tribes from the Gulf Coast of Mexico (Scarborough & Wilcox, 1991: 14-15). According to Borhegyi, a completely new group of priest-rulers came into power bringing with them a form of ancestor worship associated with the vision-serpent (mushrooms?) and the ritual ballgame (Borhegyi de, 1965: 31).

In Mesoamerica, mushrooms are also linked with warfare. Maya inscriptions tell us that the movement of the planet Venus and its position in the sky was a determining factor for waging a special kind of warfare known as Tlaloc warfare or Venus "Star Wars." These wars or raids were timed to occur during aspects of the Venus astronomical cycle, primarily to capture prisoners from neighboring cities for ceremonial sacrifice (Schele & Freidel, 1990: 130-31, 194). These wars, waged against neighboring city-states for the express purpose of taking captives for sacrifice to the gods, thus constituted a form of divinely sanctioned "holy" war.

It was the forces opposed to this mushroom holy war who provided the blood from their sacrifice to propel the sun and moon in their journeys in and out of the underworld. It was the non-initiates to Quetzalcóatl's mushroom Venus religion who were without "fire" without god, in other words, Toltec culture. In the *Popol Vuh*, the sacred book of the Quiché Maya, the Quiché secured blood sacrifices from among the non-Mexican-descended local Maya (Edmonson, 1971: 197-199).

The Mesoamerican Ballgame, Trophy Heads and Hallucinogenic Mushrooms

Many of the observations in this chapter reflect the work of Borhegyi carried out from the 1950s through 1969 and in the book *The Pre-Columbian ballgames: A Pan-Mesoamerican Tradition*, published posthumously in 1980 by the Milwaukee Public Museum where he had served as the Director.

According to Borhegyi, the ritual ballgame can only be explained as a cross-cultural phenomenon, for it transcended all linguistic barriers in Mesoamerica. "Perhaps the games channeled competition short of warfare, between villages or ceremonial centers, into the field of skill and were a means of predetermining the selection of human victims to fulfill the requirements of the cyclical, or annual ritual sacrifices" (Borhegyi de, 1980: 3).

The ballgame represented a religious function throughout Mesoamerica, and the importance of the sacred ballgame and its rituals associated with ballcourt complexes in city planning, and the game's relationship to ancestors and serpent worship, and warfare, should not be underestimated, for there are over 1200+ archaeological sites in Mesoamerica that have identified at least one ballcourt, and cities like El Tajín in Veracruz, Mexico that boast a minimum of 18 ballcourts.

Borhegyi proposed that the Olmec-influenced handball game in ancient times was probably played in open fields or open plazas, and may have used the severed heads of humans and jaguars to mark out the boundaries or as targets or goals. According to Borhegyi, depictions of ritual ballgame sacrifice by decapitation appear to be a common theme in Preclassic times (Borhegyi de, 1980: 23).

The great city of Teotihuacán's overt and disruptive presence on the Maya people during the Classic Period resulted in a suppression of Olmec-inspired rituals and cult paraphernalia, such as mushroom stones and three-pronged incense burners, commonly used during the Preclassic period. They were replaced with pottery vessels and incense burners of a Teotihuacán-type decorated with human skulls, jaguars, and such deities as Quetzalcóatl, Tlaloc and Xipe-Totec.

Sometime between the 7th and 8th century, with the fall of Teotihuacán and its influence diminished, northern and central Mexico as well as parts of highland Guatemala and most of the Yucatán Peninsula was dominated by the Toltecs, and it seems that a revival of bloody ball game rituals of Preclassic Olmec fertility rites of human decapitation once again took center stage in the great ceremonial centers of Mesoamerica.

According to Borhegyi, the Toltecs, under the influence of Topiltzin Quetzalcóatl, were responsible for a brief revival (A.D. 950-1150) throughout Mesoamerica of a trophy head cult associated with warfare and the ritual ballgame (Borhegyi de, 1980: 25). Borhegyi further proposed that the change in ballgame rituals and the switch from the Olmec handball game to the hip ball game most likely came as a result of the newly instituted Quetzalcóatl rites (Borhegyi de, 1980: 24). He believed that the ballgame and these Olmec-influenced fertility rites were linked esoterically to the use of hallucinogenic mushrooms and that these bloody rituals were banished or forced underground during the heyday of Teotihuacán (Borhegyi de, 1980: iv).

On April 8, 1954, Borhegyi wrote to Wasson noting that: "…mushroom stones follow the same pattern as the three-pronged incensarios, figurines, rimhead vessels etc. That is, they are abundant during the Preclassic, disappear from the archaeological scene completely during the Early Classic, and are revived in somewhat changed form in the Late Classic".

The apparent absence of mushroom stones in Early Classic tombs (A.D. 200-400) or within ceremonial precincts suggests that the sacred mushroom cult of Preclassic origin, proposed by Borhegyi to be ritually connected to the ballgame, was discontinued, or banished from the Teotihuacán-occupied, or influenced highland Maya ceremonial centers.

Fig. 4: Borhegyi's 1960 catalog card above of miniature mushroom stones, metates, and manos from Kaminaljuyú, Guatemala, from Karl Heinz Nottebohm's private collection in Guatemala City, dated at 1000-500 B.C. (Milwaukee Public Museum Archives).

Borhegyi's typological breakdown of mushroom stones according to their chronology and distribution, indicate that the mushroom stones that reappeared in the highland Maya area during Late Classic times were mostly the plain and tripod variety common to the lower elevations of the Pacific Coast and Piedmont area as well as in western El Salvador (Figure 4). Mushroom stones with an effigy of a human, bird, jaguar, toad and other animals, occurred earlier in time and have been mostly found at the higher elevations of the Guatemala Highlands.

Borhegyi noted the connection between the re-appearance of mushroom stones and the ritual act of decapitation, and that many Late Classic (A.D. 600-1000) stone carvings relating to the ballgame depict balls incorporating human skulls or depict human skulls in lieu of balls. He also believed that the stone heads, and later stone rings set in the walls of formal ballcourts, were symbolic replacements for the hanging of the losers' heads on walls – the trophy heads of earlier times. The hanging of human heads can be found in a passage in the *Popol Vuh*, in which one of the Hero Twins, Hunahpu, and his father Hun Hunahpu had their decapitated heads hung in a tree (Borhegyi de, 1980: 24-25).

In fact, almost all evidence of ballgame sacrifice relates to the act of ritual decapitation, both self-decapitation and by execution, which takes place metaphorically in the underworld. The mushroom and its powerful effects on the mind were the means of divine transport, thus the portal or gateway into the underworld in which one is deified and resurrected at death.

Wasson believed that the origin of ritual decapitation lay in the mushroom religion itself and terminology used in reference to mushroom parts. In a letter to Borhegyi dated June 7, 1954, he writes of the Mixe (a linguistic group of northeast Oaxaca) continuing use of the psilocybin mushroom:

> The cap of the mushroom in Mije (or Mixe) is called kobahk, the same word for head. In Kiche and Kakchiquel it is doubtless the same, and kolom ocox is not "mushroom heads", but mushroom caps, or in scientific terminology, the pileus of the mushroom. The Mije in their mushroom cult always sever the stem or stipe (in Mije, *tek* is "leg") from the cap, and the cap alone is eaten. Great insistence is laid on this separation of cap from stem. This is in accordance with the offering of "mushroom head" in the *Annals of the Cakchiqueles* and the *Popol Vuh*. The writers

had in mind the removal of the stems. The top of the cap is yellow and the rest is the color of coffee, with the gills of a color between yellow and coffee. They call this mushroom, pitpa "thread-like", the smallest, perhaps 2 horizontal fingers high, with a cap small for the height, growing everywhere in clean earth, often along the mountain trails with many in a single place. In Mije the cap of the mushroom is called the "head" "kobahk in the dialect of Mazatlán. When the "heads" are consumed, they are not chewed, but swallowed fast one after the other, in pairs.

In pre-Columbian art, ballplayers are often depicted wearing stone objects that archaeologists have called stone axes (*hachas*) and palm frond-shaped stones (*palmas*). According to ancient murals and relief sculptures, the *hachas* (Figure 5) and *palmas* (Figure 6) were part of the protective gear worn by players in the ballgame. Borhegyi believed that stone *hachas,* as well as anthropomorphic and zoomorphic vertically- and horizontally-tenoned stone heads associated with the ballgame, were symbolic of the human trophy heads of earlier times (Borhegyi de, 1980: 24-25). Stone *hachas* were depicted on ceremonial ballgame yokes worn around the ballplayer's waist, while the tenoned stone heads were set into the walls of formal ballcourts. The subject matter most frequently seen on stone yokes, *hachas* and *palmas* are decapitated heads, skulls, skeletons, trophy heads, dismembered hands, limbs and bodies, severed ears, gouged-out eyes, and outstretched

Fig. 5: This Late Classic *hacha* from Veracruz, Mexico (Figure from Whittington, 2001), represents a decapitated trophy head of a wrinkled and toothless old man wearing a cone-shaped hat that suggests the Old Fire God (Xiuhtecutli), while a closer look reveals the image of a sacred *psilocybin* mushroom encoded in the cheek and hat. The conical or cone-shaped hat is a trademark attribute of the Mexican god-king Quetzalcóatl and of his priesthood.

tongues, etc. Based on the widespread use of this ballgame paraphernalia, he proposed: "that by Middle Classic times the competitive ballgames played in formal courts from northern Mexico to as far south as Honduras and El Salvador achieved a Pan-Mesoamerican magnitude" (Borhegyi de, 1980: 3).

In Siberia, *Amanita muscaria* grows in a symbiotic relationship with the birch and pine tree, which gave rise to the World Tree within the cosmology of several Siberian tribes. An eagle is described as perched in the tree, while a serpent dwells at its base, a myth that is paralleled in both the Old World and New World (Feeney, 2013: 302; Wasson, 1968: 217). It may not be coincidental that in Mesoamerica there is a parallel belief in a World Tree with a great bird who sits on top. The cedar tree of Yucatán was called *kuche,* the "Tree of God", and was the preferred wood for idol-making. In the *Popol Vuh*, a great bird known as 7 Macaw, or *Vucub Caquix* the Principal Bird Deity, sits atop the World Tree.

There is also plenty of evidence that ballplayers from the Gulf Coast area wore knee pads with the Ahau glyph design, a symbol of God, and Maya kingship, and that 1-Ahau was another name for the Venus god of resurrection (Figure 7) (Borhegyi de, 1980: 8).

The ritual ballgame was played to commemorate the completion of period endings in the sacred calendar, such as a katun ending, a 20-year time period that always ended on the day Ahau. It was on that day, after inferior conjunction that Venus reappears as the Morning Star. According to Borhegyi, it was believed that

Fig. 6: The photograph depicts a ballgame *palma* from Veracruz, Mexico from the Late Classic Period, 600-900 C.E. (Borhegyi & Borhegyi, 1963: 51). It depicts a stylized trefoil that the author proposes is a pre-Columbian version of the Old-World Fleur de lis symbol, which signifies the divine symbol of the god-king Quetzalcóatl. The Pre-Columbian gods and kings that are crowned or encoded with the Fleur de lis symbol are also linked to a World Tree, or Tree of Life, a Trinity of gods, and a mushroom of immortality.

in order to avoid catastrophe at the end of each 52-year period which also ended on the day Ahau, man, through his priestly intermediaries, was required to enter into a new covenant with the supernatural (Borhegyi de, 1980: 8). In the meantime, he atoned for his sins and kept the precarious balance of the universe by offering uninterrupted sacrifices to the gods. The Calendar Round was considered to be so important that the world would end at the completion of 18,980 days or 52 years if sacred termination ceremonies were not performed. This was known as the New Fire Ceremony and this time cycle of 52 years was recognized by all Mesoamericans. Ritual sacrifice was a way for the ancients to nourish and sustain all the living beings of the cosmos which gave order and meaning to their world.

Mushroom Venus Warfare and the Classic Maya Collapse

To date, there are almost ninety different theories or variations of theories purporting to explain the Classic Maya Collapse which took place between A.D. 900 and A.D. 1000, when archaeologists see an abrupt halt of any new construction and that dated monuments with Long Count dates called stelae ceased to be erected. It is during this time period in the Central lowlands of Guatemala that archaeologists see a sudden decline in population or the abandonment of Maya cities.

Maya archaeologist T. Patrick Culbert explained that "the evidence all indicated that the Classic Maya had disappeared somewhere in the time-shrouded past and had left no modern descendants with even a faint touch of their glory and accomplishments" (1974: 105).

We are led to believe that some mysterious fate befell the Classic Maya, and that people just suddenly disappeared and that the once great Maya cities of the Classic Period were all

Fig. 7: Figurine holding what may be an *Amanita muscaria* mushroom in his left hand. Note its large god eyes and three Ahau icons, one on each knee and one on his ballgame yoke (Figurine from Denver Museum collection; *see also* Guzman, 2012: 93).

abandoned. There was also the deliberate abandonment of most of the Guatemala highland valley sites shortly before the close of the period. Site after site was deserted, never to be reoccupied, in spite of the fact that many of the centers had been in use for more than two millennia.

Borhegyi's theory for the Classic Maya collapse was of a Toltec invasion into the Maya region by Nahuatl-influenced Chontal Maya tribes from the Laguna de Terminos region. On March 22, 1954, Borhegyi wrote to Wasson:

> This is a completely new theory that I have recently formulated. It is quite revolutionary, and I will try to publish it as soon as possible. When you carefully check the *Annals of the Cakchiqueles* and the *Popol Vuh*, you will read that, in spite of the fact that the Quiché and Cakchiquel tribes claim origin in the legendary city of Tollán, throughout their trip until they reach the Guatemalan Highlands (they) encounter only tribes speaking a language similar to their own. The country between the Laguna de Terminos and the Usumacinta region was and still is populated by Chol Mayas. Consequently, the Quiché and Cakchiqueles must have understood this language, and therefore were also Maya speakers. When they reached Guatemala, they met the Maya and, in the *Annals*, they referred to them as "stutterers", thus implying that they spoke a language somewhat similar to their own. J. Eric Thompson, a few years ago advanced the theory that the Itzás who came to Chichén Itzá about 1000 A.D. were Mexican-influenced Chontal Maya Indians from the Laguna de Terminos region. The Yucatecan Mayas called the Itzá invaders "stutterers", or "people who speak our language brokenly". I therefore suggest that the Quichés and Cakchiqueles were equally Nahuatl-influenced Chontal Mayas. I think that the story is as follows: the priest king Quetzalcóatl/Kukulcán/Gucumatz was expelled by his enemies from Tula (Tollán), sometime around 960 A.D. He left with a small group of his followers and went to Tlapallan, that is, the Laguna de Terminos region. Here he apparently settled down. It would seem that some of the Chontal tribes accepted the mushroom cult introduced by him and after a few years, the pressure of enemy tribes forced them

> to move on, led by descendants of Quetzalcóatl and his followers. Some went northeast to Chichén Itzá; others moved southward following the Usumacinta toward Guatemala. The archaeological picture of Northern Guatemala favors this theory. Linguistically, it is far more plausible than the other. The few leaders could still refer to their homeland as Tollán, and probably continued for a while to speak Nahuatl. The great mass of followers, however, did not speak this language and therefore probably spoke Chontal Maya. The Quiché and Cakchiquel Maya are, of course, linguistically related to the Chol and Chontal Maya. Please understand, this is a completely new theory. I am in the process of gathering archaeological data, which might support it.

Borhegyi called into question the construction date of the Great Ballcourt at Chichén Itzá which is the largest in Mesoamerica. He and fellow archaeologist Lee A. Parsons believed that this Great Ballcourt was built much earlier than previously supposed, possibly Mid to Late Classic period (Borhegyi, de, 1980: 12, 25). Borhegyi also believed that the stone ballcourt rings were an Early Post-Classic addition and indicated a later change of rules in the way the game was played. He further believed the gruesome human decapitation scenes and human "skull balls" were Late Classic and were influenced by the "Tajínized Nonoalca" (Pipils) or the Olmeca-Xicallanca who spread during that period from the Gulf Coast to Yucatán and through the Petén rainforest as far as the Pacific coast of Guatemala (Borhegyi de, 1980: 25). The Nonoalca were a Nathuat-Pipil group that settled in the southern Veracruz-Tabasco area after the fall of Teotihuacán. According to the *Popol Vuh*, some continued on to Guatemala and became the forebears of the Quiché Maya. Borhegyi believed that the plain, uncarved type of mushroom stone must have been re-introduced to Guatemala and the Cotzumalhuapa area during Late Classic times, by these "Tajinized Nonoalca" Pipil groups, where the severing of human heads reached new levels during Late Classic times (Borhegyi de, 1965: 37; Borhegyi de, 1980: 25; Borhegyi letter to Wasson, November 30, 1953, Wasson Archives).

The general belief has been that the Quiché Maya and Cakchiqueles who both claimed Toltec ancestry, entered the Guatemalan highlands from the eastern lowlands after the abandonment of Chichén Itzá in Yucatán. The date in textbooks for their entry has been set between A.D. 1250-1300, using the GMT correlation

(Porter Weaver, 1981: 477). The Quiché Maya, whose traditions and history are recorded in the *Popol Vuh*, claim that their migration was led under the spiritual "guidance" of their patron god named Tohil who was associated with rain and fire and is now considered to be a variant of Quetzalcóatl (Fox, 1987: 248).

According to S.W. Miles, the archaeologist Robert Wauchope, who worked at three main sites at Gumaarcah, Iximche, and Zacualpa during the late 1940s, could not find "archaeological coordination earlier than ca. A.D. 1300, between ceramics and genealogical reckoning" (Miles, 1965: 282-283). Borhegyi questioned this date in his letter to Wauchope dated April 8, 1954 (Milwaukee Public Museum Archives), explaining:

> I will try to put down in as concise form as possible, my questions concerning Quiche archaeology.
>
> 1) As you know, Dick Woodbury found cremations in Tohil effigy jars at Zaculeu. If cremations are to be connected with the Quiche expansion under Quicab this would mean that Zaculeu was occupied by them during the Early Post-Classic period.
>
> 2) You postulated Quicab's reign in the middle of the 15th century. These lately discovered cremations at Zaculeu would infer an earlier date for this reign, i.e., around 1300. If I remember correctly, you derive the date for Quicab's reign from a passage in the Annals of the Cakchiquels, which states that the daughter-in-law of Quicab died in 1507. Can it be that this passage refers to Quicab II, and not to Quicab I? In this case, Quicab I could have reigned in 1300.
>
> 3) I think the arrival of the Quiche-Cakchiquel's to Guatemala (probably following the Usumacinta River from the Laguna de Terminos) can be correlated with the first appearance of Fine Orange X wares, Mexican onyx vases, Tohil plumbate, and effigy support tripod bowls. ... On the other hand, the Quiche expansion under the reign of King Quicab falls together with the distribution of white-on-red ware, red on buff ware, red-and-black-on-white ware, and micaceous ware. This data also suggests a reign of around 1300 for Quicab.

> 4) I have long wondered about the quick "Mayanization" of the Quiche and Cakchiquel tribes, who supposedly came from Tulan. Using Morris Swadesh's lexicostatistical system, it is quite improbable that by the time of the conquest all these tribes could have spoken Maya with practically no retention of their original language. Could it be that the Quiche and Cakchiquels, like the Itzas and Xius of Yucatan were actually Chontal speaking Mayas from the Laguna de Terminos region, who wandered southward after being influenced by Nahuatl speaking groups. I wonder if Quetzalcoatl, after leaving Tula for Tlapalan, settled among these Chontal Mayas and introduced among them a new religious cult, based on the worship of idols. Could it be that only a few of Quetzalcoatl's followers (who actually could trace their origin to Tula) led these Chontal Mayas down into Guatemala? If so, they must have arrived to the borders of Guatemala around 1000 and not, as you once postulated, around 1300. Their arrival, around 1000 AD coincides with the appearance of Fine Orange X wares, Tohil plumbate etc. (we have lately found Tohil plumbate sherds at Altar de Sacrificios and at Santa Amelia). I would appreciate very much your comments on this hypothesis and questions mentioned above. If you'd like, I could even write it up for the Research Records, amplified with the latest distributional studies of the above-mentioned wares. At any rate, I would be very much interested to know your opinion. As ever, Steve.

Since then, more recent archaeological evidence suggests that Borhegyi's original date of A.D. 1000 was right after all.

One archaeological site along the Pacific slope that provides clear evidence of both Olmec and Maya development is the archaeological site of Tak'alik Ab'aj (formerly called Abaj Takalik), a pre-Columbian archaeological site in Guatemala. This area runs along the intercontinental mountain range which was heavily influenced in Preclassic times by the powerful Olmec culture. Maya archaeologist Marion Popenoe de Hatch (2005: 1) noted that:

> According to the stratigraphic evidence and the analysis of ceramics recovered in recent excavation, it

> would seem that Tak'alik Ab'aj was conquered by K'iche groups at the beginning of the Early Postclassic period (ca. 1000 AD). This date goes a long way back from the period comprised between 1400 and 1450 AD that many ethno-historians claimed for the K'iche expansion towards the South Coast of Guatemala... The problem is when, and the Tak'alik Ab'aj information suggests that the expansion had been initiated at the beginning of the Early Postclassic period and not at the beginning of the Late Postclassic, that is to say around 1000 AD, contemporary to the dispersion of the Tihil Plumbate pottery. The chronicle states that the conquest took place in 1300 AD, but archaeological evidence shows that this happened around three centuries prior to that date, that is, around 1000 AD.

One of the early Spanish chroniclers, Diego Muñoz Camargo, recorded that the grand city of Cholula, famous for the Great Pyramid dedicated to Quetzalcóatl, was the capital of the Olmeca-Xicallanca who were from the important coastal trading center of Xicalango, located in southern Campeche. At the time of the Spanish Conquest, this was an important coastal trading center controlled by a seafaring people known as the Putún Maya who may have been related either culturally or linguistically to an earlier Olmec culture. Images of decapitated ballplayers carved on the walls of formal ballcourts at El Tajín and Chichén Itzá supports the western origin of the ballgame carried by the Putún-descended peoples when they relocated north to Chichén Itzá and south to the Guatemala Highlands.

Toltec influence can be seen throughout the Guatemala Highlands at a number of archaeological sites like Kaminaljuyú and Zacuala, and along the Pacific slope area known for its important cacao plantations, a region in which the sculptural style at sites like El Baúl, Bilbao and El Castillo is a mixture of both Maya and Mexican elements called 'Cotzumalhuapa'. Kaminaljuyú dominated the Guatemala Highlands and had complete control of earlier Olmec trade routes by establishing satellite cities, like Tak'alik Ab'aj, Chocolá, El Baúl, and Chalchuapa. According to Borhegyi, the most important Classic period settlements in Mesoamerica had at least one ballcourt, while large ceremonial centers like Kaminaljuyú had several (Borhegyi de, 1980: 3). Of the twelve recorded ballcourts at Kaminaljuyú, four are north-south in orientation, and the other eight ballcourts are oriented approximately east-west.

The great cities of Cholula, El Tajín, Xochicalco and Cacaxtla all contributed to the downfall of Teotihuacán, but Cholula may have benefited the most from the collapse of Teotihuacán where it served as a haven for its survivors after its destruction by barbaric invaders from the north. Its prosperity grew immediately, and it expanded its Great Pyramid honoring their god-king Quetzalcóatl to cover an area of over 500,000 sq. feet, making it the widest pyramid in the world. All these cities experienced a revival of ballgame rituals associated with Olmec-influenced fertility rites that included human decapitation.

In Cholula, the worship of Quetzalcóatl flourished. Spanish chronicler Friar Toribio de Benavente, affectionately called Motolinia by the Indians, wrote in his *Memoriales* that followers of Quetzalcóatl came to Cholula to give their lives in sacrifice, in return for immortality. He described the great ceremony to Quetzalcóatl which lasted eight days which, coincidentally, is the same number of days that, according to legend, Quetzalcóatl was in the underworld creating humanity by bloodletting on the bones of his father and the bones of past generations. He then emerged from the underworld resurrected as the Morning Star. Motolinia named a star Lucifer (most likely Venus) which the Indians adored "more than any other save the sun, and performed more ritual sacrifices for it than for any other creature, celestial or terrestrial" (LaFaye, 1987: 141).

Bartolome de las Casas, a Bishop of Chiapas in the mid-1500s, reported that: "after the sun, which they held as their principal god, they honored and worshiped a certain star more than any other denizen of the heavens or earth, because they held it as certain that their god Quetzalcóatl, the highest god of the Cholulans, when he died transformed into this star" (Christenson, 2007: 205). Las Casas further noted that the Indians awaited the appearance of this star in the east each day, and that when it appeared their priests offered many sacrifices, including incense and their own blood (Christenson, 2007: 205).

Archaeologists believe that Mesoamerican people may have played the ballgame to symbolize the movements of the sun, moon and the planets, most notably the planet Venus as a resurrection star, and that the ball symbolized the sun's continuous struggle to free itself at night from the clutches of the underworld. Rituals of self-sacrifice and decapitation in the underworld, allude to the sun's nightly death and subsequent resurrection from the underworld, by a pair of deities (twins or brothers) associated with the planet Venus as both the Morning Star and Evening Star. This suggests that the ballgame and its rituals are associated with the 584-day Venus cycle. As described in the Mayan book the *Dresden Codex*, the synodic revolution of Venus from Morning Star to Morning Star is 584 days, and that these revolutions were grouped by the Nahuas and Maya in fives, so that 5 x 584 equaled 2,920 days, or exactly eight solar years.

In *El Titulo de Totonicapán,* it is said that the Quiché gave thanks to the sun and moon and stars, but particularly to the star that proclaims the day, the day-bringer, referring to Venus as the Morning Star. The Sun God of the Aztecs, Tonatiuh, first found in Toltec art, is frequently paired with Quetzalcóatl in his aspect of Venus as Morning Star.

The mushroom ritual associated with warfare, and the ballgame were probably timed astronomically to the period of inferior conjunction of the planet Venus. At this time, Venus sinks below the horizon and disappears into the "underworld" for eight days. It then rises before the sun, thereby appearing to resurrect the sun from the underworld as the Morning Star. For this reason, mushroom-induced decapitation rituals were likely performed in ballcourts, a metaphor for the underworld, which was timed to a ritual calendar linked to the movements of the planet Venus as both a Morning Star and Evening Star. The mushroom experience, as well as caves and ballcourts were believed to be entrances or portals into the underworld. Dictionaries of Maya Highland languages compiled after the Spanish Conquest mention several intoxicating mushroom varieties whose names clearly indicate their ritual use. One type was called *xibalbaj okox,* «underworld mushroom», noting likely that this mushroom transported one to a supernatural realm of the underworld (Sharer, 1994: 484).

The Aztecs referred to mushrooms as flowers, and they considered their god Xochipilli to be an aspect of a young Quetzalcóatl and the patron

Fig. 8: An Aztec figurine from the collection of the National Museum in Mexico City, of the Aztec god of flowers, Xochipilli, whose name in Nahuatl, the language of the Aztecs, means "Prince of Flowers". This figurine holds what appears to be *Amanita muscaria* mushrooms in each hand.

deity of hallucinogenic plants (Figure 8). The headdress of Xochipilli contains two adornments of five plumes each – a possible reference or code to what scholars call the "fiveness" of Venus, referring to the five synodic cycles of Venus identified in the Venus Almanac of the *Dresden Codex*. Friar Diego Duran writes that war was called *xochiyaoyotl* which means "Flowery War". Death to those who died in battle was called *xochimiquiztli,* meaning "Flowery Death" or "Blissful Death" or "Fortunate Death" (Nicholson, 1967: 90).

Pottery mushrooms have been excavated at Pre-Classic Maya Lowland sites such as El Mirador and Berriozábal, and at the Olmec-influenced site of Altar de Sacrificios (Borhegyi de, 1963: 330). Pottery mushrooms dating to the middle or late Pre-Classic period have also been found with figurines of ballplayers at the archaeological sites of Tlatilco in Burial 154 (Trench 6), and at Tlapacoya in the Valley of Mexico (Borhegyi de, 1980: 2). The pottery mushroom from Tlatilco was found near a figurine of an acrobat which could support Spanish chronicler Friar Sahagún's observations that mushrooms may have been consumed to induce super-human athletic ability and agility. Sahagún noted that the Toltecs used hallucinogens before battle to enhance bravery and strength (*see* Ödman, Ch. 9). Hallucinogens taken before battle or before a ballgame likely eliminated all sense of fear and gave the combatant or ballplayer a sense of invincibility and courage to fight at the wildest levels. According to Sahagún: "This drunkenness lasted two or three days, then vanished" (Thomas, 1993: 508). A mural at Cholula discovered in 1969 by archaeologist Ponciano Salazar Ortegón, known as "The Drunkards", depicts several life-sized individuals in the act of consuming an intoxicating, possibly a mind-controlling hallucinogenic beverage.

During the Middle Classic period, ballcourts were built throughout the highlands and along the Piedmont sites at Bilbao, El Baúl, Palo Verde, Palo Gordo, Los Tarros, El Castillo, and Pantaleon, and Tonalá during a time when the area was dominated by the influence of Teotihuacán. Nowhere else in Mesoamerica does the ballgame imagery appear so gruesome. Ballgame scenes depict players, some with sacrificial knives holding trophy heads, and human sacrifice performed by were-jaguars, and heart sacrifice and dismemberment of human body parts. Some of the players wear conical hats symbolic of Quetzalcóatl, offering gifts to diving sky gods associated with the ballgame. In 1948, J. Eric Thompson wrote that this ballgame imagery in the highlands of Guatemala and the Piedmont sites suggested that the ritual ballgame was closely connected with Quetzalcóatl (Thompson, 1948: figs. 10-15).

The art style at El Tajín is reminiscent of the Cotzumulhuapa culture on the Pacific coast of Guatemala, and there was little doubt that there must have been close contact between the two regions. The Cotzumalhuapa culture depicted many Maya and Mexican elements, and represented ballplayers or ceremonies connected with the game. The great city of El Tajín, in Veracruz, Mexico, is an extensive archaeological site that boasts a minimum of 18 ballcourts, an indication that the inhabitants were likely obsessed not only with the ballgame, but with trophy heads and ritual decapitation.

According to Borhegyi: "...nowhere else in Mesoamerica does the ballgame get any more gruesome" (Borhegyi de, 1980: 16). It was in this region, that the decapitation of human heads and the dismemberment of body parts reached new levels. Borhegyi surmised that victims or captives for sacrifice were decapitated by priests or ballplayers dressed in were-jaguar attire after which the decapitated heads of both ballplayers and jaguars were hung up by ropes over ballcourts or temples. The ballgame seemed to be connected with jaguar worship and serpent symbolism associated with fertility rites. These trophy heads were venerated as sacrificial offerings, and may even have been used during certain ballgames in lieu of balls (Borhegyi de, 1980: 24-25).

In 1965, Borhegyi wrote: "...if some of the Pipil colonizers settled around the

Fig. 9: This carved image is from the coastal Piedmont of Guatemala, known for its sculptural style that includes both Maya and Mexican elements. The tenoned human head depicts the bearded Quetzalcóatl. A close look at the eyes of Quetzalcóatl show *Amanita muscaria* mushrooms, the caps being the eyelids and the stem runs down his face. (Photo by Jacques VanKirk, *from* VanKirk & Bassett-VanKirk, 1996).

Fig. 10: Above is a detail from the Cotzumalhuapa monument from El Baúl, which depicts a figure on the right wearing what appears to be a mushroom-inspired ear-plug (circled) (Photo by Ilona de Borhegyi).

area of El Tajín, Tabasco, and Campeche during the Classic period, they may have absorbed many Gulf Coast traits, and there's little doubt that El Tajín, with its specialized art style, contributed to the influence of this group" (1965: 39). He noted that Olmec archaeologist Mathew Stirling had written in 1943 that they adopted a decapitation sacrifice and trophy head cult associated with the ballgame from these coastal people.

Cotzumalhuapa is well known for its many horizontally-tenoned stone heads that represent jaguars, birds, monkeys and serpent heads, as well as human skulls and human faces, some peering out of the beaks of birds or serpent jaws. Borhegyi noted: "Some of the tenoned human heads represent rabbit-eared, snouted and fanged anthropomorphic individuals with dangling eyeballs, a feature commonly associated with the god Quetzalcóatl in his form of Éhecatl the Wind God" (Borhegyi de, 1980: 17) (Figure 9).

There is still great diversity of opinion among scholars concerning the date and nature of the Pipil migrations, however both Borhegyi and J. E. S. Thompson who excavated the sites of El Baúl and Bilbao, believe that most of the Cotzumalhuapa stone sculptures (Figure 10) are of the Late Classic period (Borhegyi de, 1965: 36, 39).

Borhegyi made an important connection; he noted that carved stone yokes worn by ballplayers are rare in Guatemala and those found depict ether serpent heads or death-heads (Borhegyi de, 1980: 7). According to Borhegyi (1965: 36), yokes, *hachas*, and *palmas* most likely originated on the Gulf coast of Mexico,

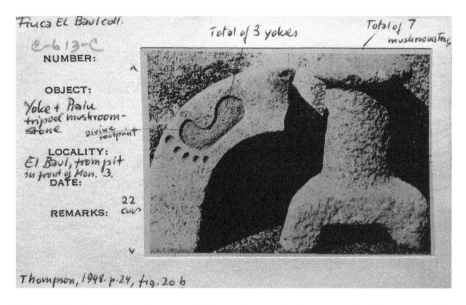

Fig. 11: Borhegyi's catalog card above shows a ballgame yoke fragment with footprint that was excavated in 1948 by J. Eric S. Thompson along with a tripod mushroom stone (Type D) from a pit in front of Monument 3 at the Pacific coastal site of El Baúl in Guatemala (Milwaukee Public Museum collection). Type D tripod mushroom stones (plain and effigy) were frequent in the Pacific Coast and Piedmont area as well as in western El Salvador (Borhegyi de, 1965: 37).

where they have been found in the greatest number and variety. Stone yokes in association with stone *hachas* are known from only three other sites in Mesoamerica, at Bilbao and Patulul, Guatemala, and at Viejon in Veracruz, Mexico (Figure 11).

Borhegyi noted that ballplayers depicted on Monument 27 at El Baúl wear tight-fitting monkey head helmets, and hand-gloves that represents either the local survival of the Olmec-influenced Preclassic handball game, or a late Classic revival of the game in the area (Borhegyi de, 1980: 16). He adds that: "These zones were once influenced by the Olmecs and later by 'warlike' Mexican Gulf Coast groups. One wonders if these grisly sacrificial activities are native to this area or are Pre-Classic survivals of a game once played with human heads with long, flowing hair in the Tajín and La Venta areas and in parts of Oaxaca" (p. 16).

Quetzalcóatl as Tohil was the Quiché patron deity. However, Borhegyi noted to Wasson in his letter of March 3, 1954:

> Unfortunately, we do not know what the stone idols of Tohil, and Avilix looked like. Tohil is also referred to in the *Annals* as Gucumatz, which is "feathered

serpent" and we might assume that he was so represented. What type of god Avilix was is still a question? (It would be more than pleasant to think of him as a mushroom God!) I think they prove beyond any doubt that at least some sort of a mushroom cult must have existed among the Quiché and Cakchiquel Mayas.

It's interesting to note that the Quiché tried to overthrow the Cakchiquel Maya at their capital of Iximche, but the Quiché were defeated, and that Cakchiquel warriors stole from the Quiché the divine image of Tohil (Figure 12), and that this deprived the Quiché of their divine power, and the Quiché did not dare attack the Cakchiquel again on their home ground (Sachse, 2001: 363).

According to testimony recorded in 1554 in the Colonial document entitled *El Titulo de Totonicapán* (Land Title of Totonicapán), the Quiché Maya revered "mushroom stones" as symbols of power and rulership: "The lords used these symbols of rule, which came from where the sun rises, to pierce and cut up their bodies for the blood sacrifice" (Sachse, 2001: 363).

There is a passage in the *Popol Vuh* in which the Quiché gods were carried on their back in pack frames:

> We have found that for which we have searched, they said. The first god to go out was Tohil, carried in his pack frame by Balam Quitze [Figure 13]. Then the god Auilix was carried out by Balam Acab, followed by Hacavitz, the name of the god received by Mahucutah... At the suggestion of Tohil, the Quichés leave Tollán. They sacrifice their own blood to him, passing cords through their ears and elbows,

Fig. 12: Here is a (Type A) effigy mushroom stone, the only one in the museum at the archaeological site of Iximché, the Cakchiquel capital, in the western highlands of Guatemala. Although the effigy mushroom stone bears the image of the Mexican goggle-eyed god known as Tlaloc, it's tempting to think that this mushroom stone may represent the divine image of the Quiche god Tohil, that the Cakchiquel warriors stole from the Quiché Maya (Photo by Carl de Borhegyi).

> and they sing a song called 'The Blame is Ours', lamenting the fact that they will not be in Tollán when the times comes for the first dawn. Packing their gods on their backs and watching continuously for the appearance of the Morning Star, they began a long migration. (Christenson, 2007: 198)

According to the *Popol Vuh*, the founders of the Quichéan lineages traveled a great distance eastward "across the sea" to the Toltec city called Tulan Zuyva where they received their gods "whom they then carried home in bundles on their backs" (Christenson, 2007: 198) (Figure 14).

Friar Sahagún (in book 9 of 12) refers to mushrooms with a group of traveling merchants known as the *pochtecas,* meaning merchants who lead because they were followers of Lord Quetzalcóatl, who they worshipped under the patron name *Yiacatecuhtli* or *Yacateuctli*, Lord of the Vanguard. He describes the mushroom's effects and their use in several passages of his *Florentine Codex* ("Historia General de las Cosas de Nueva España"). He records how the merchants celebrated the return from a successful business trip with a wild mushroom party. The *pochteca* journeyed in all directions from Central Mexico, carrying merchandise as well as spreading the mushroom religion of Quetzalcóatl (Sahagún, Book 9 chapter viii; *Florentine Codex*, fol 3 Ir-3 Iv).

Another passage from Friar Sahagún in his *Florentine Codex* reads:

> The eating of mushrooms was sometimes also part of a longer ceremony performed by merchants returning from a trading expedition to the coast lands. The merchants, who arrived on a day of favorable aspect, organized a feast and ceremony of thanksgiving also on a day of favorable aspect. As a prelude to the ceremony of eating mushrooms, they sacrificed a quail, offered incense to the four directions, and made offerings to the gods of flowers and fragrant herbs. The eating of mushrooms took place in the earlier part of the evening. At midnight a feast followed, and toward dawn the various offerings to the gods, or the remains of them, were ceremonially buried. (Sahagún, Book 9 chapter viii; fol 3 Ir-3 Iv)

In my examination of pre-Columbian art, I found that storm gods, rain gods,

Fig. 13: This Type D tripod mushroom stone from Guatemala has a human effigy on the stem (Late Classic, A.D. 600-900). The figure wears a traditional *mecapal* strapped around his forehead (tumpline) to carry what appears to be a giant mushroom on his back, or is it the god Tohil? (Photo by Stan Czolowski, A Brief History of Magic Mushrooms in BC [2019], Vancouver Mycological Society: www.vanmyco.org/about-mushrooms/psychedelic/brief-history-magic-mushrooms-bc/)

and lightning gods are all esoterically connected with psychotropic mushrooms. The idea of mushroom spores traveling to earth on lightning bolts hurled down from the sky by powerful gods appears to be a common theme in both the New World and the Old World. The ancients believed that this phenomenon was related to a magical alliance with the mushroom.

The followers of Quetzalcóatl came to the conviction very early on that, under the influence of the sacred mushroom, a divine force actually entered into their body – a state described as "god within". Because mushrooms appeared to spring magically over night from the underworld, apparently sparked by the powers of lightning, wind and rain, it would have been easy for these ancients to conclude that sacred mushrooms were divine gifts brought to them by the wind god Ehecatl-Quetzalcóatl, and the rain god Tlaloc, both of whom are avatars of the planet Venus (Figure 15).

In the *Popol Vuh*, numerous passages reveal obscure connections between Maya creation myths, the ballgame, mushrooms, ritual decapitation and self-decapitation. In another letter to Wasson dated March 3, 1954, Borhegyi wrote:

> I discovered two interesting sentences relating to mushrooms from Indian Chronicles, written around 1554 by natives: 'and when they found the young of the birds and the deer, they went at once to place the blood of the deer and of the birds in the mouth of the stones that were Tohil, and Avilix. As soon as the blood had been drunk by the gods, the stones spoke, when the priest and the sacrificers came, when they came to bring their offerings. And they did the same

Fig. 14: This image is from a Late Classic period Maya vase K4932 from the Justin Kerr Database (Photo by Justin Kerr). The author proposes that the bundles depicted on this vase may in fact be filled with *Amanita muscaria* mushrooms.

before their symbols, burning pericon (?) [probably *Tagetes lucida*] and holom-ocox (the head of the mushroom), holom=head, and ocox=mushroom". I think this section definitely indicates that the Quiché used mushrooms in connection with their religious ceremonies. I even wonder what made the stones speak?

Anthropologist Dennis Tedlock who translated the *Popol Vuh* into English in 1985, identified five episodes in the *Popol Vuh* involving underworld decapitation and self-decapitation. In one episode, the ball playing Hero Twins decapitate themselves in the underworld in order to come back to life. Tedlock explained that, based on evidence discovered by Borhegyi, he does not rule out the presence of

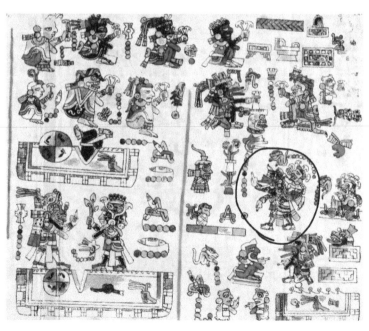

Fig. 15: Here, from page 24 of the *Codex Vindobonensis Mexicanus* is a close-up image of Quetzalcóatl as the Wind God with what can only be a mushroom inspired god strapped to his back. Quetzalcóatl is signaling with both hands pointing fingers up and down, to the god Tlaloc in front of him to open a mushroom portal of Venus resurrection (Photo from The British Museum).

an *Amanita muscaria* mushroom cult in the *Popol Vuh* (Tedlock, 1985: 250). Tedlock further mentioned that the principal gods among the Quiché Maya are listed "again and again" as Tohil, Auilix, and Hacauitz and called these three gods «the three Thunderbolts», their names being, Thunderbolt Hurican, Newborn Thunderbolt, and Raw Thunderbolt, alluding to a Trinity of Maya gods. In the *Annals of the Cakchiqueles*, the Quichés are called "thunderers" because they worship this god, and were given the name Tohohils.

Tedlock (1985: 251) specifically noted:

> The single most suggestive bit of evidence for the mushroom theory lies in the fact that a later *Popol Vuh* passage gives Newborn Thunderbolt and Raw Thunderbolt two further names: Newborn Nanahuac and Raw Nanahuac. …Nanahuac would appear to be the same as the Aztec deity Nanahuatl (or Nanahuatzin), who throws a thunderbolt to open the mountain containing the first corn. Nanahuatl means "warts" in Nahua, which suggests the appearance of the *muscaria* when the remnants of its veil still fleck the cap.

The etymology of the name of the Thunderbolt god Hurican, or *Juraqun* means "one leg", and there is plenty of evidence that the belief in a one-legged god was widespread throughout Mesoamerica (Christenson, 2007: 60).

Evidence of a trinity of gods among the ancient Maya was also supplied by ethno-mycologist Bernard Lowy, who linked sacred mushrooms with lightning and a creation myth, and a trinity of creator gods associated with divine rulership. He reported that *cakulha* was not only the Quiché term for thunderbolt but is also the Quiché Maya name for the *Amanita muscaria* mushroom (Lowy, 1974: 189). The ancient Mayan word for stone, *cauac,* comes from the word for lightning. It may be that mushroom stones were placed in sacred spots where lightning struck the ground, and or mushrooms sprouted from the ground, suggesting that it was lightning that provided the conceptual link between the sky (heaven) and Earth.

A few years later, Lowy (1980: 99) wrote:

> During a visit to Guatemala in the summer of 1978, I stayed in the village of Santiago de Atitlan, a community where Tzutuhil [Mayan] is spoken and where

ancient traditions and folkways are still maintained. There I learned that in Tzutuhil legend mushrooms are intimately associated with the creation myth. In the Quiché Maya pantheon the god Cakulja, he of the lightning bolt, one of a trilogy of supreme gods, is revered above all others, and in the Popol Vuh, the sacred book in which the traditions of the Quiché people are recorded, his position of ascendancy is made clear.

Borhegyi further identified Type B mushroom stones of the Early and Late Preclassic periods (1000 B.C.-A.D. 200; *see* Figure 16) with a circular groove around the base of the cap. According to Borhegyi, the custom of circularly grooving the base of the mushroom stone caps was discontinued after the Early Preclassic period. The Late Preclassic (500 B.C.-A.D. 200) and Classic (A.D. 200-900) period carved effigy, plain, and tripod mushroom stones have only plain caps (Borhegyi de, 1961: 499).

Archaeological evidence of a trinity of creator gods among the ancient Maya, appear at numerous archaeological sites including, El Mirador, Palenque, Cerros, Uaxactum, Caracol and at Tikal, during the Early Classic Period A.D. 250-400.

In a guide for missionaries written between 1571 and 1576 by Dr. Francisco Hernández, physician to the king of Spain, Hernández notes that the Aztecs at the time of the Spanish Conquest revered three different kinds of mushrooms (Wasson & Pau, 1962: 36). Hernández studied the natives and concluded that the Indians already believed in the Trinity. He sent a letter to Bishop

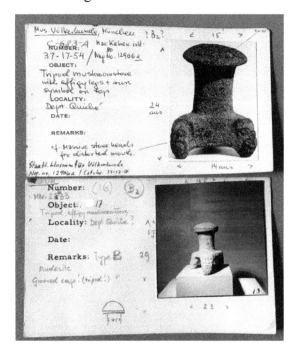

Fig. 16: Borhegyi's catalog cards of Type B mushroom stones. (Milwaukee Public Museum Archives).

Fig. 17: In the photo above, a three-sided figurine discovered in the Gulf Coast of Veracruz, Mexico, is evidence of a three-in-one deity, or Trinity of Gods in Mesoamerica (Ekholm, 1968: 378).

Bartolome de las Casas who reported what Hernández wrote:

> They knew and believed in God who was in heaven; that that God was the Father, the Son and the Holy Ghost. That the Father is called by them Icona and that he had created man and all things. The Son's name was Bakab who was born from a maiden who had ever remained a virgin, whose name was Chibirias, and who is in heaven with God. The Holy Ghost they called Echuac. (Wirth, 2012)

Conclusion

Among the ancient Maya, and Nahua, the ballgame and human sacrifice and the ritual of decapitation were believed necessary to save mankind from calamity and the cosmos from collapse. Since the greatest gift one could offer the gods was one's own life, the purpose of human sacrifice was to preserve life rather than destroy it. I believe strongly that this concept of life from death via ritual decapitation could only have been mushroom-inspired.

Although the hallucinogenic mushroom cult has survived to this day among certain tribes like the Zapotec, Chinantec, and Mazatec Indians of Mexico, there has been little of its use reported among the present-day Maya. However, on a recent visit to Guatemala, I found that the Maya Indians of the Guatemala Highlands were selling tiny *Amanita muscaria* mushroom toys in the markets (Figure 18), which all have a quetzal bird sitting in a tree painted on the stem. Although the seller informed me that the Maya did eat this variety of mushroom, it is possible she may have been referring to the non-hallucinogenic *Amanita caesarea,* commonly sold in markets in Mexico and Guatemala and much appreciated for its delicate flavor.

Although clearly a child's toy produced for the tourist trade, they bear symbolism of great antiquity. In Mesoamerican mythology, the World Tree, with its roots in the underworld and its branches in the heavens, represents the *axis mundi* or center of the world. The branches represent the four cardinal directions. Each of the directions was associated with a different color while the color green represented the central place. A bird, known as the celestial bird or Principal Bird Deity, usually sits atop the tree. The trunk, which connects the two planes, was seen as a portal to the underworld. The Quetzal, now the national bird of modern Guatemala, was considered sacred because of its green plumage.

I believe that the *Amanita muscaria* mushroom cult may still survive in remote areas of Highland Guatemala, where the mushroom grows in abundance. I also believe there is now clear evidence that the *Amanita muscaria* mushroom is a symbol of equal antiquity.

Figs 18a, 18b: *Amanita muscaria* figurines purchased in the Guatemalan Highlands featuring the green Quetzal bird on the stem (Photo by Carl de Borhegyi).

Acknowledgements: The author thanks his mother Dr. Suzanne de Borhegyi-Forrest and sister Ilona de Borhegyi for their contributions to his research.

Part IV:
Diet & Cuisine

Chapter 19

The Fly Amanita

Frederick Vernon Coville (1898)

It is well known that in some parts of Europe the fly amanita, after the removal of the poison by treatment with vinegar, is a common article of food. It was interesting to discover not long since that among some of our own people a similar practice prevails. Though most of the colored women of the markets [in Washington, D.C.] look upon the species with horror, one of them recited in detail how she was in the habit of cooking it. She prepared the stem by scraping, the cap by removing the gills and peeling the upper surface. Thus dressed the mushrooms were first boiled in salt and water, and afterwards steeped in vinegar. They were then washed in clear water, cooked in gravy like ordinary mushrooms, and served with beefsteak. This is an exceedingly interesting operation from the fact that although its author was wholly ignorant of the chemistry of mushroom poisons, she had nevertheless been employing a process for the removal of these poisons which was scientifically correct. The gills, according to various pharmacological researches [now out of date], are the chief seat of the poisonous principles in this plant and their removal at once takes away a large part of the poison. The salt and water would remove phallin or any other toxalbumin the mushroom contained, and although the presence of phallin or any of this class of poisons has not been demonstrated in *Amanita muscaria*, there is a strong suspicion that it may occur in slight amount.[2] The vinegar, secondly, removes the alkaloid poison, muscarine, and the mushroom after the two treatments is free from poisons. This process is cited, not to recommend its wider use, but as a matter of general interest.[3] The writer's recommendation is that a mushroom containing such a deadly poison should not be used for food in any form, particularly at a season when excellent non-poisonous species may be had in abundance.

It is surprising that cases of poisoning are not more frequent. At Takoma Park, D.C., on November 9, of last year [1897], a lady who has a thorough knowledge of edible and poisonous mushrooms met a family, consisting of a man, woman, and two children, who had just completed the gathering of a basketful of the fly amanita and the death cup [*Amanita phalloides*]... which they were taking home to eat. In reply to questions the woman stated that they had often eaten this kind purchased dry at an Italian store, but that they had never gathered fresh ones before. Of course they had mistaken the species, or possibly the dried ones were fly amanitas from which the poison had been removed by treatment with vinegar. After considerable persuasion the people consented to throw the lot away.[4]

It is impossible to say what amount of the fly amanita would prove fatal,[5] but in this connection it is of interest to note the custom reported by Krasheninnikoff, a Russian who traveled in Siberia and Kamchatka from 1733 to 1743, namely, that the natives of the latter country, particularly the Koraks, used the fly amanita as an intoxicant, three or four specimens constituting a moderate dose for one habituated to its use, but ten being required for a thorough drunk. The same observations, with varied details, have been made by others, particularly by Langsdorff, who traveled around the world with the Russian navigator Krusenstern from 1803 to 1806, and in more recent times by Kennan in his first Siberian journey of 1865-67.

The plant may be taken fresh, but its taste is so disagreeable[6] that only with great difficulty can a sufficient amount be eaten to produce the intoxicating effect. The Koraks have two principal methods of taking it: First, by swallowing pieces of the dried caps without chewing them; second, by boiling the dry caps in water and then drinking the liquor thus produced mixed with the juice of berries or herbs to disguise the taste.[7] The intensity of the poisonous character of the fly amanita undoubtedly varies at different ages, with different individuals, and with different methods of preparation. The amount of the poison that can be taken into the system with impunity varies, too, with the person who takes it. The fact that a Korak, who has long used the plant as an intoxicant, can eat ten specimens and merely become drunk does not prove that a similar number would not be fatal to an American who had never eaten it before.

Notes

1. This chapter is excerpted from an 1898 circular published by the United States Department of Agriculture, titled "Observations on Recent Cases of Mushroom Poisoning in the District of Columbia."
2. Ed Note: Here, Coville is suggesting that *Amanita muscaria* may contain phallotoxins similar to those found in the Death Cap (*Amanita phalloides*). Currently, it is understood that the deadly principles found in the Death Cap and other deadly *Amanitas* are amatoxins, not phallotoxins. In any case, no amatoxins or phallotoxins are known to occur in *Amanita muscaria* or its isoxazole-containing relatives.
3. Ed Note: This is typical of how *Amanita muscaria* has generally been regarded in traditionally mycophobic cultures, and this perception continues today with some drawing hyperbolic parallels between *Amanita muscaria* consumption and *fugu*, the Japanese delicacy made from poisonous blowfish, consumption of which has led to fatalities when not properly prepared (Viess, 2012).
4. Ed Note: This is a curious anecdote, and it seems most likely that this family simply did not know what they were looking for or picking. There is no known practice of drying *Amanita muscaria* for culinary use, particularly since this is apt to destroy its fine flavor, therefore it is unlikely that this is what they had been buying and consuming.
5. Ed Note: Coville wrote this piece less than a year after the highly publicized death of Count Achilles de Vecchj, an Italian diplomat living in D.C., who reportedly died after eating more than twenty fly agaric mushrooms which had been mistaken for *Amanita caesarea*, a popular and choice edible (Rose, 2006). *Amanita muscaria* very rarely leads to fatalities, and typically when cause of death has been attributed to *Amanita muscaria* it has been under circumstances where the actual cause of death is uncertain. However, fatalities have occurred, and they appear to occur most typically in people whose health is compromised, such as with the poor Count himself.
6. Ed Note: Coville's text must be understood within the context of the time, when there was an increased interest in mushrooms in the United States, as well as within his overall purpose, which was to sow seeds of caution in the public (Rose, 2006). Based on Coville's description of the fly agaric as having a disagreeable flavor, one can presume that he never sampled any properly prepared specimens.
7. Ed Note: While the taste of a fly agaric decoction leaves something to be desired, there may be a pharmacological reason for adding the juice of berries other than enhancing or altering its flavor. Some studies have shown that ibotenic acid when exposed to acidic conditions more readily converts (decarboxylates) to the more potent muscimol (Tsunoda et al., 1993c). The addition of berry juice may, in fact, increase the potency of the decoction.

Amanitas in the Family:
"Brownie Seats for Dinner… Again?"

Danny Curry

My name is Danny Curry, I'm 55 years old and I've been eating *Amanita muscaria* since I was five.[2] I know how that must sound… like some kids eat paste or crayons, you may imagine me, huddled under a bush, compulsively nibbling on the classic, bright red, fairy tale mushroom. The truth is much more mundane, if not commonplace.

I guess I should offer a bit of family history, in an effort to head off any further misconceptions. My family wasn't Russian, or Ukrainian. We were not wild eyed, shamanic Siberian reindeer herders (although my uncle Ernie was a bit of a wingnut with an esoteric bent), and I didn't grow up on a hippy commune (although, in my teens, I frequently fantasized about running away to join one).

I come from a mix of Scots/Irish Appalachian immigrants and educated English Quakers. The family farm I grew up on was in south central Ohio and was founded in the late 1790s-early 1800s. Initially, it included several hundred acres, but by the time my parents acquired it, it had been whittled down to one hundred or so. It was mostly farmland but there was a fair amount of woodlands, a mix of conifers, beech, and oaks. More importantly, it was adjacent to nearly one thousand acres of state park land, much of which had originally been family property. Part of the property had also been donated to The Society of Friends, as a Quaker retreat. We retained access to that as well.

I don't really recall eating *muscaria* when I was five, but my mother told me that I refused to eat any mushroom prior to that. You see, it was around five that I was first invited on the hunt! Being out in the woods, finding mushrooms on my own must have triggered some basic hunter-gatherer instinct, because from then on… I relished them.

We ate most of the common edible fungi: morels, chanterelles, puffballs. But in the fall, the hunt was primarily focused on "Brownie Seats"!! That's what my family called *Amanita muscaria* var. *guessowii*. Brownie was a reference to the elf of Scottish lore, not the color brown. One of my sisters would later insist on calling them "Pixie Seats," as she felt it was much cuter.

The color of our *muscaria* was mostly a variety of bright yellow/orange/red. Upon the rare occasions we would find an all red specimen, it was held in high regard, accompanied with lots of "ooohs" and "aaahs." It had a very mysterious quality to it, for us kids, even long before we had any real exposure to its renown history and lore. As we grew older and began to see colored illustrations and photos in books and paintings depicting the red, Eurasia *Amanita muscaria*, it was a thrill to recognize "our" mushroom and learn that it held a special interest for people all over the world.

My mom (and my grandma, a school teacher), were the first people I ever heard use the word "*muscaria*" in reference to our Brownie Seats. My dad would occasionally just call them "muskies." To the best of my grandma's and great grandma's recollections, the family had always hunted and eaten "Brownie Seats." There's no written historical confirmation, but they suspected that the family had been doing so for several generations, at least.[3]

By the time I had reached high school, I was slowly becoming aware that eating *muscaria* was not a common practice, at least not in our locale. I talked about them to my other mushroom hunting buddies, I asked them if they picked them too. "No way! My dad says they're poisonous!" was the typical answer.

One day I overheard a couple of girls talking about me, one said to the other, "He's one of those mushroom people!" I went home and told my mom about it… my grandma, sitting in the other room, overheard the story and shouted out "mycophobes!" Some casual explanation followed.

My initial embarrassment soon gave way to a sort of secret, exclusive pride: "Mushroom People." I felt exotic and exulted, they might as well have called me an Egyptian Pharaoh. I wanted a special hat. That feeling has stuck with me since.

Preparing *Amanita muscaria* to be eaten was almost as much fun a ritual as hunting them. Mom would spread out a plastic tablecloth on the big farm table in the kitchen and we would begin to clean and slice them up as we waited on two big, black and white speckled, graniteware canners full of water to boil. The mushrooms were then loaded up in one of the canners, boiled hard for around 10 minutes, poured off into colanders, rinsed and loaded back in the second canner to repeat the process.

It was always clear to us kids that they couldn't be eaten raw, and, as farm kids, that seemed perfectly natural. There were lots of food items we couldn't eat raw, from pokeweed to frog's legs! There was never any big to-do made of it, though… no dire warnings that we'd drop dead if we did, nothing like that. Outside of puffballs, we never picked or ate white mushrooms, period. It was the only thing we were really told to avoid (We didn't hunt any *Galerina* look-alikes, at the time, so they weren't an issue either).

After the boils and rinses were complete, the bounty was placed back on the table on towels to drain more and air dry just a little. Then they would be summarily breaded and fried, our preferred way of eating them (we ate a lot of things breaded and fried!).

A couple of pounds would be placed in bags and put in the fridge for more meals in the next week. The surplus would be frozen. Often by the end of the season we would freeze 50+ lbs of them. Mom would try to mix things up by making casseroles and mushroom gravy and biscuits, etc. But by the second week we had had our fill and the exclamation would be "Brownie Seats for dinner… again?"

Notes

1. This chapter was originally published in 2015 in *Mushroom: The Journal of Wild Mushrooming*, issue 114.
2. Ed. Note: since approximately 1965.
3. Ed. Note: If these recollections hold then the family would have been consuming *Amanita muscaria* at, or before, the turn of the 20th century, around the same time Frederick Coville (*see* Ch. 19) reported on the culinary use of this mushroom among some African American families in Washington, D.C.

Chapter 21

Cooking with Fly Agaric

Kevin Feeney

In the preceding chapters we have learned that use of *Amanita muscaria* as an edible is not a new phenomenon, nor even a historically or geographically isolated practice. Examples of culinary use can be found in Asia, Europe, and North America; however, it does not appear that use of this mushroom has ever been widespread, with most examples limited to use within small communities or individual families. Those incredulous that anyone would eat *A. muscaria* suggest that it must have been used only in the most desperate times, or by the most impoverished of people (Viess, 2012). In certain instances, this may be true. It is possible, for example, that the African American market woman interviewed by Coville (*see* Chapter 19) sold mushrooms that were "prized" by consumers at the market while reserving *A. muscaria* for use at home. However, dismissing the foods of poor people as illegitimate or undesirable is not only elitist it also ignores the ingenuity of peoples throughout the world and throughout history to discover and create delicious and nutritious cuisine within the limitations of their environment. And besides, why would one not want to eat a mushroom as flavorful and delicate as the fly agaric?

While all mushrooms have a distinct "mushroom-like" flavor, this is experienced as a mild accent with the fly agaric. This may make it potentially appealing to those who are not otherwise enamored with mushrooms, but also makes it quite versatile. The flavor of the fly agaric is best described as mild and delicate, often carrying and enhancing the flavors of foods that it is cooked with. The flavor might best be described as *umami* – savory or delicious. *Umami* constitutes a fifth flavor category, along with sweet, salty, sour and bitter, but is one that has historically been ignored or gone unrecognized in Western cuisine. As is typical of Western society, it took the discovery of *umami*-specific taste receptors before this fifth flavor category came to be seen as distinct and legitimate (Lindemann et al., 2002).

The classic example of *umami* comes from MSG (monosodium glutamate), which is added to foods to enhance flavor and impart an *umami* accent. Ibotenic acid, one of the psychoactive components of fly agaric, was actually patented as a potential flavor enhancer in 1969, though it was never taken to market (Takemoto, 1969). While one needs to leach out ibotenic acid from fly agaric mushrooms before they can be consumed as a culinary, small fractions of this compound may remain after preparation, sufficient to contribute to the flavor of fly agaric but insufficient to produce psychoactive effects. There may also be other amino acids present that act on *umami* taste receptors and thus contribute to the fly agarics distinct flavor.

The caps of the fly agaric might distantly be compared to crab as the texture is soft and meaty and the flavor mild. A quick sauté with butter enhances the natural richness of this mushroom. The stems of fly agaric can also be eaten, though the texture is more fibrous, ranging from meaty to crunchy depending on the age and condition of the stem.

In addition to its rich and delicate flavor the fly agaric is also notable for its nutritional profile. Nutritionally, fly agaric is comparable to other prized wild mushrooms, such as chanterelles and morels, and is a great source of protein, fiber, and Vitamin D. Due to its reputation little has been done to examine this mushroom's nutritional benefits, so I was compelled to seek out an analysis on my own.

One autumn day in Washington state I collected around a pound of fly agaric, specifically *A. muscaria* subsp. *flavivolvata*, cleaned and boiled them for 15 minutes to remove the psychoactive prin-

Table 1: Nutritional profile of Fly Agaric (*Amanita muscaria* subsp. *flavivolvata*).

ciples and any toxins, and rinsed them well. The deep dark color of the cap fades during the boil and the resulting mushrooms are typically alabaster in appearance, though the stems may take on a dark brownish tone. Once boiled and rinsed the mushrooms will be about 2/3rds of their original weight, so in this case I was left with 2/3^{rds} of a pound, or about 10.5 ounces. Since fly agarics must be detoxified before they are eaten this measure was necessary before submitting the sample for analysis to ensure the results reflected the nutritional profile of the mushroom as it would be consumed. The results of this analysis are presented in Table 1.

What really struck me from this analysis was the high Vitamin D content, specifically Vitamin D_2 – which is what mushrooms contain. The Vitamin D_2 content in fly agaric dwarfs the content of your typical store-bought button mushroom, which provides about 1% of the daily recommended value of Vitamin D (per 100g cooked mushrooms), and exceeds the content found in favored wild edibles, such as chanterelles (*Cantherallus cibarius*), oyster mushrooms (*Pleurotus ostreatus*), and morels (*Morchella esculenta*) (Table 2). One of the reasons that wild mush-

100 g Raw	Fly Agaric: *A. muscaria*	Button: *A. bisporus*	Chanterelle: *C. cibarius*	Morel: *M. esculenta*	Oyster: *P. ostreatus*	Shiitake: *L. edodes*
Calories	42	22	32	31	33	34
Total Fat	2.36 g	0.34 g	0.53 g	0.57 g	0.41 g	0.49 g
Protein	1.94 g	3.09 g	1.49 g	3.12 g	3.31 g	2.24 g
Carbohydrates	3.16 g	3.26 g	6.86 g	5.1 g	6.09 g	6.79 g
Dietary Fiber	2.62 g	1.0 g	3.8 g	2.8 g	2.3 g	2.5 g
Total Sugars	0.0 g	1.98 g	1.16 g	0.6 g	1.11 g	2.38 g
Minerals						
Sodium	2.34 mg	5 mg	9 mg	21 mg	18 mg	9 mg
Potassium	50.9 mg	318 mg	506 mg	411 mg	420 mg	304 mg
Calcium	6.36 mg	3 mg	15 mg	43 mg	3 mg	2 mg
Iron	0.41 mg	0.5 mg	3.47 mg	12.18 mg	1.33 mg	0.41 mg
Vitamins						
Vitamin D	**56% DV** **11.1 mcg**	1 % DV 0.2 mcg	27% DV 5.3 mcg	26% DV 5.1 mcg	4% DV 0.7 mcg	2 % DV 0.4 mcg
Provitamin D2	**770 mg**	56 mg	61 mg	26 mg	64 mg	85 mg

Table 2: Comparison of the nutritional profile of *A. muscaria* with five popular edibles (USDA, 2019). Note the stark contrast in levels of Vitamin D and Provitamin D2. (Raw nutritional values of *A. muscaria* were calculated based on values determined for cooked mushrooms [63% of raw weight] and dried mushrooms [10% of raw weight; Maciejczyk et al., 2012]).

rooms have a significantly higher Vitamin D content than store bought mushrooms is that wild mushrooms rely on sunlight to convert Provitamin D_2 (ergosterol) to Vitamin D_2 (ergocalciferol). Button mushrooms are typically cultivated in the dark, thus the low Vitamin D content. To counteract this, some businesses have begun to treat their cultivated mushrooms with ultraviolet light to increase Vitamin D_2 levels and increase the nutritional value of their product.

Because sunlight is a factor in producing Vitamin D it seemed that the exceptionally high levels of Vitamin D found in my sample may have been something of a fluke, or at the high range of natural variation. Further investigation, however, revealed that Provitamin D_2 levels in Polish specimens of *A. muscaria* var. *muscaria* were off the charts in comparison to other edible mushrooms. A Polish study found Provitamin D_2 levels of 77mg/g by dry weight in their fly agaric specimens (Maciejczyk et al., 2012), an exceptionally high return considering that most edible mushrooms have been found with Provitamin D_2 levels of 8-9mg/g or less by dry weight (Phillips et al., 2011). While these results came from European specimens of *A. muscaria* var. *muscaria* the results from my own sample of *A. muscaria* subsp. *flavivolvata* correspond with these results, suggesting that exceptionally high levels of Provitamin and Vitamin D_2 *may* be characteristic of *Amanita muscaria* and its different varieties and subspecies.

In any case, I can state without reservation that the fly agaric mushroom is both flavorful and nutritious! The important part is properly preparing it for the table so that one can enjoy a meal of this mushroom without becoming sick or inebriated.

Preparing Fly Agaric for the Table

Generally, mushrooms should not be eaten raw. The cell walls of mushrooms are made up of a substance known as chitin, which is indigestible unless cooked. While individuals may enjoy raw sliced mushrooms on their green salads, these mushrooms provide only flavor and texture since chitin prevents the body from accessing the many nutrients that mushrooms contain. Further, a number of otherwise edible mushrooms are known to cause stomach upsets if not cooked properly. The fly agaric, however, requires additional steps to remove psychoactive and potentially toxic components of the mushroom. Once this is done the mushroom can be added to your favorite dishes or prepared in some of the recipes that follow. Below, two methods are provided for making fly agaric table-ready: The Basic Method and the Multipurpose Method.

Basic Method

    1lb fly agaric mushrooms, sliced
    1 gallon water
    1-2 tsps. salt
    2-3 cloves garlic (optional)
    1 bay leaf (optional)

First, clean mushrooms of dirt and debris. Don't worry if some dirt or pine needles can't be removed, they will come off in the boil and can be removed in the rinsing stage. Slice mushroom caps and stalks into 1/8-inch pieces. Add mushrooms, salt, garlic, and bay leaf to 1 gallon of boiling water and boil for 15 minutes. Strain mushrooms and rinse thoroughly. Now your mushrooms are ready to be cooked or added to your favorite recipes (Rubel & Arora, 2008).

The above recipe was arrived at following "judicious experimentation" by William Rubel and David Arora (2008: p. 242). If one chooses to veer from, or only loosely follow the above recipe, it is advisable to proceed with caution. As a personal anecdote, I once boiled several pounds of mushrooms in a gallon of water (far exceeding the recommended 1lb/gallon) for the prescribed time before cooking and consuming my mushrooms. Within a couple hours of my meal I began to feel slightly euphoric and proceeded to have a relaxing evening at home. While this type of meal certainly has its appeal, one should never serve another person a meal that may lead to inebriation (or illness) without that individual's knowledge or consent, and one should not consume such a meal if one is responsible for driving anywhere afterwards.

As an alternative, and perhaps more conservative preparation method, some people will boil their mushrooms for 10 minutes, drain and rinse the mushrooms and then boil for another 10 minutes before a final rinse (*see* Ch. 20). It is not yet clear whether both methods are equally effective, but the cautious cook may prefer the double-boil method. This method should also be preferred if one wishes to prepare a heartier cut (ie: larger than 1/8-inch) of their mushrooms. In any case, when cooking with fly agaric be mindful of your mushroom to water ratio. As you become familiar with the mushroom you might experiment with different cooking procedures in order to arrive at the texture and flavor, and perhaps the effect, you prefer.

> **Note:** Many *fungophiles* like to keep the water that cooks off mushrooms, whether it be water that cooks off in the frying pan, water that has been used to rehydrate mushrooms, or water that has been used to boil mushrooms, so that it can be added to soups, gravies and other dishes later. Unfortunately, the water you boil your fly agaric mushrooms in is not edible and not suitable to be added to any food dishes. Toss it.

Multipurpose Method

This method is for those who are interested in culinary use of the fly agaric but also in its potential medicinal and/or psychoactive applications.

1lb fly agaric mushrooms, sliced
1 gallon water

Clean mushrooms of dirt and debris. Try to get the surface of the cap as clean as possible. The pellicle, or red skin of the cap, can be removed by pinching a piece of the pellicle on the margin of the cap and peeling it back like a sticker. Completely remove the red skin and set it aside for dehydration. If you are unable to peel the cap, proceed with slicing the mushroom caps into long 1/8-inch strips. Using a knife remove the red skin and yellowish top layer of the mushroom from the mushroom strips. Set these aside for dehydration.

Once the pellicle has been removed and the mushrooms sliced, add your mushroom pieces to a boiling pot of water. Do not add salt, spices or any other food item. Boil for 15 minutes, then drain and rinse thoroughly.

At the end you should have three products: boiled mushrooms, fly agaric pellicles, and mushroom water. The boiled mushrooms are ready to be used as food. The pellicles should be dried and saved for later. Any dirt or mushroom pieces should be strained out of the mushroom water; the water can then be boiled down until you have about a pint of liquid left. The pellicle and mushroom decoction both contain the psychoactive principles of the mushroom and can be used in medicinal or psychoactive preparations (*see* chapters 27 and 29). The dried pellicle can be stored but the decoction will need to be used or frozen for later use within several days.

> **Note:** With any wild mushroom you are eating for the first time it is best to sample about a tablespoon of the mushroom before adding it to any dishes. Wait several hours before consuming any more to make sure you don't have any allergic or other adverse reactions. If you experience any euphoria or nausea you will need to do an additional boil before the mushrooms are table-ready.

Storage

A number of good edibles like morels and porcinis are dried and stored for later use. For some mushrooms this can enhance the flavor of the mushroom, and at the very least does nothing to diminish the quality of the flavor. However, this is not a great approach for *Amanita muscaria*. Once dried the flavor is fundamentally augmented and the tenderness of the mushroom cannot be restored through rehydration. Drying is an essential step when preparing this mushroom as a medicinal or a psychoactive, but generally eliminates its value as an edible. If you have an abundance of fly agaric the best way to store it for future meals is to prepare it following one of the two methods above and then freeze for later use. One can also prepare a mushroom paste (recipe below) which can also be frozen for later use.

RECIPES

Mushroom Paste

     2 cups boiled mushrooms
     1 cup stock (chicken, veggie, or mushroom)

Place mushrooms in a blender or food processor. Slowly add stock while blending until a smooth thick paste is produced. This paste can be used as the base for a soup or can be added to white sauces, gravies, salad dressings or other fare. The paste should be used within a few days or frozen for later use.

Butter & Garlic

>1 cup boiled mushrooms chopped
>1-2 cloves garlic chopped or diced
>Butter

Frying with butter and garlic is a classic way to cook any mushroom. Fry until golden brown and then serve on crackers or bread. Once prepared these mushrooms can also be added to soups, used as a topping for pasta, pizza, or burgers, or cooked inside an omelet or pot pie.

Breaded & Fried

>1 cup boiled mushrooms
>1 egg white
>½ cup flour
>½ cup bread crumbs
>1-2 Tbsps. Butter

Place flour on a plate. Coat mushroom pieces in flour. Place egg white in a bowl and dip mushroom pieces in the egg white until moist. Next, roll mushroom pieces in bread crumbs. Once coated with bread crumbs, set aside. Heat up a skillet and add butter. Once butter is melted add breaded mushroom pieces. Cook for a couple minutes, or until brown then flip over. Mushrooms are done when browned on both sides.

These are delicious and buttery and can be enjoyed plain or dipped in your favorite dipping sauce.

Pickled Fly Agaric

>4 parts boiled mushrooms
>1 part kosher salt

This is a traditional recipe that comes from Nagano Prefecture in Japan, potentially dating back to the 14th century, and is considered a local delicacy (Phipps, 2000). Mushrooms are pickled in the fall and traditionally served during New Years' festivities but may be consumed at any time as an *hors d'oeuvre* or as part of a side dish.

To make fly agaric pickles take 4 parts (by weight) of the boiled mushroom and express any excess water. Combine in a bowl with 1 part (by weight) kosher salt and mix thoroughly. Firmly pack into a glass jar until it is close to full and make sure to spoon all of the salt into the container. Once you have packed your mixture into the jar you may find that a small amount of liquid has accumulated on top. There is no need to pour this off. Firmly seal your jar and set in a dark location for a month. Mushrooms should be stored between 62°F and 72°F degrees. After one-month mushrooms should be strained and rinsed and left to soak overnight. Strain and rinse again and your pickles will be ready to eat.

Seasoning with Fly Agaric

This is another practice that comes from Nagano Prefecture and is, perhaps, the one exception to using dried fly agarics for culinary purposes. Dried fly agarics that have not been detoxified retain a strong and distinct *umami* flavor. Individuals who have choked down several dried caps of this mushroom (for non-culinary purposes) are not likely to comment kindly on the flavor, but when used in small amounts, sprinkled into rice, stir fry, or other dishes, it acts as a flavor enhancer. For seasoning purposes, dried caps of the fly agaric should be broken into small flakes, about the size of red pepper flakes. When used lightly as a seasoning (perhaps a ¼ tsp) one will not experience the psychoactive effects that are produced by larger doses, and the flavor should enhance rather than overpower the seasoned food.

Salmon with Mayo-Agaric Sauce

    1 ½ lb. salmon fillet
    3 Tbsps. Mayonnaise
    1 – 1 ½ Tbsps. Mushroom paste
    Garlic powder
    Oregano
    Salt
    Parmesan, shredded

In a small bowl mix together mayonnaise, mushroom paste, garlic powder, and salt. Lay out fillet on a cooking sheet and salt lightly. Cover fillet with a generous layer of the mayonnaise-mushroom mixture. Lightly season with oregano. Cook for 15 minutes at 400°F. Check fish with a knife. If close to done sprinkle with parmesan and return to oven to broil for 2 – 3 minutes. Serve with bread and fresh salad.

Fly Agaric Sushi Roll

    Boiled mushroom caps, dry sautéed
    Sticky rice
    Avocado
    Lemon juice
    Cream cheese (cube or loaf)
    Nori (seaweed sheets)

Cooked fly agaric has a delicate texture and flavor that is reminiscent of crab meat and can be used in various sushi recipes as an alternate to real or imitation crab. The boiled mushrooms are too moist for this purpose and frying the mushrooms in butter gives it a richness that is generally incompatible with the style and flavors of sushi. To prep your mushrooms, you will need to dry sauté them, which simply means that you sauté the mushrooms without butter or oil. This will help reduce the moisture and also gives the mushroom a firmer texture. Use only caps and dry sauté for several minutes or until mushrooms are either dry to the touch or slightly browned. Set mushrooms aside to cool.

    Next, peel and pit your avocado and slice into ¼ inch pieces. Sprinkle with lemon juice to prevent browning. Then slice the cream cheese into long sections that are ¼ inch in diameter. Lay out your nori on either a bamboo rolling mat or a piece of parchment paper and spoon a small amount of sticky rice on top. Flatten the rice with a wooden spoon. Continue to add and flatten rice until your sheet of nori is fully covered. Place fly agaric pieces, avocado, and cream cheese in a line on top of the rice. Using your bamboo mat (or parchment paper) roll the prepared nori tightly and firmly. Use a sharp knife to cut your sushi roll into pieces. Enjoy with soy sauce, wasabi, or another garnish.

Hopefully, the above recipes have given you a sense of the culinary versatility of the fly agaric. The rest is up to your imagination!

Part V:
Pharmacology & Physiological Effects

Chapter 22

Amanita Muscaria Chemistry:
The Mystery Demystified (?)

Ewa Maciejczyk

Institute of Natural Products and Cosmetics
Faculty of Biotechnology and Food Sciences
Lodz University of Technology, Poland

With its bright red cap and white spots, the fly agaric (*Amanita muscaria*) is probably the most conspicuous member of subgenus *Amanita*. It is also one of the most striking mushrooms in general. It is a common species in forests all around the world – it grows everywhere where there are birches and conifers. Against this background, the fact that it may have been the earliest hallucinogenic substance used for religious or shamanic purposes (dating back over 10,000 years) should not be surprising. The human fascination with altered states of consciousness caused by various drugs has not changed over the centuries. However, the "magical power" of hallucinogenic mushrooms is subject to scientific rationalization today. While chemical investigation of fly agaric toxins can be traced to 1869, it wasn't until the 1960s when two primary compounds, ibotenic acid and muscimol, were found in *A. muscaria* at pharmacologically active levels and were determined to be responsible, at least in part, for the fly agaric's psychoactivity. Further research has shown that the hallucinogenic effect following fly agaric consumption is greater than after ingestion of an equivalent amount of the active compounds (Ott, 1993). Thus, the possibility that other unidentified hallucinogenic substances are produced by these mushrooms has been taken into consideration (Barceloux, 2008; Catalfomo & Eugster, 1970; Eugster, 1967; Feeney, 2010; Festi & Bianchi, 1991; Liu, 2005; Matsushima, Eguchi, Kikukawa, & Matsuda, 2009; Michelot & Melendez-Howell, 2003; Schultes, 1977; Wasser, 1967).

Although no new substance with psychoactive action has been confirmed since then, further research has resulted in a better understanding of the chemical composition of the fly agaric. This mushroom, as with many fungi, has been found to be an important source of compounds with significant medical, chemical, and pharmacological potential (as its traditional application suggests: shamans of different tribes used it as a 'miracle cure' for many diseases). Consequently, this review surveys current literature dealing with the isolation, structure elucidation, and biological activities of natural products from the *A. muscaria* fruiting body.

The Search for Fly Agaric's Inebriating Compound(s)

Plants, fungi, and other living organisms that have been used in traditional medicine or rituals have always attracted great interest among scientists. The search for active/toxic substances in *Amanita muscaria* mushrooms dates to 1869 when the properties of muscarine isolated from *Amanita muscaria* were first described (Schmiedeberg & Koppe, 1869). For nearly a hundred years it was thought that muscarine was the main toxic component of the red toadstool – despite the significant differences between the effects of the toadstool and of muscarine intoxication. First of all, muscarine is not psychoactive, but it causes abundant salivation, tearing and hyperhidrosis (excessive sweating) through stimulation of the autonomic nervous system. Such symptoms are rarely found after fly agaric intoxication. This seems to be a consequence of the extremely low concentration of muscarine in *A. muscaria* tissues (Festi & Bianchi, 1991; Ott, 1993; Schultes, 1977; for a different perspective *see* Feeney & Stijve, Ch. 23).

There have also been reports of the isolation of *L*-hyoscyamine and bufotenine from fly agaric, however, subsequent studies on the chemical composition of *A. muscaria* were unable to confirm the presence of these compounds. Substances responsible for the psychoactive properties of fly agaric were not identified until 1964. Interestingly, this success was achieved almost simultaneously by three independent research groups – in Japan, England and Switzerland (Michelot & Melendez-Howell, 2003; Ott, 1993). From the chemical point of view, the most interesting discovery was that the identified substances contained an isoxazole ring (five-membered ring containing three carbon atoms and nitrogen and oxygen atoms next to each other), which rarely occurs in natural substances. In 1967 an international agreement was reached on the nomenclature and the identified substances were named: ibotenic acid and muscimol. Ibotenic acid derives its name from the Japanese name for fly agaric – ibo-tengu-take (long-nosed goblin mushroom) (Hatanaka, 1992). Soon after, both compounds were isolated from other species of toadstools: *A. strobiliformis*, *A. pantherina*, *A. cothurnata*, *A. gemmata*, *A. regalis* and *A. muscaria* var. *alba* and *formosa* (Ott, 1993).

N-containing active substances

Ibotenic acid (Figure 1) is α-amino-3-hydroxy-5-isoxazoleacetic acid, which under standard conditions is a colorless crystalline substance, well soluble in cold water. The largest amount (548 nmol/g f.w.) of this compound are found in the red skin of the cap and the yellow flesh below it (Michelot & Melendez-Howell, 2003). Ibotenic acid is an amino acid that stimulates the glutamic acid receptors by altering membrane permeability and resting potential (Barceloux, 2008). In 1985 it was discovered that the substance called pantherine, identified from *A. pantherina*, was in fact ibotenic acid. Heating and drying of the compound leads to its decarboxylation, and muscimol is formed as a result. Given that ibotenic acid readily decarboxylates when exposed to heat and acidic conditions it is possible that, after eating cooked or dried mushrooms (or even after the digestion process itself), only muscimol reaches the brain, causing the mushroom's famed psychotropic effects (Barceloux, 2008; C. Li & Oberlies, 2005; Michelot & Melendez-Howell, 2003).

Muscimol (Fig. 1) is a colorless, highly water-soluble crystalline substance. Fresh fly agaric contains small amounts of this compound (compared to the ibotenic acid content; *see* Chart 1), though it is believed that muscimol found in fresh mushrooms may simply be an artifact formed during the isolation procedure. The presence of muscimol in older specimens appears to be the result of decarboxylation of the naturally occurring ibotenic acid (Hatanaka, 1992). Muscimol binds to γ-aminobutyric acid (GABA) receptors, being their agonist, and blocks the neuronal and glial uptake of GABA, resulting in increased serotonin and acetylcholine levels, and lowered norepinephrine. The effects of muscimol on the brain are most apparent in the cortex, hippocampus and cerebellum. It is assumed that muscimol is responsible for the majority of physiological effects after ingestion of fungi containing both ibotenic acid and muscimol, much as psilocin is considered primarily responsible for the psychoactive effects of fungi containing both psilocybin and psilocin (Barceloux, 2008; Hatanaka, 1992; C. Li & Oberlies, 2005; Michelot &

Melendez-Howell, 2003).

Muscazone (Fig. 1), α-amino-2,3-dihydro-2-oxo-5-oxazole acetic acid, is a colorless crystalline compound, also isolated from the discussed species; its structure has been confirmed by synthesis. This cyclic amide, or lactam, is formed by the photochemical conversion of ibotenic acid and is produced mainly during the extraction of other compounds from fly agaric. Muscazone has always been isolated as a racemate and does not undergo decarboxylation or exchange hydrogen easily. These facts suggest that muscazone is an artifact of isomerization of ibotenic acid. Compared with previous substances, muscazone induces weak pharmacological effects (Barceloux, 2008; C. Li & Oberlies, 2005; Michelot & Melendez-Howell, 2003).

Muscarine (Fig. 1) was thought to be the main active substance in *A. muscaria* for decades. As previously mentioned, this presumption was discredited because the effects of muscarine are distinct from the effects caused by the fly agaric. However, it should be remembered that muscarine itself is a strong poison. This substance is found in *A. muscaria* in small amounts (0.0002-0.0003 %) in comparison with *Inocybe spp.*, or *Clitocybe spp.* (0.43% in *Inocybe subdestricta*, and up to 0.15 % in *Clitocybe dealbata*) (Michelot & Melendez-Howell, 2003). *A. muscaria* should not show effects of muscarinic poisoning except, perhaps, in cases of excessive consumption (however, *see* Ch 23.). In addition to muscarine, the fly agaric also contains several related stereoisomers: (-)-(2S,3R,5R) *allo*-muscarine, (+)-(2S,3S,5S) *epi*-muscarine (Festi & Bianchi, 1991; Michelot & Melendez-Howell, 2003; Ott, 1976b).

Muscaridine (Fig. 1) has been isolated from *A. muscaria* and several other fungi. Initially, this substance was erroneously characterized as tetrachloroaurate (Pedersen & Schubert, 1993). It is an acyclic isomer of muscarine and is among the active ingredients of the fly agaric, belonging to a group of stimulant substances (Dunn, 1973).

(R)-4-hydroxypyrrolidin-2-one (Fig. 1), which has a structure similar to muscimol and ibotenic acid, is a compound characteristic of cells of some microscopic fungi which demonstrates potential biological activity (against bacteria and other fungi). This hydroxypyrrolidin derivative gives some information about the biogenesis of ibotenic acid, muscimol and muscazone – most probably all come from the same precursor, which is β-hydroxyglutamic acid (Michelot &

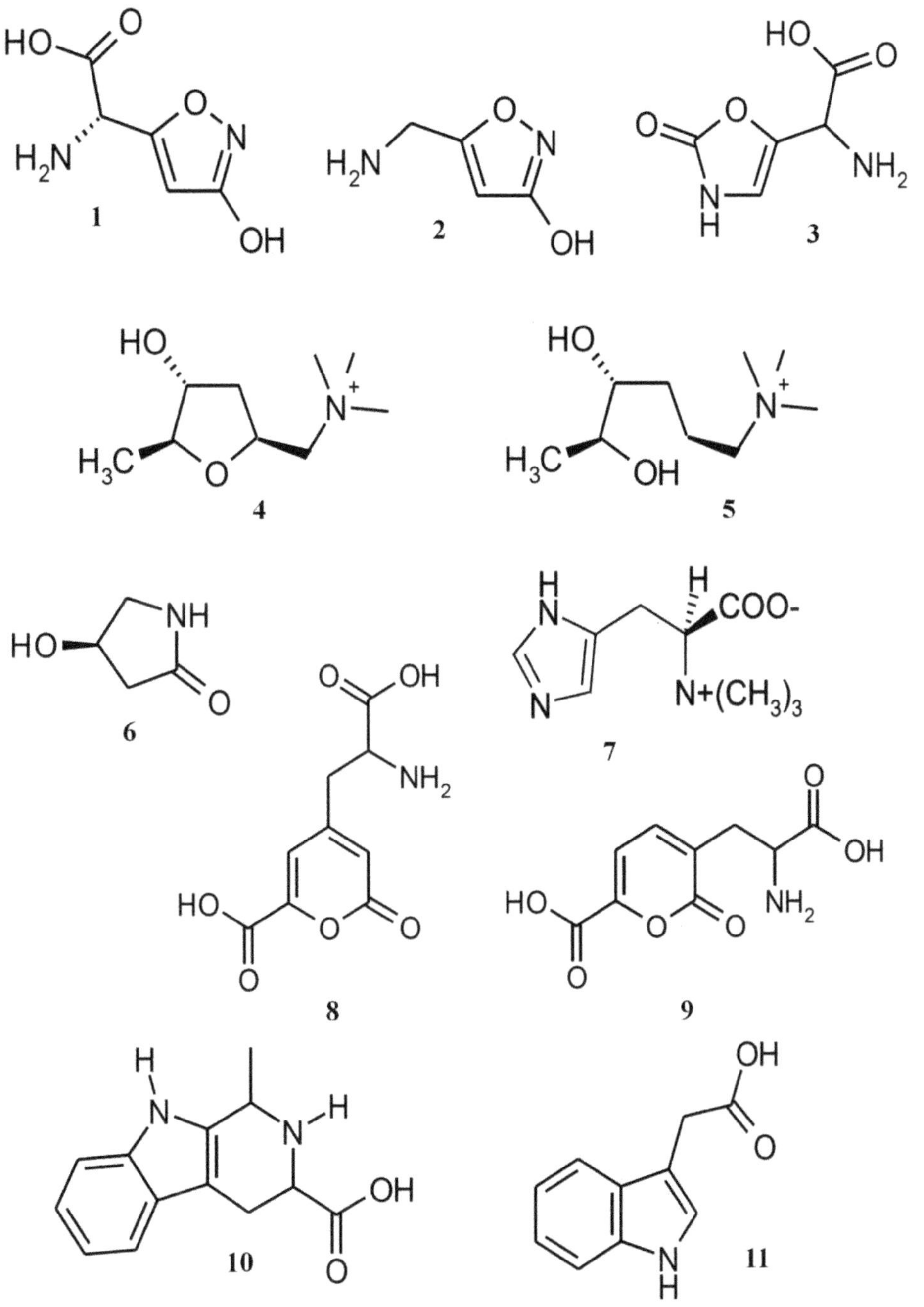

Fig. 1: Fly agaric N-containing substances: (1) ibotenic acid, (2) muscimol, (3) muscazone, (4) muscarine, (5) muscaridine, (6) (R)-4-hydroxypyrrolidin-2-one, (7) hercynine, (8) stizolobic acid, (9) stizolobinic acid, (10) 1,2,3,4-tetrahydro-1-methyl-β-carboline-3-carboxylic acid, (11) β-indoleacetic acid.

Melendez-Howell, 2003).

Hercynine (N,N,N-trimethyl-L-histidine) (Fig. 1) is an amino-acid betaine and a precursor of L-ergothioneine (L-ergothioneine has established free radical scavenging properties and antiapoptotic activity and could become of therapeutic value in Parkinson's disease) (Ming Yi Tang, Kee-Mun Cheah, Shze Keong Yew, & Halliwell, 2018). This nonproteinogenic amino acid occurs in high concentrations in certain fungi, including *A. muscaria*, and has been the subject of a substantial number of studies. However, the mechanism of its activity remains unknown (Kohlmunzer & Grzybek, 1972; Tulp & Bohlin, 2005).

Stizolobic acid and **stizolobinic acid** (Fig. 1) (α-pyrone amino acids) have been detected in small amounts in a few toadstool species in addition to *A. muscaria* (Hatanaka, 1992). Stizolobic and, to a lesser extent, stizolobinic acid have a stimulating effect on the isolated rat spinal cord. Studies on the biosynthesis pathway indicated that 3,4-dihydroxyphenylalanine (DOPA) is a precursor of both amino acids (Michelot & Melendez-Howell, 2003).

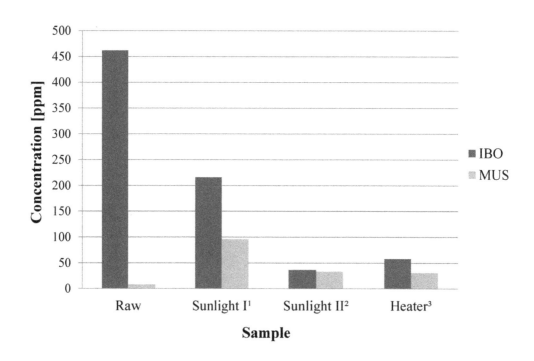

Chart 1: Mean concentration [ppm*] of ibotenic acid (IBO) and muscimol (MUS) in raw *A. muscaria* and after drying; 1 - sunning for three days; 2 - sunning for eleven days; 3 - drying near the oil heater for two days; * the value reduced to the raw weight. (Tsunoda et al., 1993).

Another compound isolated from *A. muscaria*, **1,2,3,4-tetrahydro-1-methyl-β-carboline-3-carboxylic acid** (Fig. 1) (1-methyl-2,3,4,9-tetrahydro-1*H*-pyrido[3,4-b]indole-3-carboxylic acid) is a substance of unknown pharmacology (Eugster, 1968; Ott, 1993). This compound has also been determined in various food samples, including fruit juices, jams, soy sauce and alcoholic beverages (in beer and wine). This alkaloid is formed in the condensation of Pictet-Spengler – a reaction between acetaldehyde and tryptophan, which occurs in nature or during food processing (Cao, Peng, Wang, & Xu, 2007; Cox et al., 1997). Interestingly, similar β-carboline compounds determined in Ayahuasca (*Banisteriopsis caapi*) are monoamine-oxidase (MAO) inhibitors (Ott, 1993).

β-indoleacetic acid (Fig. 1) (indole-3-acetic acid) which is also found in fly agaric tissue is a metabolic breakdown product of tryptophan (deamination) or tryptamine (decarboxylation). It is often produced by bacteria that live in the intestine. Literature provides information on endogenous production of this acid in mammalian tissues. Quite often it is determined in low concentration in urine, while in the case of patients with phenylketonuria its level increases drastically. Primarily, β-indoleacetic acid is an important plant hormone belonging to the auxin group (Prusty, Grisafi, & Fink, 2004).

Pigments

Determination of dyes responsible for the bright red color of the fly agaric cap was one of the research priorities in the 1970s. Currently, most structures of these compounds are already known. In 1930 **muscarufin** (Figure 2) was

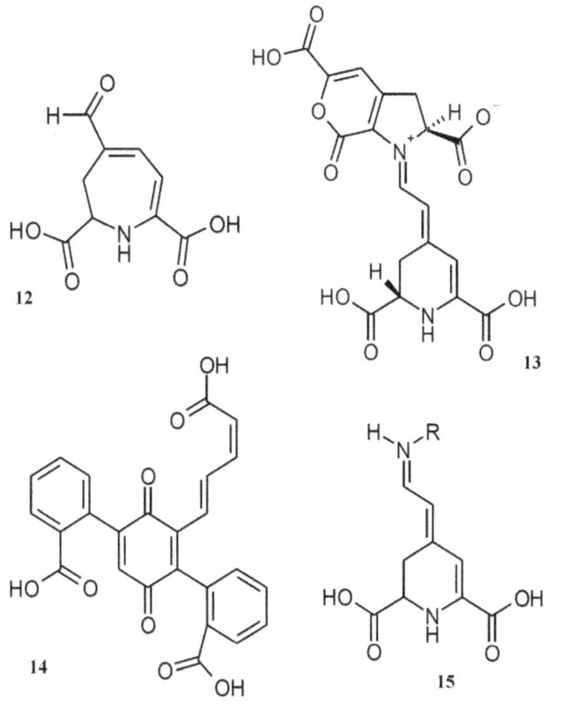

Fig. 2: Pigments of fly agaric: (12) muscaflavin, (13) muskapurpurine, (14) muscarufin, (15) general structure of musca-aurins.

proposed to be the main pigment responsible for the red color of the fruiting body but it was not confirmed later by any group (Hatanaka, 1992; Michelot & Melendez-Howell, 2003).

Structurally interesting is the yellow pigment – **muscaflavin** (Fig. 2), which is synthesized on the same path as beta-lactic, stizolobic and stizolobinic acids (Hatanaka, 1992; Michelot & Melendez-Howell, 2003).

The next group of dyes are the I-VII **musca-aurins** (Fig. 2), which belong to betalains – pigments resulting from the transformation of DOPA. It was surprising that mushrooms contain pigments which normally occur in red beet (*Beta vulgaris*) and other plants. Interestingly, ibotenic and stizolobic acid are an essential part of the structure of musca-aurine I and II, respectively (Hatanaka, 1992; Michelot & Melendez-Howell, 2003). Structurally related substances to muska-aurins are **violet muskapurpurine** (Fig. 2) and red-brown muscarubin (Michelot & Melendez-Howell, 2003; *see also* muscarubrin [Stintzing & Schliemann, 2007]).

Amavadin (Figure 3) is a light blue vanadium complex isolated originally from *A. muscaria*. In general, the accumulation of metals by organisms is designed to protect against the growing toxicity associated with increased levels of metals in the soil. However, the concentration of vanadium in some fungi of the genus *Amanita* is extremely high, often several hundred times higher compared to the concentration of this element in plants (C. Li & Oberlies, 2005; Michelot & Melendez-Howell, 2003). Amavadin is an eight-coordinate vanadium complex with 1:2 stoichiometry with (S,S)-2,2'-(hydroxyimino) dipropionic acid. This unusual molecule aroused enormous interest among chemists, due to an atypical bare vanadium (IV) complex (no V=O bond), with coordination number eight and containing two ligands, which resulted in its structure being confirmed in many crystallographic and spectroscopic studies (Da Silva, Fraústo da

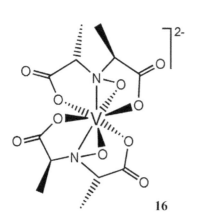

16

Fig. 3: Structure of amavadin (16) (Da Silva et al., 2013).

Silva, & Pombeiro, 2013; C. Li & Oberlies, 2005; Ooms et al., 2009). Amavadin has also been isolated from other psychoactive *Amanitas*, including *A. regalis* and *A. velatipes* (Berry et al., 1999).

Arsenic derivatives

In the 1990s, the nontoxic **arsenobetaine** (Figure 4), and related methyl arsenic derivatives, was discovered for the first time in the terrestrial environment in mushrooms accumulating arsenic, including in *A. muscaria*. However, relatively little is known about this type of chemical in connection with the terrestrial environment. In *Amanita* species, the following derivatives have been identified in addition to arsenobetaine: (18) **arsenocholine**, (19) **tetramethyl-arsenic salt**, and (20) **cacodylic acid** (dimethylsulfonic acid). Moreover, other arsenic derivatives, most likely sugar and phospholipid derivatives were found in fly agaric. Arsenobetaine and arsenocholine demonstrate virtually no toxicity. On the other hand, methyl derivatives are characterized by various activities. For example, cacodylic acid was one of the components of the Agent Blue defoliant used during the Vietnam War (Stellman & Stellman, 2018). In general, formation of organo-arsenic compounds occurring in living organisms is a result of biological detoxification of toxic arsenic (inorganic) forms (Byrne et al., 1995; Kuehnelt, Goessler, & Irgolic, 1997; Vetter, 2005).

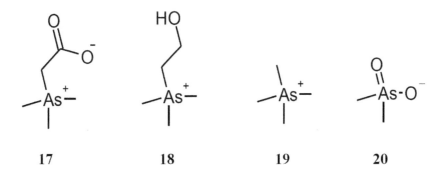

Fig. 4: Fly agaric arsenic derivatives: (17) arsenobetaine, (18) arsenocholine, (19) tetramethyl-arsenic salt, (20) cacodylic acid.

Other Compounds & Constituents

The fly agaric, like every living organism, produces many substances essential to sustain life. Some of these are typical for fungal cells, and some have special properties. Among these are proteins, lipids, sugars and their derivatives.

In the group of proteins identified in *A. muscaria*, most are enzymes. One of these is DOPA 4,5-dioxygenase, a central enzyme in the biogenesis of betalains in the species. This enzyme is an extradiol-cleaving dioxygenase and as an oligomer is composed of a varying number of identical subunits (Girod & Zryd, 1991). Another example of a fly agaric enzyme is an aspartic protease. The general aim of studies on proteases from mushrooms has been to identify enzymes with potential applications in industry and medicine (Erjavec, Kos, Ravnikar, Dreo, & Sabotič, 2012).

Although tissues of mushrooms mainly contain water, they also produce lipids. Typical constituents for this class of substances are fatty acids which occur in living organisms as esters: triacylglycerols, phospholipids, sterol esters. Fly agaric whole cell fatty acid analysis shows that the fungus' main fatty acids are: linoleic, oleic, stearic, and palmitic acids respectively. Hydroxy fatty acids were observed in very low concentration (Karliński, Ravnskov, Kieliszewska-Rokicka, & Larsen, 2007).

A representative of diacylglycerols is 1,3-diolein which was isolated from the squeezed juice of fly agaric fruiting bodies. This compound is a housefly attractant (Muto, Sugawara, & Mizoguchi, 1968). It is highly interesting that this mushroom produces an attractant and insecticide (ibotenic acid) for the same insect concomitantly. One possible explanation for this phenomenon would be that the fungus lures flies with the attractant, kills them with the toxin and then digests their bodies. Many fungal species are capable of utilizing proteinaceous materials of insect carcasses. However, this ability has not been proved in *A. muscaria*. Usually this mechanism for ensuring a supply of nitrogenous substances from insects is relatively sophisticated (Hatanaka, 1992).

The name of the mushroom: fly agaric, and latin *muscaria*, refer to its use as a flycidal. Interestingly, in many countries a common name for this species literally means "fly killer": fly agaric (English-speaking countries), amanite tue-mouche (France), fliegenpilz (Germany), muchomor (Poland), Мухомор (Russia), moscario (Italy), hongo mosquero, hongo matamoscas (Spain) (Guzman, 2001; Michelot & Melendez-Howell, 2003; Wasser, 1967). Similarly, in Japan one of the epithets used to describe this mushroom - haitori - can be successfully translated as fly agaric. In fact, the toadstool has a weak insecticidal effect. Although, according to folk sources, fresh mushrooms of this species (mixed with water or milk) were used to poison flies - the flies sipping juice fell into a state of numbness for a few hours to several days. A good housewife knew, however, that such stunned flies needed to be killed or removed from the home. Another explanation for the name fly agaric is associated with medieval and even earlier beliefs in which madness was associated with flies. In the lands of northern Eurasia, it was thought that possessed people were captured by flies, and after exorcisms, insects were supposed to exit the body through the nostrils - which was tantamount to a cure for possession (Michelot & Melendez-Howell, 2003; Wasser, 1967).

Another group of compounds obtained from fly agaric are sphingolipids, mainly ceramides and, to a lesser extent, cerebrosides (Weiss & Stiller, 1972). Ceramides are N-acyl derivatives of aliphatic amino alcohols (called sphingoid base) and cerebrosides possess an extra sugar group (in *A. muscaria* it is only glucose [Weiss & Stiller, 1972]). In general, many biological cellular processes are dependent on sphingolipids, including growth regulation, cell migration, adhesion, apoptosis, aging and inflammatory reaction (Olsen & Færgeman, 2017).

Screening with ^{31}P NMR elucidated presence of another lipid type – phospholipids – in *A. muscaria*. This research also demonstrated other forms of organic phosphorus in the fungus: monoesters, diesters, pyrophosphates, polyphosphates and phoshphonates (Koukol, Novák, & Hrabal, 2008; Maciejczyk et al., 2015). Presence of the above-mentioned substances reflects high metabolic activity in *A. muscaria* basidiocarps.

Among many different active substances separated from fungi (including *A. muscari*a), there are also sterols – effective as antiviral, antibacterial, anti-inflammatory and anticancer substances (Chen et al., 2017; Yoshino et al., 2008). Such properties are characteristic for **ergosterol** (Figure 5). Although it was first isolated more than 100 years ago from *Claviceps purpurea*, it is one of the most common fungal sterols and its therapeutic value was recognized during the search for active metabolites of medicinal mushrooms. For example, a hexane extract from the mycelium *Grifola frondosa*, containing a mixture of fatty acids and ergosterol and ergosta-4,6,8(14),22-tetraen-3-one, showed inhibitory properties towards cyclooxygenases 1 and 2 (Zhang, Mills, & Nair, 2002). Other studies report that tumor growth (sarcoma) in mice was disrupted without side effects after oral administration of ergosterol isolated from *Agaricus brasiliensis* (Zaidman, Majed, Mahajna, & Wasser, 2005). Inhibition of new blood vessel formation in Lewis lung cancer was also observed following administration of ergosterol. On the basis of *in vivo* studies, it was found that the anti-cancer effect of ergosterol is associated with the inhibition of angiogenesis caused by the growing tumor (Li et al., 2015).

In addition, it is worth mentioning that ergosterol is a precursor of vitamin D_2 (ergocalciferol). This vitamin is essential for proper bone development. It is also used in the treatment of skin diseases, secondary hyperparathyroidism, and various types of cancer (Jäpelt & Jakobsen, 2013). The *A. muscaria* fruiting body is a very rich source of ergosterol – containing up to 77 mg/g of dry mass. For comparison, concentrations of this compound in the popular edible mushroom, *Agaricus bisporus*, was more than six-fold lower (Maciejczyk et al., 2012).

Recently, polysaccharides derived from mushrooms have arisen as an important class of bioactive substances. Many medicinal and therapeutic properties are attributed to polysaccharides present in *Basidiomycetes* (Ruthes, Smiderle, & Iacomini, 2016). *A. muscaria* is not an exception – a fucomannogalactan and a (1→3),(1→6)-linked β-D-glucan were isolated from its fruiting bodies (Kiho, Katsurawaga, Nagai, Ukai, & Haga, 1992; Kiho et al., 1994; Andrea Caroline Ruthes et al., 2013). These compounds' structure determination showed that the fucomannogalactan is a heterogalactan formed by a (1→6)-linked α-D-galactopyranosyl main chain partially substituted at O-2 mainly by α-L-fucopyranose and a minor proportion of β-D-mannopyranose non-reducing end units, and the β-D-glucan is a (1→3)-linked β-D-glucan partially substituted at O-6 by mono and a few oligosaccharide side chains. Both polysaccharides were evaluated for their anti-inflammatory and antinociceptive potential, and they produced potent inhibition of inflammatory pain (Ruthes et al., 2013). Glucan obtained from *A. muscaria* by a Japanese research group also exhibited significant antitumor activity against sarcoma 180 in mice (Kiho et al., 1992).

Other interesting sugar metabolites are polyols which are obtained by substituting an aldehyde group with a hydroxyl one. One example of this type of sugar alcohol is **Mannitol** – a white, crystalline substance (Figure 5). Along with sorbitol and other polyols, it is a characteristic fungal metabolite present in *A. muscaria* and other fungi (Lewis & Smith, 1967). In addition to being a source of carbon and energy, its high content in mushrooms is related to the fact that it has a protective function in the event of water and thermal stress (especially low temperatures) (Tibbett, Sanders, & Cairney, 2002; Wingler, Guttenberger, & Hampp, 1993). Mannitol generates osmotic potential during the growth of the fruiting body (Wingler et al., 1993). This poly-alcohol is also widely used in medicine: in the control of elevated intracranial pressure, in the protection of kidneys in various types of transplants and in the treatment of rhabdomyolysis (Shawkat, Westwood, & Mortimer, 2012). Furthermore, other research showed interesting properties of mannitol to improve delivery of drugs to the human brain (opening the blood-brain barrier) (Bhattacharjee, Nagashima, Kondoh, & Tamaki, 2001; Brown, Egleton, & Davis, 2004; Ikeda, Bhattacharjee, Kondoh, Nagashima, & Tamaki, 2002). These results inspired the hypothesis that relatively high concentration of mannitol in the tissues of fly agaric (1.02 ± 0.02 g mannitol per 100 g dry weight [Reis et al., 2011]) enables more efficient transportation of ibotenic acid and muscimol into the brain and thus enhances their hallucinogenic activity (Maciejczyk & Kafarski, 2013). This hypothesis may explain why the psychoactive effects produced following ingestion of *A. muscaria* is greater than after ingestion of an equivalent amount of ibotenic acid and muscimol (Ott, 1993). However, other researchers have suggested that this greater effect may be related to other hallucinogenic substance which remain undiscovered (Barceloux, 2008; Catalfomo & Eugster, 1970; Eugster, 1967; Feeney, 2010; Festi & Bianchi, 1991; Liu, 2005; Matsushima et al., 2009; Michelot & Melendez-Howell, 2003; Schultes, 1977; Wasser, 1967).

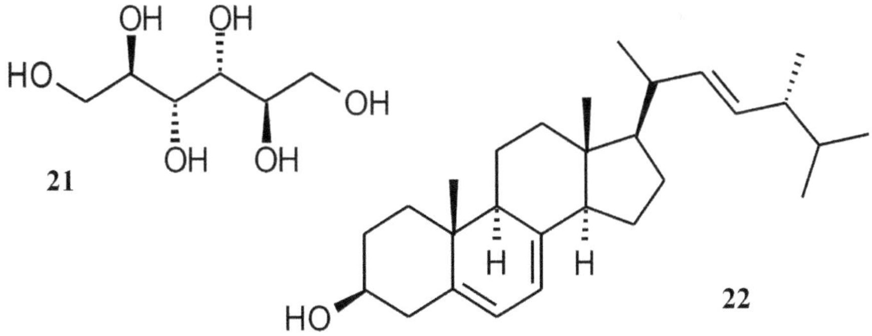

Fig. 5: Structure of mannitol (21) and ergosterol (22).

Conclusions

This chapter reviews the chemistry of *Amanita muscaria* by providing the type and number of compounds isolated from the species, and references for the description of their isolation, structure elucidation studies and relevant review papers. The data reveal that the pharmacology of *A. muscaria* remains largely unexplored. The discussion presented here focused mostly on hallucinogenic compounds, pigments, and substances with proven activity in other species. Thus, there is a lack of analysis for the total chemical composition of the fly agaric – a mushroom which has been used as a 'miracle cure' for many diseases.

The neurological symptoms following fly agaric ingestion are well known. Despite some individual differences in the psychoactivity of this species (resulting from the variable level of active substances, as well as distinctive individual reactions), some effects are characteristic: limb twitching, euphoric states, macropsia; religious feelings or insights, and sporadically - colorful visions of supernatural phenomena. Attacks of aggression and stages of a deep sleep have also been reported.

As previously mentioned, the effects following the consumption of muscimol and ibotenic acid do not explain all of the observed pharmacological effects produced by *A. muscaria* inebriation. Most likely, other substances in *A. muscaria* contribute to the mushroom's effects. Stizolobic and stizolobinic acid have proven stimulating effects (in animal experiments), as does muscaridine, and perhaps these contribute to the hyper-stimulation that is sometimes reported. There is also a possibility that some of the fly agaric's constituents exhibit synergistic effects with each other. But all this remains in the sphere of conjecture.

Additionally, the long history of fly agaric use in folk medicine should not be forgotten, uses which include reducing rheumatic inflammation, treatment of bruises and other injuries, or in alleviating the effects of insect bites. It should also be noted that the mushroom, once called 'Agaricus muscaricus', has been used for centuries in the form of tincture (35 g Amanita muscaria in 100 ml solution) to treat depression, various types of tics, epilepsy, etc., as well as having been used, in combination with Mandrake root tincture, as a treatment for Parkinson's disease.

Bearing in mind the pharmacological potential of fungi (including antibacterial, anti-inflammatory, anti-cancer, and immuno-stimulating properties, as well as their ability to synthesize atypical secondary metabolites), research on *Amanita muscaria* can be considered only through the benefits of using fungi in medicine. This approach seems to justify ethnopharmacological research. However, discovering "active substances" for the world and ways of using "holy" plants/mushrooms, in addition to the obvious benefit (from a scientific point of view), also has its negative repercussions. First of all, the modern world deprives the mentioned species of the *sacrum* element. In the past, their consumption was an element of ritual, frequently facilitated by a shaman who was the only person who knew the secret recipes and the peculiarities of the substance's effects. Today, the magical power of hallucinogenic mushrooms is subject to scientific rationalization, a perspective frequently at odds with older sacred and magical understandings of the mushroom. The only thing that has not changed over the centuries is the human fascination with different states of consciousness, which are caused by drugs of various kinds.

Re-examining the Role of Muscarine in Fly Agaric Inebriation

Kevin Feeney & Tjakko Stijve

The pharmacology of the *Amanita muscaria* mushroom has long been a puzzle for scientists, and many pieces of this puzzle remain in dispute. One recurrent dispute centers on the role of muscarine in *Amanita muscaria* inebriations and poisonings. Currently, it is widely believed that muscarine does not occur in *Amanita muscaria* in pharmacologically active levels, however, anecdotal and folkloric evidence suggest otherwise. In order to resolve this confusion, nearly 500 anecdotal accounts of *Amanita muscaria* inebriation and poisonings were collected and analyzed for evidence of muscarine-like symptoms. The findings of this study demonstrated that consumption of moderate amounts of this mushroom was occasionally reported to produce physiological effects akin to muscarine poisoning in consumers. While there are limitations to the current study, the results suggest that a more comprehensive study of *Amanita muscaria*'s pharmacology, to determine the prevalence of muscarine or muscarine-related compounds, is in order.

Confusion over Muscarine

Muscarine, first discovered in 1869, was the first compound isolated from *Amanita muscaria*, from which it derives its name. Originally muscarine was believed to be the primary active agent of this white-speckled scarlet fungus; however, instead of acting on the central nervous system as hallucinogens do, muscarine, a cholinergic agonist, primarily affects the peripheral parasympathetic nervous system. The effects of muscarine poisoning are generally marked by excessive perspiration and salivation, blurring of vision, abdominal pain, chills, nausea, vomiting, lacrimation (tearing) and diarrhea, while symptoms of *Amanita muscaria* inebriation are typically "characterized by confusion, pronounced muscle spasms, delirium, hallucinations, and disturbances of vision" (McKenny & Stuntz, 2000: 223). Despite the mismatch between the symptoms of muscarine and fly agaric poisoning atropine, a muscarine antidote, was long used as the standard treatment for fly agaric poisoning.

Interestingly, the once mysterious compounds in fly agaric that cause CNS (central nervous system) stimulation came to be known as pilzatropine, or "mushroom atropine", because of similarities between the effects of atropine and the fly agaric (Tyler, 1958). Because of these similarities' hospitals treating fly agaric poisoning with atropine ended up compounding the effects of the mushroom rather than alleviating them. It is now understood that atropine is contraindicated in cases of fly agaric poisoning and the primary inebriating agents of *Amanita muscaria* are now recognized as the isoxazole derivatives ibotenic acid and muscimol – though it is possible that some other unidentified compound(s) may contribute to the psychoactive effects of this mushroom (Catalfomo & Eugster, 1970).

The current understanding of muscarine concentrations in *Amanita muscaria* is based on a 1950s study that found a concentration of muscarine at a mere 0.0003% by fresh weight in European specimens (Ott, 1993, *citing* Eugster, 1956; Eugster, 1959; Catalfomo & Eugster, 1970). Based on this study it has been claimed that "a man would have to eat the equivalent of his own weight of *Amanita muscaria* before manifesting any muscarinic symptoms" (Benjamin, 1995: 346). However, considerable losses are generally involved in the type of isolation procedures employed, a fact that should give pause before coming to any conclusions about muscarine concentrations, not to mention that isolation procedures have almost certainly improved in the last sixty years. Another consideration is that purposeful ingestion of *Amanita muscaria* typically involves dried rather than fresh specimens (Feeney, 2010). The dehydration of mushrooms removes much of the water mass, but is unlikely to affect concentrations of muscarine, meaning that concentrations of muscarine in dried mushrooms should be about 10-fold higher than those found in fresh specimens.[2]

A little-known study conducted in the 1980s provides a very different view of muscarine concentrations in *Amanita muscaria*, a view which appears to have been overlooked in the English mycological literature. The study, using high performance thin-layer chromatographic determination methods, demonstrated concentrations of muscarine between 0.005% and 0.011% by dry weight in European specimens (*See* Table 1 [Stijve, 1981; Stijve, 1982]). These concentrations are considerably higher than those found in the previously mentioned study, which is universally cited in the English literature; and concentrations of muscarine reported as high as 0.011% place *Amanita muscaria* on par with the muscarine content of some species of *Inocybe* (Catalfomo & Eugster, 1970), a genus generally considered to be poisonous due to high concentrations of muscarine. It has also been estimated that mushrooms "must contain a minimum of 0.01% muscarine" before ingestion will lead to the appearance of clinical symptoms (Puschner, 2013: 671), which suggests that at least some *Amanita muscaria* specimens will produce clinical symptoms of muscarine poisoning if ingested in a dried state.

Table 1: Analyses by HPTLC of muscarine and muscimol: Analyses performed in autumn 1980.[3]

Mushroom species	Origin	Muscarine content	Muscimol content
Amanita muscaria	Kühlsheim, Ger., Sept. 1974	0.007	0.015
	Himmelstadt, Ger., Sept. 1974	0.005	< 0.01
	Gramschatz, Ger., Oct. 1974	0.011	< 0.01
	Neuwirtshaus, Ger., Sept. 1975	0.005	0.02
	Helmstadt, Ger., Oct. 1975	0.009	0.02
	Chamonix, France, Sept. 1979	0.009	0.15
	Lally, Switz., Sept. 1979	0.008	0.22
	Lally, Switz., Sept. 1980	0.01	0.16
Amanita pantherina[4]	Gamburg, Ger., Sept. 1977	< 0.0005	0.025
	Gamburg, Ger., Sept. 1980	< 0.0005	0.19
	Puidoux, Switz., Sept. 1980	< 0.0005	0.31

All values expressed in percentage on dry matter. (<) means lower than the limit of determination, which is somewhat dependent on the amount of co-extractives.

This study alone indicates the importance of recognizing the Eugster (1956; 1959) study as a single data point. Adding the numbers provided by Stijve (1981;

1982) provides us with a range (0.003 – 0.011%)² that gives us a more complete understanding of possible variations in muscarine concentrations in *A. muscaria* mushrooms. Additionally, anecdotal reports of *Amanita muscaria* ingestion indicate that muscarinic symptoms can occur after consumption of moderate amounts of this mushroom, which suggests that neither of the cited studies fully capture the range of muscarine concentrations in this mushroom. One individual reported, after ingesting a preparation of 15 grams of dried *Amanita muscaria*, that he "started pouring out water, outta every pore and orifice. I was like a wet rag being rungout [*sic*]" (Schiffer, 2003). Another individual, after consuming 10 grams of dried *Amanita muscaria*, had the following to say: "I could feel all the water in my body going up to my salivating glands, spilling and being swallowed, making the whole circuit again" (Neuroglider, 2000).[5]

Jonathan Ott (1976b), writing in the *Journal of Psychedelic Drugs*, was cautious when reporting about muscarinic effects he and others experienced from *Amanita muscaria* specimens picked in Washington State. In explanation, Ott speculated that "either these effects were due to muscarine in the carpophores (in which case *A. muscaria* from Washington must contain a much higher concentration of muscarine than is reported for European specimens), or they were produced by some yet-unidentified compound with muscarinic activity" (p. 32). Despite skepticism about the role of muscarine in *Amanita muscaria* inebriation, disparities between the "accepted science" and frequently reported muscarine-like symptoms, following *A. muscaria* ingestion, require reconciliation. In addition to anecdotal reports, a brief review of folklore offers another, unique perspective on the presence of muscarine in Siberian and European populations of *Amanita muscaria*.

Muscarinic Symptoms in Religion and Folklore

The folklore surrounding the use of the *Amanita muscaria* in Europe and Asia seems to suggest that cultures familiar with this mushroom recognized its muscarinic effects and considered them important components of *Amanita muscaria* inebriation. In Siberia, Koryak legend tells us that *Amanita muscaria* was created by the spittle of Vahiyinin, the God of Existence (Schultes & Hofmann, 1992). Similarly, a Croatian variation on the myth of Wotan's wild hunt describes how the *Amanita muscaria* was formed by the bloody spittle that fell from the mouth of Wotan's horse (Morgan, 1995).[6] More recently, a Koryak woman shared a story in which after eating *Amanita muscaria* she "awoke with her mouth full of saliva. The *mukhomer* (*Amanita muscaria*) told her she was the heroine of a Koryak fairy tale and that she should spit to make a big river, which she did" (Salzman et al., 1996: 45; *see also* Ch. 6).

Methods

In order to determine the prevalence of muscarine-like symptoms produced by *Amanita muscaria*, 490 accounts of inebriation and poisoning with *Amanita muscaria* were compiled for analysis.[7] Sources included journals (Anonymous, 1993; BD, 1994; BF, 1994; D.D., 1999; MN, 1996; Ott, 1976b; Pollock, 1975; Salzman et al., 1996; Tengu, 1998), newspaper articles (Associated Press, 2006), toxicology reports (Beug, 2007; Beug, 2006; Cochran, 2000; Cochran, 1999; Cochran, 1985; Trestrail, 1998; Trestrail, 1997; Trestrail, 1996; Trestrail, 1995) and various websites (Erowid, 2009; Lycaeum, 2000; Tolento, 2008) and forums (Drugs Forum, 2009; Entheogen Dot Com, 2008; Mycotopia, 2009; Shroomery, 2009). Textual analysis was used to identify the following muscarinic symptoms: abdominal pain, blurry vision, chills, diarrhea, excess perspiration, and excess salivation. Since nausea and vomiting can be caused by muscarine or ibotenic acid, incidents of nausea and vomiting were not cited as primary evidence of muscarinic symptoms. However, incidents of muscarinic symptoms coupled with either nausea or vomiting were analyzed to see if the presence of other muscarinic symptoms were related to higher frequencies of nausea and vomiting.

Results

Of the 490 reports collected, 115 reported at least one of the following muscarinic symptoms: abdominal pain, blurry vision, chills, diarrhea, excess perspiration, or excess salivation (*See* Table 2). The most common muscarinic symptoms in the sample appeared to be perspiration (69%), salivation (44%) and blurry vision (37%). While unable to contribute to the psychoactive effects of this mushroom, the blurring of vision may contribute to the visionary aspect of the experience. Abdominal pain (3.5%) and diarrhea (8%) were the least common reported symptoms. Interestingly, nearly 14% of those reporting muscarinic symptoms did not become inebriated; suggesting that in some specimens of *Amanita muscaria*, muscarine may be more active than either ibotenic acid or muscimol.

Table 2: Frequency of Individual Muscarinic Symptoms.

Muscarinic Symp.	Inebriating	Non-Inebriating	Total
Abdominal Pain	4% (4)	0% (0)	3.5% (4)
Blurry Vison	36% (36)	44% (7)	37% (43)
Chills	24% (24)	25% (4)	24% (28)
Diarrhea	6% (6)	19% (3)	8% (9)
Nausea	62% (63)	19% (3)	57% (66)
Perspiration	69% (68)	69% (11)	69% (79)
Salivation	47.5% (47)	25% (4)	44% (51)
Vomiting	27% (27)	19% (3)	26% (30)
Total	86% (99)	14% (16)	100% (115)

While the occurrence of nausea and vomiting cannot be solely attributed to muscarine, the data (Table 3) show that individuals who experienced other muscarinic symptoms were more likely to experience nausea (57%) than those who experienced no muscarinic symptoms at all (35%). This would appear to suggest that the presence of muscarine contributed to and increased the nauseating properties of *Amanita muscaria*, a notion supported by application of a chi-square test for independence which found a statistically significant association between the occurrence of muscarinic symptoms and the experience of nausea ($\chi^2 = 18.46$,

	Total	Inebriating	Nausea	Vomiting
Non-Muscarinic	76.5% (375)	75.5% (283)	35% (131)	16% (61)
Muscarinic	23.5% (115)	86% (99)	57% (66)	26% (30)
Total	100% (490)	78% (382)	40% (197)	19% (91)

Table 3: Occurrence of muscarinic symptoms in association with Nausea and Vomiting.

$p < 0.001$).[8] Based on raw numbers the rate of vomiting also increased when other muscarinic symptoms were present (from 16% to 26%), a trend similarly supported by application of the chi-square test which indicated a statistically significant association between the muscarinic and emetic effects of *Amanita muscaria* ($\chi^2 = 5.61, p < 0.05$).

Discussion

Recognizing the muscarinic-like activity of *Amanita muscaria* allows theories regarding the cultural and religious significance of *Amanita muscaria* to be revisited with new eyes. For example, a re-examination of the Rig Veda for descriptions of muscarinic symptoms associated with use of Soma may provide further support for R. Gordon Wasson's theory identifying the Vedic sacrament as the *Amanita muscaria* mushroom.

A passage from the Rig Veda, known as the Frog Hymn, suggests a potential correlation between muscarinic symptoms and use of Soma. In this passage frogs are compared to perspiring Brahmins gathered around the Soma bowl:

> *Like Brahmins at the overnight Soma-sacrifice speaking around as it were a full lake, ye celebrate that day of the year which, O Frogs, has begun the rain.*
>
> *Soma-pressing Brahmins, they have raised their voice offering their yearly prayer, Adharvu priests, heated, sweating, they appear; none of them are hidden.* (MacDonell, 2006: 145)

While it has been asserted that the Brahmins in this passage are sweating because they are gathered around a heated cauldron of milk (Doniger, 2005), a new possibility arises if Soma is indeed *Amanita muscaria*.

Another suggestive account of Soma intoxication comes from Phillipe de Félice. Referencing a story from the *Satapatha Brahmana* in 1936, de Félice provided the following description of Soma inebriation, "It happens sometimes that the inebriation is accompanied by organic disturbances, which are in reality symptoms of an acute intoxication. Men know and fear the baleful effects of the drug, and, though he was a god, Indra himself did not escape them, since one day the Soma came forth from every opening in his body" (Wasson, 1968: 135, *citing* de Félice, 1936). These effects, cited by de Félice are similar to the experience cited earlier where one individual reported that "(he) started pouring out water, outta every pore and orifice" (Schiffer, 2003). The experience of the god Indra, as described above, suggests typical muscarinic symptoms, which include: salivation, perspiration, vomiting and diarrhea. There are few orifices that are left unaffected by the properties of muscarine.

The results of this study suggest that muscarine content may vary greatly from mushroom to mushroom in much the same way that the fly agaric's psychoactive principles, ibotenic acid and muscimol, also greatly vary. Undoubtedly, some in the mycological world will worry that discussing the muscarinic effects of *Amanita muscaria* will lead down a dangerous path where doctors and hospitals re-introduce atropine as an antidote for *A. muscaria* poisoning. Such fear, however, is unwarranted. Atropine is contraindicated in cases of isoxazole poisoning and the levels of muscarine remain on the lower end (0.003 – 0.011%); here, supportive and symptomatic treatment should be sufficient. It has been estimated that a lethal dose of muscarine ranges somewhere "between 40 mg and 495 mg (Puschner, 2013: 671), and fatalities are extremely rare (Kendrick, 2000). Assuming a high concentration of 0.011% one would have to consume nearly a pound of dried *Amanita muscaria* before reaching a *potentially* fatal dose. Of course, the possibility remains that some other compound with muscarine-like activity is contributing to the reported symptoms.

Limitations

The anecdotal reports used for this study were collected from a variety of sources, and the extent of details provided varied from source to source. As a result, the data used were not entirely uniform and may have been incomplete in some instances. Additional limitations include a lack of information regarding where the ingested mushrooms were harvested as well as what specific varieties and subspecies of *Amanita muscaria* were ingested. While folkloric evidence has been offered supporting the presence of active levels of muscarine in European and Asian varieties of *Amanita muscaria*, the results of the current study do not necessarily support this contention due to the lack of geographical information on ingested specimens in the surveyed accounts; however, the results of Stijve's work indicate that significant levels of muscarine can be found in some European populations of this fungus.

A further limitation of this study is that no bioassays were performed, nor any chemical analysis available on the particular mushrooms consumed within the 490 analyzed reports. While the reported symptoms support and indicate a finding of physiologically active levels of muscarine in some *Amanita muscaria* collections, without hard data to support this finding it cannot be ruled out that some other compound with muscarine-like activity is responsible.

Conclusion

Through analysis of the collected anecdotal accounts of *Amanita muscaria* inebriation, it appears that muscarine-like symptoms, while not universal, are not infrequent. The associations between *Amanita muscaria* and salivation in Croatian and Siberian folklore also suggest that these cultures may have been familiar with the muscarinic effects of this mushroom. While the findings of this study were limited by the tools that were available, they nevertheless indicate that current understandings of muscarine concentrations in *Amanita muscaria* need to be adjusted to recognize a range of concentrations (0.003 – 0.011% [by dry weight]) and also suggest that further chemical analysis of this mushroom is required before any conclusive statements can be made about the role of muscarine in shaping the symptoms produced by ingestion of *Amanita muscaria*.

Notes

1. An earlier version of this chapter was published under the title "Re-Examining the Role of Muscarine in the Chemistry of *Amanita muscaria*," in the 2010 Spring-Summer issue of *Mushroom, the Journal*. The chapter has been revised and updated for purposes of the present volume, which you now hold in your hands.
2. Since mushrooms are over 90% water, fresh weight concentrations of muscarine pegged at 0.0003% would be approximately 0.003% for dried specimens of the same mushroom.
3. Many of the specimens were already several years old at the time the analysis was performed in 1980. An examination of the data suggests that muscarine concentrations remain stable while muscimol degrades rapidly with time.
4. The results for *Amanita pantherina* suggest that levels of muscarine are significantly lower than in *Amanita muscaria*, despite other similarities in their pharmacological profiles. This is supported by a review of fifty *A. pantherina* ingestions that was conducted for an earlier version of this paper which showed only two incidences of reported muscarinic symptoms, or 2% of the sample.
5. At muscarine concentrations of 0.011% a 10 g dose would contain approximately 1.1 mg of muscarine while a 15 g dose would contain 1.65 mg. The exact dose of muscarine required to produce clinically observable symptoms is unknown.
6. Interestingly, the name for the World Tree, Yggdrasill, translates as Ygg=Wotan, drasill=horse, or Wotan's horse; this story would thus place *Amanita muscaria* at the base of the World Tree, an interesting association given its mycorrhizal nature (*see* Leto, Ch. 8).
7. When initially published this study included reports of *Amanita pantherina* inebriations and poisonings mixed in with data on *Amanita muscaria* inebriations and poisonings. The mix of data on separate species confused the purpose and results of the study. All data points concerning *Amanita pantherina* have been removed to provide a cleaner picture of the role of muscarine in *Amanita muscaria* inebriation.
8. The chi-square test for independence is a statistical test used to determine whether or not there is an association between two variables.

Chapter 24

Agaricus Muscarius:
The use of Fly Agaric in Homeopathy

Kevin Feeney & Bill Mann

While generally considered poisonous and without medical application the fly agaric mushroom (*Amanita muscaria*) has been used as a remedy within the field of homeopathy since 1828, where it is known as Agaricus muscarius. This designation comes from Carl Linnaeus, the "father of modern taxonomy," who proposed the name in his work *Species Plantarum*, published in 1753. Linnaeus was inspired by its use in rural Sweden as a poison for flies and thus designated the mushroom "muscarius," which is derived from the Latin word for fly, *musca*. The scientific name was officially changed to *Amanita muscaria* in 1821, seven years before its incorporation into homeopathy, but the still popular Agaricus muscarius was the term used when the mushroom was introduced to the Homœopathic Pharmacopœia and remains the popular term for this treatment within homeopathy.[1]

Before a substance is introduced into homeopathic practice it must undergo a "proving." This is a process by which a dozen or more individuals take the substance over a period of time and record whatever symptoms they experience following its consumption. Homeopathy relies on the principle of *similia similibus curentur,* or "like cures like" (Hahneman, 1833), and by determining what symptoms a particular substance causes, through the process of a "proving," the homeopath discovers what ailments can be treated with that substance (or *poison*[2]). The *proving* is typically done blind and involves three primary roles: a homeopathic pharmacist, a Proving Master, and Provers (subjects). Only the pharmacist will know the identity of the substance in question, while the Master Prover handles compiling and analyzing the subjective reports of the Provers for any compelling patterns or consistencies in the symptoms reported by the Provers.[3] Once a substance has been *proved* it becomes a *remedy*.

While a *proving* is a relatively small trial, Agaricus muscarius has been the subject of at least five recorded *provings*, the most notable being the initial *proving* by Schreter and Stapf in 1828, as well as the second *proving* conducted by Samuel Hahnemann, the founder of Homeopathy, who recorded 715 symptoms in 1830 in a trial involving a dozen people (Hering, 1879). The field of homeopathy has also relied on "outside" reports on the effects of substances, including medical data, poisoning reports, historical accounts, and other sources. This type of data falls under the category of "soft provings," as it is not gleaned from the standard *proving* practices of homeopathy.[4]

In its nearly 200 years of use within homeopathy Agaricus muscarius has been used to treat a number of complaints, including chilblains and other skin afflictions, tics and trembling of the limbs, confusion, vertigo, nervous afflictions, rheumatism, cataracts, and various other conditions. When new "remedies" are added to the Homœopathic Pharmacopœia following a *proving*, they are typically given a grade of "1" for any symptoms that are identified. This is the lowest grade a remedy can have in relation to the treatment of particular symptoms or afflictions but indicates its potential use as a remedy. As more *provings* are completed and successful case studies are reported, the grade for a remedy in treating a particular ailment can rise to a 2, 3 or 4. This grading system gives the homeopathic doctor a gauge for how historically reliable and effective a remedy is in treating particular symptoms. While Agaricus muscarius has not been a historically popular treatment in homeopathy, growing interest in the remedy over the past twenty years has increased its "grade" (reputation) for treating a number of different symptoms.

To give a broader view of Agaricus muscarius, and its use within homeopathy, a case study is provided below involving the treatment of a young man with Agaricus muscarius by homeopath Bill Mann.

CASE STUDY:

Feb. 1996
Patient: "Doug"
Age: 30 years old
Sex: Male

PATIENT COMPLAINT:

Patient has intense anxiety and a neurotic obsession with his health, including fears of having or developing Multiple Sclerosis or Cancer. He also reports intense sexual obsessions that impede his ability to form meaningful intimate relationships. Additionally, patient experiences vesiculations and twitching of muscle groups, especially upper arms and thighs, and of facial muscles –most notable in upper lip. Symptoms have been relatively constant for the past two years. He experiences extreme stiffness after increased levels of twitching and tingling. He has had swelling of lymph nodes mostly below the jaw. Frequent tingling and numbing in the face, especially in the morning. His lip recurrently cracks in the middle, especially the lower lip. He often bites his tongue involuntarily. He reports recurring problems with fatigue and muscle stiffness over the last three years, though he improves with rest.

PATIENT PROFILE:

Patient describes himself as a sex addict and has been in a recovery program. He remembers being relatively content as a child until high school when he had to deal with his emerging sexuality:

> I was so intensely sexual, just the thought of a female and I was totally distracted. I masturbated constantly. When a female teacher or student would come into the classroom and I could see any of her breast or nipple I would ejaculate. It was so embarrassing. I never had a girlfriend in high school; I was a geek type and felt odd. By my first year of college I grew physically, and women found me attractive. I then became sex-crazed and had many superficial sexual adventures. My world became about conquer and pillage and always in search of the most physically attractive female I could find. I was into the ecstasy, and [would] then move onto the next one I could find.

> I liked to party, sing and get crazy with my friends, and then hunt for sex. This was a weekly weekend adventure, and I was happy.

Patient is obsessed with sex shops and strip joints. If a strip club is close by, he is compelled to go. The charge of watching women perform is intense and exciting to the point that his heart is pounding with great exhilaration and desire. He will masturbate and then feels worse physically, followed by enormous remorse and guilt. Though he knows this is the result he cannot stop himself. He drives in seedy parts of town to find where prostitutes do business just so he can feel the rush and excitement of being in their proximity. He rarely sees prostitutes but has on occasion. He tells me these things as if *confessing*. This behavior has been intense for many years.

Patient has had many unsuccessful relationships, mostly because he becomes overwhelmed and he "devours" his partners sexually. They all leave within a few months, typically complaining that he is too detached emotionally and they feel used as an object. He knows they are right:

> I cannot seem to get close with a woman, never have. I want sex and that is it. It is like I am an animal. I love women's bodies and that is as far as it goes. I am always looking for the most perfect body.

Patient confessed that he knows how selfish and superficial he is, but he has not been close to a woman ever. He says that therapy groups have been helpful to him and he has stayed abstinent for several months.

After college he started his own business taking groups of people on extreme adventures. He and an assistant take these groups on 10 – 20 day wilderness excursions in the most extreme environments and terrains, i.e.: the Alaskan Hinterlands, Canadian wilderness, the Rockies, etc. Dangerous and intense hiking, climbing, rafting, canoeing, and hunting excursions. Every year he plans more difficult and challenging trips; he always needs something bigger and more intense the next time. He studies and prepares intensely for months by reading maps, reading about terrain, rivers, and climbing routes, possible dangers, wildlife and where it is going to be the most challenging for him and his groups. He says the more challenging and exotic the better.

Patient loves the extreme adventure and has been able to make a good living for the past 7 years, working primarily during the spring and summer months. He says that he always loved maps, even as a boy, and constantly fantasized about wild adventures in the wilderness, imagining himself as a daring explorer.

Patient had his ten-year reunion a few years back and the whole preceding year he planned his most dangerous adventure (he did a ten-day rafting journey in the Rockies), so he could tell all of his old friends about his exploits, daring's and doings. He wants to be known as a fearless adventurer.

When he's working he feels his best; he experiences less anxiety and concern regarding his body and cannot act upon his sexual obsessions, though fears of Cancer and other diseases continue to plague him.

Before his current symptoms developed patient was in Alaska on several back to back trips. He met a woman he reported having strong feelings for, a rarity, but an old boyfriend returned into her life and she dropped him. He says he was intensely hurt and wanted to kill this man. He broke apart the cabin he was in, smashed a chair and basically went *berserk*. He was jealous, hurt, and outraged. His ex-girlfriend was unsympathetic. Afterwards he came back to California and has not been well since.

I asked him to describe his feelings about the break-up:

> I went insane and went into a full-blown rage attack. I could have killed this other man. Though I had never even seen him I was in constant fantasy about an attack on this man, I wanted revenge. I hated her and hated him with a frenzy, I went *berserk*. It freaked me out and really, I think this crushed me.

It is during the mid-fall through the winter months that his troubles typically arise. He reports becoming wildly preoccupied with sex and with his health when he is at home. He has an obsession that he has some incurable illness, typically Cancer, Multiple Sclerosis, or some incurable neurological disease that no one can find (he mentioned that his dad had Multiple Sclerosis in his 20s but went into a complete remission about 10 years later). Patient has gone to many physicians, even oncologists, and not once have they found anything. When he becomes particularly perturbed, he will visit medical libraries and research diseases that he believes he might have. It is a day-in and day-out obsession. He has even gone to physicians in

Mexico where he could get a C.A.T. scan for a fraction of the cost in the U.S. But to no avail. No one has ever found any pathology. Still, he continues with an intense conviction that he has Cancer and is absolutely terrified of the implications.

His search for medical explanations has left him emotionally and financially exhausted: He has a twinge in the kidney region and finds himself in a new physician's office. He experiences mild diarrhea that lasts two days and he is certain it is the beginning of his demise; he finds another physician to examine him. He is convinced that his glandular swelling is definitely Cancer and he becomes crazed. He begs and pleads with the physician to get to the bottom of it. He says he knows that he sounds insane, but the conviction drives him, and he cannot stop thinking about it. He said that he drives his family crazy with his constant complaining and never-ending obsession with his health.

DREAMS

Dreams often provide additional information about a patient's symptoms and overall presentation and can be helpful in determining the appropriate remedy. I asked about his dreams:

Dream #1: He said that he dreams often of running from something gigantic and then finds that he can fly and experiences great exhilaration and joy, pure pleasure, and no fear at all. Then wakes up sweating with his heart pounding. I asked what feelings this provokes. He said it is a remarkable feeling of power with no fear, an adrenalin rush and then *whoosh* "I am floating and then flying with the earth below."

Dream #2: This is a recurring dream from childhood of a huge ominous being. He can only ever see its face, not its body. He fights against it with all his might. He is not afraid but wants to kill this thing before it kills him. He rages, fights, kicks, and screams. It is a very heavy being and lies upon him as he struggles. He cannot move an inch, he cannot breathe, and he begins to suffocate unbearably, and wakes gasping for air. PRIMARY FEELING FOR HIM: the worst part is the suffocation. He wants to fight this thing but cannot strike its body as he cannot see it. He feels small and defeated.

PATIENT SYMPTOMS
- Fear of cancer/Anxiety about His Health
- Satyriasis (uncontrollable sexual desire)
- Twitching muscles

PATIENT TRAITS
- Pathological Boldness
    - audacious
    - seeks extreme adventure and challenges
    - conquest of women and nature
- Fearful & Submissive
    - fear of Cancer, Multiple Sclerosis
    - anxiety about health
    - unable to control urges/obsessions
    - need for reassurance
    - hatred and revenge overwhelming

DREAM THEMES
- Flying
- Exhilaration
- Being Chased
- Giant Beings
- Struggle
- Suffocation
- Defeat

HOMEOPATHIC RUBRICS

In the field of homeopathy there is a large catalog of Rubrics that practitioners may consult to determine the appropriate remedy for a patient. Often finding the correct remedy requires a degree of triangulation within a patient's presenting symptoms. Below is the rubric for "Fear of Cancer," one of the patient's primary presenting symptoms, which lists Agaricus muscarius as "4AGAR". The number 4 indicates that Agaricus muscarius is one of the higher graded remedies for this treatment, but other remedies are also graded with a 4, and may be similarly or more effective, depending on the patient.

> **mind; FEAR; cancer, of (51):** acan-pl., **4<u>AGAR.</u>**, 4<u>ARS.</u>, aur-m-n., bac., bamb-a., bar-c., beryl., cadm., 4<u>CALC.</u>, calc-f., calc-p., 3Carc., chin-ar., coco-n., electr., falco-p., 3Fl-ac., ign., 3Kali-ar., lac-c., lac-

dr., lac-eq., lac-h., 3Latex, 3Lob., 4MANC., **3Med.**, methylp-h., **nat-m.**, 3Nit-ac., onc-t., pele-o., **phos.**, plac., **4PLAT.**, 3Plut-n., 2posit., **4PSOR.**, ruta, sabad., scol., scorp., sep., soph-m., spect., tax., 2tax-br., thuj., 2uran-n., verat. (Zandvoort, 2013)

In order to pick the most appropriate treatment it is necessary to cross-reference symptoms to see which remedies are cross-listed, which can indicate a superior or preferred treatment. Cross-referencing with *Hypochondriacal Anxiety* and *Satyriasis* show that Agaricus muscarius is indicated for these symptoms with a grade 4 and grade 1 rating, respectively.

mind; ANXIETY; hypochondriacal (74): 3Acon., **4AGAR.**, agn., aloe, 2alum., 2am-c., anac., arg-n., 4ARN., 4ARS., 2asaf., asar., bar-c., 3Bell., bry., 2calad., 4CALC., calc-sil., calen., calop-s., cann-i., 2canth., caust., 2cham., cinis-p., 3Con., conch., 2cupr., 2dros., ferr-p., geoc-c., graph., 3Grat., 3Haliae-lc., hyos., ign., 3Iod., kali-ar., 3Kali-c., kali-chl., kali-m., kali-p., lach., lat-h., 4LEC., lob., 3Lyc., 2m-arct., **med.**, 4MOSCH., nat-c., **4NAT-M.**, 4NIT-AC., 3Nux-v., ol-an., oro-ac., ox-ac., 2ph-ac., **4PHOS.**, **plat.**, pras-f., **psor.**, 4PULS., 3Raph., 4RHUS-T., 3Sep., spect., squil., 3Staph., sulph., 4SYPH., tarent., thuj., 2valer.

mind; SATYRIASIS (52) : agar., agn., anac., 3Anan., androc., 3Apis, 3Aster., 4BAR-M., 3Bell., 3Calc-p., 3Camph., 3Cann-i., cann-s., canna-i., 4CANTH., chir-f., coca, con., cyna., 3Fl-ac., 3Gels., 4GRAPH., grat., 3Hyos., 3Kali-br., kali-c., lac-cpr., 4LYC., 3Lyss., **med.**, menth., 4MERC., **3Nat-m.**, 4NUX-V., oryc-c., pen., **4PHOS.**, 3Pic-ac., **4PLAT.**, **psor.**, sabin., 3Salx-n., saroth., 4SIL., 4STRAM., 3Sulph., 3Tarent., thymu., ust., 3Verat., zinc., zinc-pic. (Zandvoort, 2013).

Examining the three rubrics above we find only six remedies (in bold) that are common to all three symptoms. These include:

AGAR.   (Agaricus muscarius)
NAT-M. (Natrum muriaticum)

PHOS.   (Phosphorous)
PLAT.   (Platinum metallicum)
PSOR.   (Psorinum)
MED.   (Medorrhinum)

The remaining remedies can thus be eliminated as treatments, but a full six potential treatments remain. After this elimination I found myself leaning towards Agaricus muscarius because of its association with the fearless battle-frenzy of the Berserkers, from which we get the phrase "going berserk." The frenetic energy and audacity that led the Berserkers in conquest appeared to mirror symptoms and behaviors of the patient, but another round of elimination was required before I could be positive that Agaricus muscarius was the correct treatment.

While the patient's mental and emotional state seemed to predominate in his presentation, he also complained of physical symptoms, including twitching of muscles in the legs, arms, and face. Rubrics for twitching in the face and extremities were examined next for further eliminations.

> **face; TWITCHING; lips; upper (24)**: adam., **3Agar.**, 3Ars., bacch-a., 4CARB-V., digin., gink., 3Graph., hep., 2irid., lap-be-e., latex, lute-o., methylp-h., nat-c., nicc., onc-t., ozone, **plat.**, plut-n., sabad., stront-c., 3Thuj., 3Zinc.

> **extremities; TWITCHING; upper arms (67)**: **agar.**, am-c., ant-c., arg-s., 2arn., 2asaf., bamb-a., bapt., bell., brachy-s-p., bry., 3Calc., camph., caust., chin., 3Cina, clem., 2cocc., coloc., croto-t., cupr., dig., dulc., hell., 2ign., inul., junc., kali-bi., kali-c., kali-n., lap-be-e., 3Lyc., m-arct., mag-m., mang., mang-acet., 3Meny., merc., mez., mur-ac., **nat-m.**, nit-ac., olnd., ox-ac., petr., ph-ac., **phos.**, phyt., 4PIC-AC., plb., plut-n., pter-a., 3Ran-b., seneg., 3Sep., sil., spig., squil., stann., sulph., 3Tarax., 3Tax., 3Teucr., thuj., uro-h., valer., zing. (Zandvoort, 2013)

When all five of the rubrics above are combined, only one remedy remained: Agaricus muscarius; and this was selected for the treatment of the profiled patient.

PATIENT TREATMENT

On intake patient presented with severe Hypochondriacal neurosis, rating a 10 on a scale from 1-10. Patient was treated over a period of 9 months with Agaricus muscarius with doses ranging from 200C and 1M.[5] Patient received treatment between 5 and 10 times and began to show improvement after 2-3 months. After three months his anxiety levels came down significantly, ranging between a 2 and 5. After 9 months anxiety levels were consistently down to a 2, with periodic upward oscillations. Patient's improvement was confirmed by his sister, also a patient, who reported that his health neuroses had significantly diminished, to the delight of his family, and that his troubling pattern of sexual conquest had also subsided.

Several years after completion of treatment I ran into the patient, who was performing in a theater production I had attended. Patient reported that he had moved on from adventure tourism, had become a teacher, and had also taken up acting. Patient presented well and appeared to be without neuroses or tics.

Notes

1. *Amanita pantherina* is also used within homeopathy under the name Agaricus pantherinus, the name by which it was originally described by Swiss botanist Augustin Pyramus de Candolle in 1815. Agaricus pantherinus is rarely used within homeopathy but has been a recognized homeopathic remedy since at least the early-1870s. According to the *Encyclopedia of Pure Materia Medica Index* (Allen, 1874: 125) Agaricus pantherinus is indicated for the following symptoms:

   Mind. – Delirium. – Maniacal disposition to rave. – Loss of memory. – State of consciousness resembling coma.
   Head. – Great heaviness of the head.
   Mouth. – Lips tremble.
   Throat. – Difficult deglutition.
   Stomach. – Some loss of appetite.
   Stool and Anus. – Slight diarrhoea.
   Upper Extremities. – Trembling of the hands.
   Generalities. – General overpowering sense of fatigue. – Extreme lassitude and torpor. – Loss of power of co-ordinating muscular movements. –Convulsive movements.
   Sleep and Dreams. – Invincible drowsiness. – Stupor. – During sleep, respiration embarrassed; face congested and of a livid hue; pulse rather slow.

2. It should be noted that substances/remedies used in homeopathy are typically considered poisons, and the poisoning symptoms caused by the substance indicate the types of symptoms it can be used to treat homeopathically, e.g.: *similia similibus curentur*.

3. To give the reader a better idea of the nature of a *proving*, several *proving* reports, drawn from *A Cyclopædia of Pathogensey* (Hughes & Dake, 1886), are provided below (minor edits have been made to the original text for purposes of readability):

   > FRANZ KRAUS, medical student, æt. 23, sanguineous-choleric temperament, of robust frame, with the exception of an ague never seriously ill.
   > -30th Oct., morning, took some drops of tincture agaricus [muscarius] in water. 11 a.m., slight transient heat in head, combined with mental fatigue as after long-continued intellectual labour, at same time slight pricks in left upper eyelid, obtuse pains in eyeball similar to the pain caused by pressing on the eyeball with the hand. Frequent call to urinate, the quantity of urine passed much greater than usual. This last symptom lasted all day, as also the occasional recurrence of heat.
   > -31st. Same dose. Same symptoms, with the exception of the eye pains; call to urinate again present, in addition, shooting pain in almost all the joints, especially well

marked in left knee joint and head-joint; the right knee-joint was very painful on going upstairs. (p. 162)

OHLHAUTH took a few drops of 1_x dilution: after ¼ hour, while standing reading the paper, he noticed the paper moving with the beat of his heart; pulsations were as distinctly felt and violent as in climbing a mountain. Next day, after 7 drops of 1st centesimal [1C], same symptoms recurred.(p. 138)

SAMUEL MAX, medical student, æt. 20, strong constitution, sanguine temperament, in good health.
-21st Oct., morning, 5 drops tincture agaricus [muscarius] in water. In a few minutes, heaviness in feet and head, very transient.
-22nd, 10 drops, some confusion of head, heaviness and coldness of extremities, quivering of left upper eyelid, repeated at noon.
-23rd, 15 drops, confusion of head and quivering of upper eyelid.
-31st, 5 drops. Soon confusion of head and coldness of lower extremities. All right after breakfast.
-4th Nov., 10 drops. In 10 minutes, coldness of lower extremities. Heaviness of head and inclination to vomit, with watery eructations.
-9th, 10 drops. Soon coldness in lower limbs and inclination to vomit. Transient headache. (p. 169)

4. A *soft proving* is any account of "poisoning" that falls outside the parameters of a homeopathic *proving* that is also considered sufficiently informative to be retained for reference within the field of homeopathy. The following is an example of a *soft proving* from *A Cyclopædia of Pathogensey* (Hughes & Dake, 1886):

> Three half-starved soldiers in the retreat from Moscow in 1812 made a meal of agaricus muscarius, which they roasted on the coals, with a little butter and salt, and ate without bread. One ate four, the two others three each. They then lay down to sleep. At 10pm, the first soldier commenced to speak nonsense, passing from one subject to another in a gay delirium with great loquacity. Some time afterwards he had violent convulsions. Seen soon after he was found with convulsive movements of muscles of face and extremities, the jaws were firmly clenched, he could not be got to take anything; he always wanted to talk, but could hardly articulate. His eyes rolled

in their orbits, and were sometimes quite turned up. The agitation was extreme, the lower limbs strongly retracted, and the arms so agitated that his pulse could not be felt. Cold sweat on face, neck, and chest. Tip of nose and lips bluish. Some froth at the commissure of lips; respiration oppressed and noisy; breath had a sickly and sour odour. In an instant he became more calm, and the jaws were loosened. An emetic was given. He threw up much phlegm, which smelt sour, together with fragments of the fungus. He then got vinegar and water to drink. He passed the rest of the day alternately convulsed and in stupor. The beginning of the night was restless, but he grew quiet, and in the morning was as usual.

The second was attacked by convulsions soon after the first with great anxiety and pain at epigastrium, which was relieved spontaneously by vomiting. He then developed much strength with gay delirium; he sang and talked, but gave no replies to questions. He imagined himself an officer commanding at drill, and various manœuvres which he thought he was conducting. The convulsions ceased for a short time, but soon recommenced; irregular and hurried movements of the upper extremities; restlessness of hands, which he pressed together as though he was rolling a soft body between his hands in order to make it round. Speaking with volubility and animation to his father and mother, as though he were beside them, giving no replies to questions, he sang and lamented alternately, embracing his comrades and kissing their hands. All this took place in the midst of a general spasm resembling trembling rather than convulsions. Half an hour later he fell into a faint, which did not last long, but left him in a deep stupor. Extreme alteration of physiognomy, general prostration, with cold clammy sweat all over the body. The scene ended in a soporous state, from which he gradually roused and recovered, but remembered nothing of what had happened since eating the fungi.

The third was seized with some pains which he referred to the stomach, with great oppression, then convulsive movement that lasted but a short time, but was violent, and was succeeded by a yellow tinge of the whole body, a kind of jaundice, most apparent on the face, neck and chest. Soon he vomited spontaneously a great quantity of fetid matter, and had some stools which relieved him greatly. When seen he was very feeble, pulse scarcely perceptible. The rest of the day was found in a state of stupor and spasmodic agitation, but he slept well at night, and next day was well. The yellow colour disappeared in a few days. (p. 191-192)

5. The potency and dosing of homeopathic remedies tends to be a point of confusion for those outside of the field of homeopathy and may appear counter-intuitive to those trained in and familiar with Western medicine. Doses of homeopathic remedies are typically infinitesimal and follow a philosophy that considers the most dilute preparations to be the most therapeutically potent. Potencies are generally prepared at the centesimal level, where 1C represents a preparation that is one-part of the "mother tincture" and 99 parts water or alcohol. Standard potencies include 30C (diluted 30 times), 200C (diluted 200 times), and 1M (diluted 1,000 times).

Chapter 25

Agaricus Muscarius

Horace P. Holmes, MD, Omaha, NE (1894)

This remedy, though not appearing in Hahnemann's *Materia Medica Pura* [1846], is the initial remedy given by him in his *Chronic Diseases* [1828], and it is there classed by Hahnemann as an antipsoric [remedy for itchy/irritated skin]. The remedy was first proved by Shreter and Stapf, later by Hahnemann and his students. Apelt followed with a still better proving and Hartlaub added the provings of Drs. Woost and Seidel. From this collection of material Hahnemann gathered the 715 symptoms which forms the article above referred to and to be found in the 1845 American edition of *Chronic Diseases*.

From a criticism published in Clotar Muller's *Quarterly* in 1859, many of the symptoms of the provings were deemed unreliable and stricken out. But the re-proving of the remedy by the Vienna Society confirmed the symptoms which had been questioned and they were reinstated.

Agaricus Muscarius is the name used by Hahnemann in his *Chronic Diseases* and by Hering in his *Condensed Materia Medica* [1877]. The latter author, however, in his *Guiding Symptoms* [1879], adopts the title Amanita, but does not state his reason for changing the name. Dr. T. F. Allen [1874], in his *Cyclopeodia of Pure Materia Medica*, incorporates provings of nine different members of this family and gives to Agaricus Muscarius the simple title of Agaricus. In this latter article are 2,496 symptoms gleaned from the authorities to date and 48 references given.

In the opinion of the writer, Agaricus Muscarius is a remedy but little used by the great mass of our homœopathic physicians. It seems to be seldom thought of in the many diseased conditions to which it is applicable. Again, physicians too often mentally limit the field of a remedy to the few affections to which they personally know it to be applicable. In my own experience I limited this remedy, for several years, to those dyspeptic troubles in which I found the symptom "*relief from eating*" a prominent characteristic. I had my attention called to Agaricus Muscarius in a case of atonic dyspepsia where there seemed a strong suspicion of cancer. The patient was a man of nearly forty years, was pale, haggard, anaemic, lean and lank. There was nausea, poor appetite, loss of spirits, irritable disposition, with faintness and languor in the forenoon equal to Sulphur. Over all these symptoms the immediate *relief from eating*—even a cracker or a crust of bread—was prominent. After several remedies, fairly well indicated, had been prescribed with little benefit, Agaricus Muscarius was prescribed in the 3^x and the relief was something almost magical.[2] The remedy repeated at infrequent intervals seemed to effect a perfect cure. At least the patient is still living, now ten years since I prescribed for him, and in good health as far as his stomach is concerned.

In this case I would say the analogues of Agaricus appeared to be Arsenicum and Sulphur, and in dyspetic cases I would rank it with those two remedies and Nux Vomica and Lycopodium. It certainly is one of the grandest dyspepsia tonics we have. The peculiar symptom "much hunger but no appetite" occurs for you to wrestle over. The "all gone" feeling of Sulphur. The "soon satisfied" feeling of Lycopodium and also the "sleepiness after eating" of Lycopodium. The nausea, vomiting, burning in the stomach and thirst of Arsenicum. The vertigo, eructations and constipation of Nux Vomica.

My next use for Agaricus was in twitching of the eyelids. So many times have I used this remedy successfully in blepharospasm that I seldom think of any other, though the symptom is common to many remedies, especially Cicuta and Belladonna. Twitching of muscles is a characteristic of Agaricus and it makes little difference if the offending muscle be in the eyelid or elsewhere, the remedy is to be thought of.

One of my greatest successes with Agaricus was in a case of a little girl of seven years. It was about as complicated a case as I was ever called upon to treat. There had been periodic attacks of asthma; hay fever came annually and with it chorea. At one time there was the most serious endocarditis I have ever met with. This was brought under control by my friend, Dr. Hawkes, most beautifully with Lycopodium 1^m. Later on the chorea remained very troublesome when I found Agaricus to cover the case thoroughly. It was given in the 1^m and the result was marvelous. Since then there have been threatenings of the malady to return but it has been kept off by a few doses of the 200th [200C]. The father of the child has demonstrated positively that the 30th [30C] potency aggravates so that it is worse than useless in the case.

Chorea is one of the principal affections calling for Agaricus. Probably its nearest analogue in this trouble is *Mygale lasiodors*. The Cuban black spider. This latter remedy should be very carefully studied as it will be found very useful in nervous affections and especially those of choreic type.

The skin symptoms of Agaricus are often called upon to differentiate the remedy in nervous affections. These are the affections typical of frost-bite. There is burning, itching, redness, swelling. In many cases the terrible discomfort from chilblains I have relieved the trouble by applying Agaricus locally. This practice I find is more called for in rural districts than in the city as our city people do not seem to get frost bitten so frequently. Intense itching of the skin is likely to call for Agaricus in any of the skin diseases. There may be miliary eruptions or hard nodules, sebaceous tumors or carbuncles, phagedenic or carious ulcers. Some years ago I had a horse taken sick with a disease new to the veterinarians and termed by them scarlatina. It was two years after the "pink eye" epidemic in Illinois. I find the prescription quite accurately given in two symptoms of Agaricus in *Guiding Symptoms under Skin*: "Small nodules deep in skin, with cough, especially when eyes are also affected (horses)." The provings of Agaricus certainly show that it would be a grand remedy in the pink-eye and scarlatina of horses.

In perhaps no class of affections does Agaricus prove more tonic under homœopathic treatment than in sexual difficulties and especially loss of virility in the male. As the immediate drug effects are wildly stimulating and intoxicating, so the reverse effect is true—complete lassitude and languor. Under homœopathic prescribing one can expect from Agaricus as great benefit in sexual stimulation as the old school claim for Damiana. In the sexual sphere of woman there does not seem to be as many indications, probably for the reason that most all the provings have come from the male sex.

What might have been first spoken of, are the mind symptoms; but I preferred to deal with this remedy first in the line of my personal experiences. Agaricus is one of the wildest remedies to be thought of in the mind symptoms. Remembering the wildest symptoms of delirium tremens, and we have a possible picture of Agaricus. The Russians[3] make a drink from this variety of toad-stool and the intoxication is rapid and intense. So potent are the effects and so great the craving for this stimulant that men have been known to drink the urine of those intoxicated with Agaricus in order to gain the stimulant where the supply has been exhausted.[4] The stimulating effect of such urine seems about as potent as the original draught. The intoxication is wild, gay, dancing, loquacious, with prodigious strength and a general magnifying of distances and objects. The subject will jump high to get over a small object or far to get over a small hole that appears to him to be a frightful chasm. The intoxication is followed by a deep sleep that leaves the subject greatly depressed. In delirium with constant raving and efforts to get out of bed, this remedy will do good work. When the nervous system is affected by diseases so there are twitching and jerking of muscles with convulsions threatened or in fact, Agaricus is indicated. It is here analogous to Belladonna, Cina, Stramonium and Hyoscyamus. It has the delirium of the above remedies, the poisonous symptoms of Lachesis and Tarentula, the cold, icy feelings of Calcarea carb. and Veratrum album, and the pains of Pulsatilla and Rhus.

Dr. Th. Ruckert wrote an essay comparing the symptoms of Agaricus with those of incipient tuberculosis. It would be well for us to keep in mind these symptoms and to carefully compare them with those of Cetraria Islandica—both of which remedies promise much in tuberculosis.

In closing, I would say to bear Agaricus in mind especially in nervous troubles, deliriums (whether from disease or intoxicants), dyspeptic difficulties, especially of an atonic character, and tubercular affections of the lungs in the earlier stages.

Notes

1. This chapter is excerpted from an entry in Vol. XXXI of *The Medical Advance: A Monthly Magazine of Homoeopathic Medicine*, published 1894 in Chicago, Illinois.
2. Ed Note: Roman numerals are used in homeopathy to indicate the degree of dilution used for a particular remedy. An "X" preparation contains one-part of the homeopathic substance to nine-parts water or alcohol, a 1/9 ratio, whereas a "C" preparation is diluted by a 1/99 ratio and an "M" preparation diluted by a 1/999 ratio.
3. Ed Note: Here the author refers to the well-known use of *Amanita muscaria* among some Siberian peoples, not to Russians generally.
4. Ed Note: The author appears to interpret the practice of urine-drinking as an act of desperation brought on by the addictive qualities of the mushroom. It is understandable that someone from a culture that regards urine with disgust might come to this conclusion rather than considering other explanations, such as scarcity, an advanced understanding of the mushroom's pharmacology, or even the existence of different cultural attitudes towards urine and its consumption. Contrary to the author's suggestion, *Amanita muscaria* is not known to produce physical or psychological dependence.

Chapter 26

Fly Agaric as Medicine:
From Traditional to Modern Use

Kevin Feeney

The fly agaric (*Amanita muscaria*) mushroom has been used medicinally for hundreds of years among tribal peoples in Siberia, as well as in parts of Scandinavia, Eastern Europe and Russia, where it has been used both topically and internally for its analgesic, anti-inflammatory, anxiolytic, and stimulant properties. Recent research on the mushroom's pharmacology supports these traditional uses of this mushroom and has also demonstrated that certain compounds exhibit anti-tumor and memory-protecting activities. While these are promising developments in understanding the ethnomycological uses of this mushroom, most of the world considers *Amanita muscaria* to be poisonous. Despite this general view, anecdotes of self-medication with this mushroom have been cropping up online in topical Facebook groups, discussion boards, blogs, and other social media platforms, with individuals claiming to treat symptoms related to Lyme Disease (fatigue, cognitive deficits), tinnitus, substance dependence/withdrawal, depression and other conditions. While these accounts are not prolific, they suggest an increased public interest in this mushroom and its potential therapeutic applications.

This chapter seeks to examine some of the emerging therapeutic uses of this mushroom, as well as how and why it is being used, and whether it is perceived as effective. To this end, thirty surveys and five interviews have been conducted with individuals reporting therapeutic use of this mushroom. These reports will also be evaluated to determine whether self-treatment practices parallel known traditional therapeutic uses or are otherwise supported by current scientific literature on the properties of *Amanita muscaria* and its pharmacological constituents. The findings of this study suggest that some applications of the fly agaric may be therapeutically

A Brief History of Medicinal Use

The fly agaric is a mycorrhizal mushroom that grows in association with birch, pine, and other trees throughout the northern hemisphere. Famous for its bright red cap and snow-white spots, it is likely that humans took early notice of this mushroom when appearing in their environment. Many may have moved on from this mushroom after experiencing nausea, vomiting, or bizarre cognitive and perceptual changes following its ingestion, but others found therapeutic value in small, measured doses. Evidence for this use can be found in different parts of Europe, Russia, and Siberia, though traditions of fly agaric use in Siberia are the most well-known.

Fantastic stories of mushroom inebriation began filtering out of Siberia in the 17th and 18th centuries from explorers, soldiers, and prisoners of war (*see* Wasson, 1968), but the more subtle uses of the fly agaric did not receive the same types of attention. Some of the primary therapeutic uses of fly agaric observed in Siberia are its use as a stimulant, an analgesic, an anti-inflammatory, an anxiolytic, and as a sleep aid. One of the more interesting uses of this mushroom is as a stimulant, providing the user with stamina and staving off fatigue. Here, it is used primarily as an aid to work, perhaps similar to the use of coffee in the West, though with the added benefit of making mundane tasks more interesting and tolerable. Siberian reindeer herders use dried fly agaric for strength and energy to help them keep up with the herd (Lincoff, 2005; Salzman et al., 1996; *see also* Ch. 6). A Koryak man interviewed by Adolph Erman (*see* Wasson, 1968: 53) in the mid-1800s explained "[i]n harvesting hay… I can do the work of three men from morning to nightfall without any trouble, if I have eaten a mushroom." A Chuckchee woman interviewed by Takashi Irimoto (2004: p. 49) explained that she and the other women "could progress quickly with their work if they ate [fly agaric] before or while they were tanning reindeer hides." Dried fly agarics are also used by the Khanty to treat psychophysical fatigue (Saar, 1991b: 176). While this practice appears to have been common among some Siberian peoples of the Russian far-East similar uses in different regions of the world are unknown. In contrast to its use as a stimulant, the fly agaric has also been used in Siberia to aid sleep and treat insomnia (Irimoto, 2004; Kopec, 1837; Lincoff, 2005), for which its effectiveness has been attested

Fig. 1: Fly agaric cream used to treat joint pain and inflammation. Produced in the Temple of the Prophet Elijah, in Krasnodar, Russia.

(Cossack, 1998).

The Evensk and Koryak are known to use fly agaric as a poultice to treat pain and inflammation (Lincoff, 2005; Salzman et al., 1996; *see* Ch. 6). Rather than using a poultice, Chuckchee women eat the dried mushroom to relieve muscle pain and soreness from long hours of tanning hides (Irimoto, 2004: 227). Anxiolytic uses of the mushroom are found among the Khanty and Koryak, who use the mushroom to relieve anxiety and embolden the user. Among the Khanty there is a tradition of performing "heroic epic songs" and singers would frequently consume the fly agaric to help animate their performance and reduce performance anxiety (Saar, 1991a: 164). This practice was also present among the Koryak, as Russian anthropologist Waldemar Jochelson recounts: "Once I asked a Reindeer Koryak, who was reputed to be an excellent singer, to sing into the phonograph. Several times he attempted, but without success… After eating two fungi, he began to sing in a loud voice, gesticulating with his hands" (Saar, 1991a: 164). Jochelson's account further suggests that the fly agaric either induces courage, reduces inhibition, or both.

This ability to impart courage (or reduce anxiety) also played an important role in the shamanic use of this fungi. Among the Khanty, bears have been observed eating the fly agaric "during the rutting season in order 'not to fear,'" a practice that has been compared with the shaman "who eats [fly agaric] not only for foretelling the future after having dreamed frightening events but also in order to encourage himself before meeting several spirits" (Saar, 1991a: 163, *quoting* Kulemzin, 1984). In this way, the fly agaric not only provides the shaman with the means for his otherworldly travels, but also imparts the courage necessary to embark on the journey in the first place.

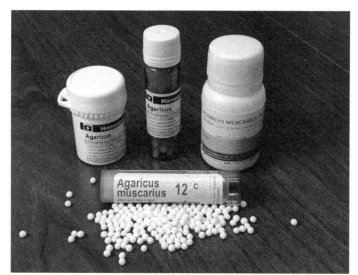

Fig. 2: A number of homeopathic preparations of fly agaric (Agaricus muscarius) are available for the treatment of a variety of conditions, including confusion, vertigo, and chilblains, among others.

In Europe and Russia, fly agaric used for medicinal purposes was typically applied topically, as a tincture, or in homeopathic remedies, though there are some exceptions. In the 1980s, Russians and Ukrainians from the Sukhodol River Valley were still using home preparations of fly agaric topically to treat joint ailments (Moskalenko, 1987: 236). Fly agaric is still popular in parts of Russia as a topical remedy, though now a variety of creams and ointments can be bought for home use (Figure 1). In the 1990s, ecologist and mycologist Marja Härkönen (1998) discovered that fly agaric was still used by the Karelian people of northwestern Russia as a home remedy. The Karelians would remove the red skin of the cap and soak it in alcohol. This could be used topically to treat bruises or other pains or taken internally in small amounts to treat headache or stomachache. In parts of Eastern Europe and Russia an infusion of fly agaric has also been used to treat rheumatic pains and there are also reports of its use to treat epilepsy and various nervous disorders (Dunn, 1973; Rolfe & Rolfe, 1974). Within homeopathy the fly agaric has been used to treat a variety of ailments, including those related to pain and insomnia (for more on the use of fly agaric in homeopathy *see* Feeney & Mann, Ch. 24 *and* Holmes, Ch. 25; Figure 2).

Fly Agaric Pharmacology

While cultural symbols, rituals, and belief systems all play an important role in the healing process (Feeney, 2014) a plant's pharmacological make-up is what typically draws the interest of western scientists. Although pharmacology cannot fully explain the processes of healing, it is important to understand what the active pharmacological constituents of a plant or fungus are as well as understanding how they work within the human body. For the fly agaric, the primary constituents that concern us are muscimol, ibotenic acid, muscarine, fucomannogalactan, β-d-glucan, and Vitamin D (*see* Maciejczyk, Ch. 22).

Muscimol and ibotenic acid are the principal components responsible for the inebriating effects of the fly agaric. Of these, muscimol, a GABA-A agonist, appears to be the component of primary therapeutic importance. Muscimol has been shown to have several potential therapeutic applications, however, studies with this compound have typically been limited to experiments on mice and rats where muscimol is injected as a pure compound rather than being ingested as one component of a pharmacologically complex mushroom. Still, the findings are interesting and appear to support some of the historical and cultural uses of this mushroom described above. One study, for example, found that muscimol injected into the spinal canal of rats significantly reduced levels of neuropathic pain, though effects were short acting and diminished over the course of three hours (Hosseini et al., 2014). Another study found that muscimol, when injected into the body-cavity of rats, produced anti-anxiety effects "of similar magnitude to that of diazepam" (Corbett et al., 1991: 312). This anti-anxiety effect is supported by another study which found that muscimol reduced the fear or "freezing" response in rats (Muller et al., 1997).

Muscimol also appears to enhance and protect memory by reducing brain inflammation and by reducing rates of acetylcholine metabolism in the brain (Pilipenko et al., 2018). Acetylcholine is recognized for its role in enhancing memory; slowing the rate at which the brain metabolizes this neurotransmitter can help ensure that adequate amounts of acetylcholine are available. Finally, another study has demonstrated that the incidence of gastric cancers in rats artificially inducted with gastric carcinogenesis is reduced by prolonged treatment with muscimol (Tatsuta et al., 1992).

Interestingly, muscarine may also play a neuroprotective role. While the symptoms of muscarine are frequently the more unpleasant (though inconsistent) effects produced by the fly agaric, including symptoms such as excessive perspiration or salivation, nausea and vomiting, a recent study has shown that stimulation of the muscarinic receptors in the brain "provides substantial protection from DNA damage, oxidative stress, and mitochondrial impairment" (Sarno et al., 2003:

11086). It may be that small amounts of muscarine provide necessary amounts of stimulation to muscarinic receptors, thereby preventing neuronal loss or the development of neurodegenerative diseases.

The next major pharmacological constituents to consider are fucomannogalactan and β-d-glucan. These constituents haven't been studied as widely as muscimol but do show some overlap with muscimol in terms of potential therapeutic applications. A recent study demonstrated that both fucomannogalactan and β-d-glucan "produced potent inhibition of inflammatory pain" and found that β-d-glucan effectively reduces neurogenic pain (Ruthes et al., 2013: 761). Another study on the medical properties of β-d-glucan extracted from *Amanita muscaria* found that β-d-glucan "exhibited significant antitumor activity against Sarcoma 180 in mice" (Kiho et al., 1992: 237).

The final constituent of fly agaric that might be considered medically or therapeutically important is Vitamin D. While not normally regarded by the public as a drug or a medical treatment, Vitamin D is an essential component of the human diet and confers significant health benefits. A deficiency in Vitamin D levels can lead to fatigue, depression, back pain, muscle weakness and an increased vulnerability to illness or infection. Adequate amounts of Vitamin D in the diet can prevent these deficiency-related symptoms and help optimize muscle strength and prevent muscle and bone pain (Grant & Holick, 2005). There is also significant evidence that "[s]ufficient vitamin D levels in adulthood may significantly reduce the risk for many types of cancer" (Grant & Holick, 2005: 97). The high levels of Vitamin D in fly agaric mushrooms may contribute to its use and importance among Siberian tribes and others with traditional uses of this mushroom who live above the 60th parallel where sunlight is limited, and Vitamin D becomes a necessary dietary nutrient.

Based on the traditional and cultural uses of fly agaric as a medicine and tonic as well as the brief overview of the mushroom's pharmacological constituents provided above, there is plenty to suggest that this mushroom *might* be used efficaciously to treat some medical complaints.

Methods

Data was collected for this study using online surveys that were advertised in Facebook groups and online discussion boards devoted to the topics of mushrooms, mycology, entheogens and psychedelics. Surveys were conducted online using Qualtrics and included question categories related to demographics, health, *Amanita muscaria* use, healthcare, and others. A total of thirty surveys were completed and deemed acceptable for purposes of this study. Of the thirty survey respondents, five were interviewed with the goal of adding qualitative depth to the collected survey results.

Results

Survey respondents were asked to identify any conditions they had treated with fly agaric, or any conditions for which they experienced relief or an alleviation of symptoms following use of fly agaric. Respondents were *not* provided a list of conditions to choose from and instead were required to self-identify the conditions they treated. The following conditions were the most frequently identified by respondents:

- **Pain** — **53%** (16/30)
- **Depression** — **50%** (15/30)
- **Anxiety** — **40%** (12/30)
- **Insomnia** — **20%** (6/30)
- **Fatigue** — **17%** (5/30)
- **Addiction/Withdrawal** — **13%** (4/30)

Many respondents reported treating several different conditions which explains why the numbers above add up to over the number of total participants. Several other conditions were also identified, including epilepsy, muscle spasms, and dementia/cognitive dysfunction, but were only identified by one or two subjects each.

Respondents were next asked to comment on the effectiveness of their therapeutic use and, in order to acquire a more objective measure, were also asked whether they had been able to reduce or eliminate any medications they had previously been taking to treat their identified condition. The bar graph provided below (Figure 3) illustrates the differences between the numbers of individuals who treated a particular condition, those who felt the treatment was effective, as well as the number of respondents who reported reducing or eliminating the use of other medications as a result of their *Amanita muscaria* use.

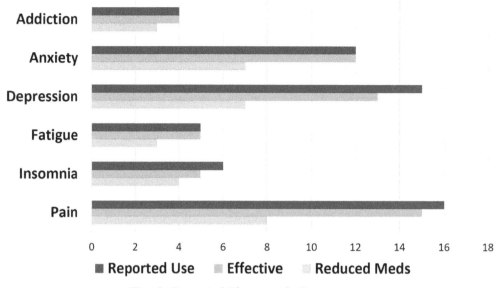

Fig. 3: Reported Therapeutic Purpose.

While the sample size of the study is admittedly small it is interesting to observe the high number of respondents who reported that their treatment was effective. And though the number who reported reducing medications is less than the numbers reporting effectiveness, the numbers remain impressive with between 47% and 75% of respondents reporting reductions, for an average of 55% across conditions. If these results could be repeated on a larger scale it would be remarkable, to say the least.

Demographically, the diversity of respondents was limited. Over 80% of respondents were white and 70% identified as male. Americans comprised over 60% of respondents, but participants hailed from a variety of other locales, including the United Kingdom (10%), Australia (7%), and one each from Canada, Ecuador, Finland, Portugal, and Russia. The group of respondents was also highly educated, with 57% reporting a college degree or higher, and up to 87% reporting at least some college completion. Finally, the median and modal age for this group was individuals aged from 35 to 44 years old.

One of my interests as a researcher was to determine what factors inspired people to seek out *Amanita muscaria* for use in a therapeutic capacity. Respondents were given a list of ten possible options and could select multiple options. The most commonly identified factors are shown in Figure 4.

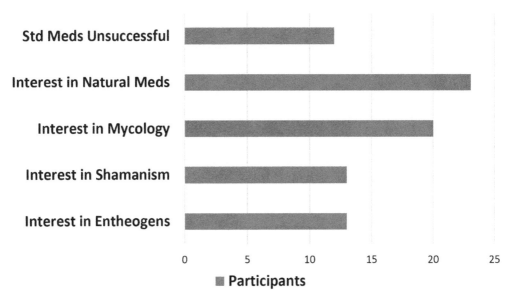

Fig. 4: Inspiration for Therapeutic Use.

Several other questions were asked of respondents to determine whether subjects would respond in a manner corroborating the factors identified above as inspiring or influencing their decision to use *Amanita muscaria* in a therapeutic way. These questions included the following: (1) *Do you, or have you ever picked mushrooms from the wild?* (2) *If so, what types of mushrooms do you typically pick?* (3) *Where did you acquire Amanita muscaria mushrooms?* (4) *Have you had any luck treating your condition or symptoms with modern medicines?* (5) *Do you use any other mushrooms medicinally?* (6) *Do you use any other herbal remedies?* And (7) *do you rely more on modern or alternative and complementary medicines?* The responses to these questions appear to conform to and support the factors that respondents identified as influencing their choice to use *Amanita muscaria*.

Approximately 87% (26/30) of respondents reported a history of mushroom collection, with 92% (24/26) of *these* respondents reporting collection of edibles, 81% (21/26) reporting collection of medicinals, and 38% (10/26) reporting collection of psychoactive *Psilocybes*. Additionally, 73% (22/30) of all respondents reported picking their own *Amanita muscaria* mushrooms. The remainder of these questions focused on health and medical care, with 73% (22/30) of all respondents reporting little to no success treating their symptoms and conditions with standard medical treatments. Sixty-three percent (19/30) of respondents reported use of other medicinal mushrooms while 87% (26/30) reported using other herbal remedies.

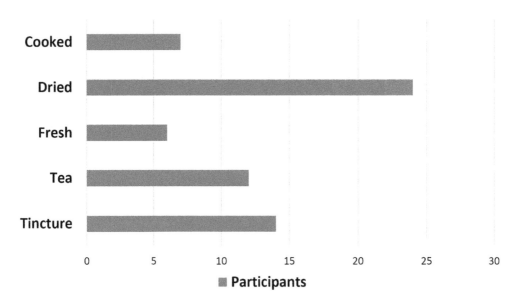

Fig. 5: Preparation of Fly Agaric for Therapeutic Use.

Of all the respondents, 83% (25/30) reported using alternative and complementary medicines alone or in combination with modern medicines to meet their healthcare needs.

The final category of results to address before moving on to the discussion concerns how the mushroom was prepared and used. Respondents were asked to select the preparation type they used from the following categories: *Fresh*, *Cooked*, *Tea*, *Tincture*, *Dried* (Figure 5). The primary methods of preparation employed by subjects was to use the mushroom in its dried state (38%), to make a tea (19%), or to ingest it or apply topically as a tincture (22%). While "smoking" was not provided as an option, two participants also mentioned using the mushroom in this manner.

The dosage and frequency of use of the mushroom also varied significantly. The dose of dried mushrooms reported varied from a piece of dried cap the size of a grain of rice up to 17 g of dried material, an extremely high dose which was reported in the treatment of severe pain. While there was great variance there also appeared to be some consensus. Seven of the participants identified doses between 1 and 5 g (or less) of dried material to be sufficient. While the potency of fly agaric mushrooms varies from mushroom to mushroom, 5 g dried is typically considered a threshold psychoactive dose, and doses above this amount could lead to impairment or otherwise interfere in daily activities. The largest reported doses were 10 to 17 g for pain, and 5 to 15 g for insomnia. These are clear outliers and represent doses that tend to produce mild to strong psychoactive effects in most people.

Of the remaining twenty-one participants, eight reported that they were unable to determine a consistently appropriate or effective dose; three offered no response; seven reported using tincture; one reported using small amounts topically; one reported smoking one to two puffs; and the final participant reported adjusting the dose until it "felt right." Individuals using tincture reported using several drops up to 2 ml, though most did not report specifically on the potency of their tinctures. However, one participant reported using up to 2 ml of a tincture with a 100 mg to 1 ml ratio, and another reported using up to 1 ml of a tincture prepared from 1-part mushroom and 1-part 80 proof vodka.

In regard to frequency, some reported daily use, others used two to three times a week, while others would only use the mushroom once every few months. Several participants reported using the mushroom regularly until their supplies would run out. Other participants expressed uneasiness about using the mushroom regularly due to concerns about potential poisoning or unknown long-term consequences of regular use.

Discussion

While there is a lot of data to sift through the small number of participants significantly limits the value of quantitative analysis. Fortunately, there is a fair amount of qualitative data as well as a fair degree of scientific research that can be drawn upon to evaluate the reported therapeutic applications of *Amanita muscaria*, as collected in this study. To begin the discussion, I will be examining the different reported applications of this mushroom, exploring the effectiveness of the reported treatments, and reviewing the scientific literature for potential corroboration of reported therapeutic outcomes. This will be followed by a discussion of the research subjects and their motivations for using *Amanita muscaria*.

Addiction / Withdrawal

Four study participants reported using the fly agaric to help treat substance dependence and to treat or reduce withdrawal symptoms. The substances identified by these four subjects included alcohol (1), benzodiazepines (1), and opioids (2). All four participants reported that use of fly agaric was effective in reducing withdrawal symptoms associated with the above substances, and three of the four subjects reported that their fly agaric use allowed them to wean and eventually discontinue their problematic use of these substances. One participant reported that "[A]fter 20+ years and multiple attempts to quit [alcohol], *Amanita muscaria* was the tool that was able to minimize cravings and mitigate… withdrawal symptoms." Another participant who had been using fly agaric to treat pain and injuries resulting from a motorcycle wreck reported that prior to using fly agaric he had been on "a monthly prescription of hydrocodone [10 mg daily], Percocet [30mg daily], methadone [10mg daily] and dilaudid [4mg daily]. $240 a month. I quit them all for obvious reasons." Interestingly, another subject who reported use of fly agaric for unrelated issues reported "I no longer felt I needed to use as much Cannabis or alcohol" after using fly agaric daily for six weeks. Another subject, who had been using fly agaric for insomnia, reported that taking the mushroom "helped me to limit alcohol consumption. Something about the effects of muscimol kills my cravings for alcohol, an unexpected side effect."

While there is no known use of fly agaric, traditional or otherwise, to treat cravings or withdrawal symptoms associated with substance dependence, the unique pharmacology of muscimol, a GABA-A agonist, may help explain the successes reported here. GABA (*gamma*-Aminobutyric acid) is an important neurotransmitter in the brain and has an inhibitory effect on the central nervous system. Drugs that are GABA agonists, meaning they bind to various GABA receptors in the brain, typically produce sedative, anxiolytic, anti-convulsant, and muscle relaxing effects. While alcohol and benzodiazepines are not GABA agonists they are both *GABA-A receptor positive allosteric modulators*, which means that when ingested they increase the activity of GABA-A receptors in the brain which ultimately depresses the central nervous system, easing anxiety and producing a sedative type effect (Caputo & Bernardi, 2010). Benzodiazepines are frequently used to treat alcohol withdrawal but are similarly addictive (Caputo & Bernardi, 2010). Muscimol, on the other hand, has the advantage of being a GABA-A agonist that is non-addictive. In this way, muscimol can activate the brain's GABA system which might ease the symptoms of alcohol or benzodiazepine withdrawal while simultaneously reducing the body's dependence on these substances. While the sample size here is small, each subject reported that fly agaric eased withdrawal symptoms, and with what we know about this mushroom's pharmacology, further investigation into its potential use to treat withdrawal symptoms seems warranted.

Anxiety

Twelve of the study's participants reported using fly agaric to treat and ease anxiety. Each of these subjects reported that this treatment was effective in reducing their anxiety and seven (58%) reported that use of the fly agaric had allowed them to decrease other medications. One subject reported "I don't need to take Benzodiazepines to control panic attacks as often due to having a natural alternative." Another explained his use of fly agaric as being "[e]xtremely effective as a natural benzodiazepine alternative to help alleviate Generalized Anxiety Disorder. Taken on a PRN [as needed] basis when anxiety levels are high and need bringing down to more manageable levels." A third subject reported that fly agaric "helped me taper off of Levetiracetam [seizure medication] and Diazepam [a benzodiazepine] …" The ability of these individuals to reduce their use of habit-forming benzodiazepines and simultaneously treat excessive anxiety is also likely connected to the nature of muscimol as a GABA-A agonist. As stated before, GABA-A agonists, like muscimol, bind to GABA receptors in the brain and produce effects that limit and reduce anxiety. Because muscimol is non-addictive it would be worth further investigating its potential as an anxiolytic for treatment of anxiety.

Depression

Depression was one of the most common reported conditions that participants sought to treat using the fly agaric mushroom. Of the fifteen participants that reported using fly agaric for depression thirteen reported that it was effective in allaying depressive symptoms, though only seven (47%) reported that fly agaric use enabled them to reduce the use of other medications for treating depression. One participant reported that after using a low daily dose of fly agaric for nine months he experienced his "[f]irst time free of depression, depressive behavior. First time since the age of four having no suicidal plans whatsoever." The same individual reported that "[t]raditional SSRIs and trintellix were ineffective" in treating his severe depression. Another individual explained "I sometimes have to take SSRI medications (usually more so in winter months) supplementing with *amanita* helps me keep the dose of pharma meds low."

Depression is one of the more interesting conditions that presented itself in this study because there is little in the scientific or historical literature that corroborates this use. Vitamin D deficiency can contribute to depression and it is notable that one of the above subjects mentioned using SSRIs more heavily in the wintertime, which may be an indication of low Vitamin D levels. While the fly agaric has exceptional levels of Vitamin D (*see* Feeney, Ch. 21), there is little to suggest that Vitamin D could solely explain the results of this study. Researchers are investigating the relationship between Vitamin D and depression, but results concerning Vitamin D's ability to alleviate symptoms of depression have been inconclusive (Gowda et al., 2015; Parker et al., 2017). For the moment, the role the fly agaric might play in treating depression is a mystery, but a notable and interesting one.

Fatigue

Five participants reported using fly agaric to treat issues surrounding fatigue, though treatment of fatigue was typically secondary to treatment of other more prominent conditions, such as pain or depression. Despite being a secondary consideration, all five respondents reported that their fly agaric use was successful in alleviating fatigue and three reported reducing or eliminating related medications. For some respondents there was a correlation between fatigue and mood or depression. One participant noted that the fly agaric was "very effective for mood upliftment, and energy" while another explained how fly agaric was effective at "removing years of lethargy and dispair [*sic*]." Fatigue is often a symptom of depression which makes the application and success of using fly agaric to treat fatigue a little unclear. Are the reported results due to the alleviation of depressive symptoms? Or is fly agaric effective at treating these conditions independently? The traditional use of fly agaric as a stimulant in Siberia, however, would suggest that the effectiveness of the fly agaric in treating fatigue is not necessarily limited to treating fatigue rooted in depressive conditions.

Unfortunately, there is little about the known pharmacology of the fly agaric that would explain its use as a stimulant except for speculation that ibotenic acid, the pre-cursor to muscimol, might have stimulating properties. It is also possible that some participants suffered from Vitamin D deficiency, which can lead to fatigue. Use of the fly agaric could have remedied this deficiency, thereby increasing energy levels. While the exact mechanism of this effect is unclear, there appears to be plenty in terms of ethnographic evidence to support this use of the fly agaric.

Insomnia

A total of six participants reported using fly agaric to treat issues related to insomnia and sleeplessness. Of these six, five reported that it was an effective treatment and four (67%) reported reducing their intake of related medications. Regarding the effectiveness of fly agaric in treating insomnia one subject reported that "[i]n small doses of chewed and swallowed dry caps I was able to fall asleep faster and stay asleep longer. It increases my sleep pattern from the typical four to six hours, to a full seven to nine hours of what I consider to be good sleep." The same subject also reported that use of fly agaric "has allowed me to avoid OTC [over-the-counter] sleep aid and helped to limit alcohol consumption." Another subject when prompted about the effectiveness of fly agaric reported that she falls asleep "within five minutes of micro-*dozing* [*sic*]." The final participant who reported some benefit from using fly agaric for insomnia characterized it as "mildly effective."

Based on traditional use of the fly agaric in Siberia to treat insomnia I was somewhat surprised that the results here were not stronger. Of course, six individuals make an extremely small sample. It appears that muscimol, as a GABA-A agonist, is the most likely source of the sleep-inducing effects of fly agaric. Of the participants reporting successful use of fly agaric as a sleep aid the dose was typically determined by trial and error. One participant reported eating up to one or two caps before determining that "a tiny flake the size of a rice grain is perfect." Another reported using up to 10g, when unpleasant side-effects caused him to reduce his dosage to 4.5 to 5g. The participant who characterized the fly agaric as mildly effective as a sleep aid reported using between 1 and 5g. One participant, reporting reduced reliance on sleep medications, reported using fly agaric as a sleep aid for over two years, while another reported use over a six-month time period. The consistent and long-term use by these participants suggests a successful track-record of using fly agaric as a sleep aid.

Pain

Treatment of pain was the most frequently identified therapeutic use of the fly agaric. Sixteen participants reported using fly agaric for this purpose with fifteen reporting some success treating pain. While fifteen reported this treatment as effective only eight (50%) reported reductions in use of other medications and medical treatments. This category of treatment differs from those previously discussed in that participants reported both internal and external use of the fly agaric to treat pain. Interestingly, those who reported external use of the fly agaric, usually in the form of a tincture or ointment, reported the most striking and consistent results. Five of the sixteen subjects in this category reported using fly agaric externally and each of these five reported that it was effective and allowed them to reduce or limit other medications and medical treatments. One subject stated that "[m]assaging Fly Agaric ointment into affected areas [back pain] provided relief and eased tension the following morning. It's the best remedy I've found thus far." As a result of his use, this subject explained "I take less ibuprofen and get fewer massages." Another participant explained that topical use of fly agaric tincture was "Astonishing for [treating pain related to] sciatica, the problem is gone like *poof* within minutes." Another participant's description of its effects was less enthusiastic, stating that topical use provided "some pain relief" when applied to treat joint pains, muscle strain, and a knee injury. Though her perception of its effectiveness was moderate she also stated that she was "taking less ibuprofen and using less cbd [cannabidiol]" as a result. A sixth subject reported "temporary relief" from arthritic pain when using fly agaric, but did not clearly state whether this use was topical or internal.

The use of fly agaric to treat pain is one of the more common traditional uses of this fungus that we see in Europe, Russia, and Siberia. Most of these traditions also appear to favor topical application of fly agaric in the form of a poultice, ointment, or direct application of tincture. Creams and ointments made from fly agaric can currently be purchased in different parts of Russia and, relatedly, an individual recently approached me after a talk I gave on "Fly Agaric as Medicine" to share his experience of being treated with one of these ointments during his travels in Russia. He expressed surprise that such a product existed but was more surprised still by the relief from back pain that it provided. Interestingly, one of the study participants also had a Russian connection. She shared that "I have a Russian friend who introduced me to mushroom foraging. She mentioned that her mother also had bad knees and that she made a tincture out of *Amanita muscaria* and vodka for the pain." After learning of this use, the participant was inspired and stated "[w]hen I encountered a bloom of the mushrooms I figured I'd give it a try."

From a pharmacological point of view there appears to be strong support for the use of fly agaric to treat pain. Studies have demonstrated a potential role for muscimol in treating pain (Hosseini et al., 2014). Fucomannogalactan & β-d-glucan have also demonstrated analgesic and anti-inflammatory properties (Ruthes et al., 2013), and even Vitamin D can prevent and alleviate achy muscles and bone pain (Grant & Holick, 2005).

Other Reported Conditions

Participants mentioned a handful of other conditions they treated with fly agaric, including ADD, epilepsy, IBS, TBI and bulimia, but the only other condition that was cited by multiple participants was dementia and mental clarity, which I will briefly explore here. Two participants reported using fly agaric to treat issues related to confusion, dementia, or what one subject termed "brain fog." Both individuals described the effects of fly agaric in ways that suggest it was nothing short of life changing. One subject described his use of fly agaric as "very efficacious and successful in helping to rebuild [and] restore cognitive facilities... by pushing me beyond a cognitive delay plateau that is the result of more than twelve lifetime concussions." This individual reported that though he has used fly agaric daily for up to a week at a time that he will "typically eat a food dish with the [fly agaric] powder once or twice a month." The other subject reported struggling with dementia and cognitive dysfunction associated with chronic Lyme Disease. She reported that a few drops of fly agaric tincture under the tongue would quickly provide mental clarity and energy (chronic fatigue is also a symptom of Lyme Disease), and that she would periodically take additional doses during the day to maintain stamina and mental clarity. This subject reported that after going twenty-four hours without the tincture that her thinking noticeably slowed, she became more of an "observer" in social situations and felt that her mind would downshift to "autopilot." After resuming her therapeutic regimen, she felt back in control, and better able to communicate and engage with others.

Muscimol is known to reduce inflammation in the brain, a condition which can lead to memory loss and cognitive decline (Walker et al., 2019). Muscimol also reduces rates of acetylcholine metabolism, a neurotransmitter essential to memory (Pilipenko et al., 2018). These two functions of muscimol might explain the benefits experienced by the study participants described above. Muscarine may also play a role by stimulating muscarinic receptors in the brain. Regular stimulation of the muscarinic receptors may help prevent neuronal loss and the development of neurodegenerative diseases (Sarno et al., 2003).

Side-Effects

The accounts described above suggest that fly agaric may have several potentially important therapeutic applications, however, ingestion of the fly agaric is frequently accompanied by unpleasant side-effects, including nausea, vomiting, and excessive perspiration and salivation. Of course, as an inebriant, higher doses might produce additional side-effects, including euphoria, hallucinations, dizziness or lack of coordination. Depending on the person and the context these might be seen as positive or negative, though these symptoms probably indicate that the dose is too high to be therapeutically appropriate. The most common side-effect reported was euphoria (19%) followed by unsteadiness and excessive perspiration (14%), dizziness (13%), nausea (11%), and hallucinations (10%). Vomiting was the least common side-effect reported accounting for only 3% of all reported side-effects. For perspective, it should be remembered that many over-the-counter and prescription drugs also produce many uncomfortable and problematic side-effects. Because side-effects can determine whether an individual continues to use a certain treatment or not, thus affecting its overall usefulness or effectiveness, participants were asked to address whether these side-effects had an impact on their decision to continue to use fly agaric as a therapeutic treatment.

Participants had a wide variety of responses to the reported side-effects and dealt with these in different ways. Some felt that the side-effects were acceptable, such as the following subject who stated "[i]ncreased perspiration can be an annoyance, however, it's a small price to pay for relief from anxiety symptoms." Another explained "Only the perspiration was an issue. This happened only a few times, in social occasions when I took too many microdoses... I realized it was probably too much muscarine, and that I had just taken too much. It didn't put me off it." One participant stated, "nausea only lasts for a half hour and it's not that strong." Others adjusted the time of day they used fly agaric to account for side-effects, including one participant who explained "I only use at bedtime, so I wouldn't notice [the] dizziness." Still others found that certain preparations seemed to reduce or eliminate side-effects. Referencing these side-effects one participant noted that "[t]hese symptoms were never from the tincture I used, only from larger doses of dried mushrooms or tea." Here, the subject found that limiting his use to tincture was sufficient to eliminate side-effects.

Perhaps the most interesting response I received on the question of side-effects came from the following subject who explained that her therapeutic fly agaric use had "impacted what I eat, as I am now vegetarian, and those meals tend to make me feel less nauseous." For most participants the side-effects did not appear to be a deterrent; with subjects using trial-and-error to determine the right dose and optimal preparation (ie: tincture vs. dried), and others adjusting their medication schedule to account for potential side-effects.

What inspired study participants to use fly agaric therapeutically?

Study participants were provided a list of options to choose that most closely related to their reasons or inspirations for using fly agaric therapeutically. Here, as with other questions, participants could select multiple options. One of my interests with this question was to see how much of the therapeutic interest in this fungus was being driven by a primary or underlying interest in its hallucinatory or inebriating properties. For better or worse, the legitimacy of the fly agaric's therapeutic uses is likely to be judged by other researchers and by the public-at-large by the degree to which it is associated with or disconnected from recreational "drug" use. Consequently, I felt that it was important to put a finger on the pulse of participants to determine what was driving or "inspiring" the therapeutic uses reported in this study.

As reported in the results section, a significant portion of participants (43%) identified an interest in "entheogens" as a major inspiration and 33% of all participants reported a history of picking *Psilocybe* mushrooms, which would corroborate this interest. However, the fly agaric's psychoactive properties did not appear to be a primary motivating factor. Interest in natural and herbal medicines (77%) and in mycology (67%) far outweighed interest in entheogens among participants as a motivating factor. Further, between the thirty study participants a total of 105 "inspirations" were selected from the ten available options, and when accounting for every selection made "interest in entheogens" fell to just 12% compared to 22% for natural medicines and 19% for mycology.

Several interesting stories arose following this line of questioning, including one individual who discovered the fly agaric's properties through a foraging mishap. Having misidentified several fly agarics in the button stage as puffballs this participant experienced the full-blown effects of the fly agaric after cooking them up in a soup. While the experience was overwhelming, and resulted in a trip to the emergency room, the subject reported an elevated mood and alleviation of depressive symptoms lasting two months following his accidental ingestion. Another subject reported that her inspiration came from a dream. This individual explained that she had not been dreaming for awhile but one morning woke up with the phrase "red top" stuck in her head. She did not know what it meant but later recognized the "red top" when she spotted some fly agarics while on a walk. After some research she decided to give it a try and found it to be effective in treating conditions that she had had little success treating with standard medications. One individual indicated that his therapeutic use arose out of family tradition, but unfortunately, did not provide further details about the scope or length of this tradition. In any case, while some 43% of participants cited an interest in the mushroom's psychoactive properties, the primary inspirations driving therapeutic use of the fly agaric appeared to arise from elsewhere.

Limitations

There were several significant limitations to the present study, with the primary limitation being the small sample size. While the sample size is not sufficiently large to conduct any significant type of quantitative analysis, the qualitative data gathered was sufficiently interesting and informative to suggest that further research is warranted. Another limitation has to do with how participants were recruited. The targeting of topical groups in online discussion boards, forums and Facebook groups resulted in a narrow demographic make-up, both by interest and by racial and socio-economic status. While it is possible that individuals interested in and practicing therapeutic uses of the fly agaric are predominantly white, middle-class, and highly educated, the method of sampling used for this study may be a better explanation of the ultimate demographic make-up of study participants.

Participant self-selection might be another limitation. People that have something to say or who have had a positive experience using the fly agaric therapeutically might be more likely to participate in a study like this than those who tried fly agaric but found it to be ineffective for their condition or symptoms. If individuals with negative or unsuccessful experiences using fly agaric therapeutically are less inclined or are disinclined to participate in a study on potential therapeutic applications of this mushroom, then the results will be skewed.

Conclusion

Pharmacologically, the fly agaric has an intriguing profile, one that is suggestive of its therapeutic potential. Studies that have been conducted on various chemical constituents of the fly agaric in animal experiments suggest anxiolytic (anti-anxiety), anti-inflammatory, analgesic (pain killing), and anti-tumor properties. There are limits, however, in the conclusions that such studies allow us to draw. For one, the studies are conducted with pure compounds, which are typically injected and, more importantly, these studies have been conducted on rodents not humans. Nevertheless, participants in the present study report that they have used various preparations of the fly agaric mushroom and have experienced relief from anxiety, pain, and depression, among other conditions. These uses are also supported by traditional uses found in different parts of Europe, Russia, and Siberia. The fact that these uses, and reported successes, are corroborated by laboratory experiments on rats and mice is significant. At the very least, it suggests that the pure isolated compounds could play an important role in modern medicine.

Problematically, there is little financial incentive for pharmaceutical companies to produce and sell naturally-occurring compounds and medicines, which means any medical application of this mushroom or its compounds is likely to only occur within the realm of natural and home remedies. Although the fly agaric is known to be temperamental in terms of dose and to cause some potentially unpleasant side-effects, the participants of this study appeared to be unhindered by these obstacles, and many were able to arrive at appropriate doses through trial-and-error. Some reported that making a tincture or powdering large batches of dried mushrooms allowed them to establish a uniform dose that could be used until supplies ran out. New harvests of mushrooms might vary in potency, but this same process of creating a uniform potency can be used again, though trial-and-error may be required to determine the appropriate dose in any new batch.

While the results of this study are far from conclusive the results tend to corroborate traditional therapeutic uses of this mushroom as well as what is currently known about the pharmacological properties of the fly agaric's various constituents. A larger data set could add further clarity to these findings, and further study of this mushroom's therapeutic potential should be encouraged. Nevertheless, traditional and emerging therapeutic uses of this mushroom should not be ignored, and hopefully continued research in this area will shed new light on the fly agaric's therapeutic potential as well as help reduce the stigma surrounding the fly agaric as simply a "poisonous" mushroom. While the stigma of "poisonous mushroom" is unlikely to fully disappear, the present study should help to demonstrate the veracity of the old maxim "what is one man's poison is another man's medicine."

Chapter 27

Preparing Fly Agaric as Medicine

As covered in the previous chapter, the fly agaric has a significant history of use as a therapeutic agent and has demonstrated potential in the treatment of various conditions and ailments. While the therapeutically active compounds in the fly agaric are unlikely to ever be introduced into modern medicine – in part due to low-profit potential – the mushroom itself can be easily gathered and prepared for home use.

While the potency can vary significantly from mushroom to mushroom individuals using the mushroom therapeutically have found that creating a homogenous mixture of dried mushrooms by powdering large batches, or by making a tincture, can help to average out the potency of the mushrooms. Essentially, homogenization allows for a greater degree of measurability and repeatability of dose. Further, participants used trial-and-error to arrive at a dosage that was effective for treating their individual symptoms, one that was also manageable in terms of repeatability and without producing over-powering or otherwise unwanted effects. Although doses reported in the study ranged from a dried piece of cap the size of a grain of rice up to 17g of dried material, a narrower set of guidelines for dosage can be constructed based on these study results as well as

Fig. 1: Dried *Amanita muscaria* in a grinding mortar (Photo by Inge Geurts).

upon anecdotal and other data regarding the potency of muscarioid *Amanitas*.

The consensus dose arising from the study in the previous chapter falls between one and five grams of dried material. Five grams is typically considered a threshold psychoactive dose for muscarioid *Amanitas* and holds the potential to interfere with daily activities such as driving. The highest doses recorded in the study were reported by individuals treating insomnia and severe pain. For individuals treating insomnia a psychoactive dose might be acceptable since the psychoactive effects are unlikely to interfere with daily activities. For those dealing with debilitating pain, higher doses might be appropriate, particularly where pain is the primary barrier to engaging in normal daily activities. For most, the sweet spot is likely to fall between one and three grams of dried material, which should be sufficient to produce a therapeutic effect but not so high a dose that one needs to cancel their plans for the day. In any event, some trial and error will be required for each individual to determine the optimal dose for treating their symptoms or condition. Start with no more than ½ gram and work your way up (or down) ½ gram at a time until you discover your optimal dose. If you reach five grams without noticeable therapeutic benefit it is time to consider whether this mushroom will be an appropriate remedy for you since the benefit of higher doses may be offset by the psychoactive effects of the mushroom.

As noted previously, the primary methods for using the fly agaric therapeutically include using the mushroom dried, as an infusion or decoction, taking it in tincture, and applying extractions of the mushroom topically to treat pain and inflammation. Below I will discuss several of these methods in more depth as well as providing some guidelines for using these different preparations therapeutically.

Infusions & Decoctions

Taking the fly agaric as a "tea" is almost as easy as simply eating the dried mushrooms, but there are important differences between making an *infusion* and a *decoction*. An infusion is made by steeping your mushroom material in hot water, just as one would brew a cup of *tea*. The hot water will help release the active principles of the mushroom into the tea, but the final beverage will not be as potent as a decoction. An infusion may be preferred for reasons of "taste," and where a mild dose is preferred.

A decoction is made when your dried material is brought to a boil and boiled for a period of time. Boiling your mushrooms for 15 to 20 minutes should be sufficient to release nearly all of the desired compounds into your drink. In either case, you will want to strain the mushroom solids from your drink and discard them. Consuming left-over mushroom solids will contribute nothing to the therapeutic effects of the beverage and may actually increase the potential for gastrointestinal distress (Feeney, 2010). In any case, here is what you will need:

- 1 – 3 g dried mushrooms (homogenized)
- 1 cup water

If making tea (infusion), carefully measure your crushed or powdered mushrooms and place in a muslin bag or a tea ball. Pour one cup of boiling water over your mushrooms and allow to steep for 5 to 10 minutes. Remove the tea bag and drink.

If making a decoction, bring 1 cup of water to a boil. Measure and add your crumbled mushrooms to the boiling water. Boil for 15 to 20 minutes. Add a quarter cup of water if the water level gets too low during the boil. Strain and let sit for 5 to 10 minutes. Drink.

Tincture

Making tincture is a fairly straightforward process, though it requires about a month to produce a good quality tincture. The wait is worth it, however, since a tincture will be a homogenous solution, making doses easy to measure, and a couple drops can be taken straight or in a glass of water anywhere you go. Here's what you will need:

- 1 – 2 oz dried mushrooms
- Vodka (80 or 100 proof)

Fill a one-cup jar to the top with dried mushrooms. Pour vodka over the top until the jar is filled. Replace the lid and store in a dark cupboard overnight. Open your jar the next day and top off with vodka until mostly full. Close and place in a dark and cool cupboard for 1 month. After one month strain out the mushroom pieces and pour your tincture into a clean jar or dropper bottle. With tincture you will want to start with a few drops and work your way up until you find a dose that works for you. Be sure to shake or stir your tincture before use.

An alternative method is to begin with left-over cooking water. If you plan to detox your mushrooms for cooking (*see* Ch. 21), set aside any water you used to boil and detox your mushrooms and combine in a large pot. This water should not have been salted or had any flavorings added to it in the cooking process; it should be plain mushroom water and be free of any dirt or mushroom pieces. Boil your leftover water until you have about a cup of liquid left. Combine this in equal parts with the highest proof alcohol you can find; 190 proof is preferred. After combining you will be left with a tincture that is between 80 and 100 proof. Store in a cool and dark place.

Topical Use

One of the most popular and seemingly effective uses of this mushroom is as a topical for treating pain and inflammation. For topical applications, it is best to first prepare a tincture, as described above. Once you have prepared your tincture, several drops can be applied directly to any problem areas. Alternately, tincture can be added to massage oil (or any number of oils) for a more thorough application.

To create a more potent and faster acting topical one can add fly agaric tincture to a pre-made DMSO topical cream. DMSO facilitates absorption through the skin which will speed up the action of the tincture and DMSO also facilitates the decarboxylation of ibotenic acid to muscimol, which may result in a more therapeutically active cream (Filer et al., 2005; Filer, 2018). It is unclear whether DMSO that is already suspended in a topical application will effectively convert ibotenic acid to muscimol, but it might. To maximize the potential conversion, add the desired amount of tincture to your DMSO cream, mix and allow to sit at room temperature for at least a week. An added benefit is that DMSO has also been shown to assist with inflammation and pain relief. A note of caution, however. When using DMSO the site of application needs to be immaculately clean. If your skin is covered in lotion, perfume, insect repellent, soap, or anything else DMSO will cause these substances to pass through your skin and possibly enter your bloodstream. As long as you are mindful, topical applications containing DMSO should be safe.

Insecticide

While use of fly agaric as an insecticide falls outside of the main topic of this chapter it is nevertheless a home remedy and given the mushroom's history and moniker a brief discussion of this folk method of pest control is warranted.

Recently, the efficacy of the fly agaric as an insecticide has come into dispute (Bunyard, 2018). There may be several reasons for this changing view. One potential factor is that the fly agaric cannot be effectively used against every type of insect. While preparations of the fly agaric have been used effectively to control house flies (*Musca domestica*; Carapeto et al., 2017) and against the southern house mosquito (*Culex quinquefasciatus*; Cárcamo et al., 2016) it is ineffective against fruit flies (*Drosophila melanogaster*; Mier et al., 1996). The insecticidal properties of the fly agaric are also dose dependent (Cárcamo et al., 2016; Carapeto et al., 2017), which could affect the perceived efficacy of fly agaric as a natural insecticide. One can see how one might conclude that the fly agaric is an ineffective remedy if they are using poor or inactive material or if they are targeting insects for which fly agaric is an inappropriate remedy. Finally, the fly agaric is thought by some to be "the most preferred host of mushroom-consuming flies in North America, and globally" (Bunyard, 2018: 41). Of course, this is neither here nor there if mycophagous insects, such as fungus gnats or scuttle flies, are not the target of one's fly trap.

Nevertheless, fly agaric has a long history of use to control and kill house flies in Europe and current studies suggest that there is a scientific basis for this use (Cárcamo et al., 2016; Carapeto et al., 2017; Lumpert & Kreft, 2016; Muto & Sugawara, 1970; Takemoto et al., 1964). The pharmacology of the fly agaric is particularly interesting since it contains both a fly attractant (1, 3-diolein) as well as flycidal compounds (ibotenic acid, muscimol). The combination is really a perfect match for creating an effective fly trap.

Typically, the fresh or dried cap of the mushroom is placed in a saucer and partially covered with either milk or water, to which sugar or honey might be added (Harkonen, 1998; Lumpert & Kreft, 2016). Because ibotenic acid and muscimol are the fly killing agents of the mushroom the higher the percentage of these compounds in the milk or water solutions the more effective they will be. Some traditional recipes call for bringing the mushrooms to a boil in either water or milk before setting the solution out for catching flies (Lumpert & Kreft, 2016). It does not appear to matter much whether water or milk is used. Use of dried caps will also produce a more potent solution, though dried caps may lack a sufficient quantity of the fly attractant (1, 3-diolein) and thus require sugar or honey as a lure. In any event, the fly agaric can be used to lure and kill house flies and may also be effective against some varieties of mosquito.

Chapter 28

The Experience

The categorization of *Amanita muscaria* as a psychedelic or hallucinogenic mushroom is problematic on several levels. While the terms "hallucinogenic" and "psychedelic" are problematic in their own right, the application of these terms to *A. muscaria* may give the impression that the effects of this mushroom are similar to the more well-known effects of *Psilocybe* mushrooms, or other more common psychedelics, like LSD. This would be a mistake. While there are similarities – both alter mood and perception and both can lead to ego-loss and death-rebirth experiences – the differences can be striking and for those unprepared, unsettling.

What have come to be known as the "classic" psychedelics, such as LSD, psilocybin, and mescaline, are typically divided into three categories: ergolines (LSD and relatives), phenethylamines (mescaline, MDMA, etc…) and tryptamines (DMT, psilocybin, etc…). The ergolines and tryptamines are primarily serotonergic, meaning they act upon and disrupt the various serotonin receptors in the brain. The phenethylamines are structurally similar to the neurotransmitters dopamine and noradrenaline and tend to mimic their function in the brain, though phenethylamines are also highly active within the brain's serotonergic system. As a result, these "classic" psychedelics tend to produce similar symptoms and effects, effects that have generally come to be recognized as "psychedelic."

Other substances have broadly been included in the category of "hallucinogens," including *Amanita muscaria*, *Salvia divinorum*, *Datura* and other tropane-containing plants, but these do not specifically act on the brain's serotonergic system and do not produce effects that are typically recognized as "psychedelic." Individuals with a penchant for the *classic* psychedelics are frequently surprised that these substances are not "truly" psychedelic and often describe their experiences as *weird* and tend not to repeat them. Of the three substances mentioned

Fig. 1: *Stargazing* by Joy Muller. Mixed media: clay, paint, wood, lichen, dried poppy pods and bark.

above, however, *A. muscaria* most resembles the classic psychedelics in effects, though in high doses loose-comparisons might be made to the dreamy delirium of tropane-containing plants like *Datura*.

There are a clear set of effects that distinguish the isoxazole-containing mushrooms, like *A. muscaria*, from *Psilocybes* or other classic psychedelics. Muscimol, the primary psychoactive agent in the fly agaric is GABAergic, meaning that it resembles and imitates the inhibitory neurotransmitter GABA (*gamma*-Aminobutyric acid) and acts upon the brain's GABA receptors. This is the same system that plays a role in alcohol inebriation, so it should not be too surprising that some of the symptoms of fly agaric inebriation include loss of coordination and memory lapses or blackouts. Ibotenic acid, the other psychoactive compound in fly agaric, resembles the neurotransmitter glutamate and binds with the brain's glutamate receptors. Because large portions of ibotenic acid metabolize into muscimol upon ingestion it is unclear what it specifically contributes to the overall effects of the fly agaric, though it has been speculated that it produces stimulant-like effects (Chilton, 1975; Ott, 1976b; Theobold et al., 1968). The fact that different neuronal pathways are affected largely explains the differences in effect between the fly agaric and "classic" psychedelics, though the presence of muscarine (or muscarine-like compounds) in fly agaric and some of its relatives, also accounts for some significant differences in symptoms.

One of the primary differences between fly agaric and the classic psychedelics is that the effects produced by the classic psychedelics are fairly consistent and predictable in comparison to the fly agaric, which is not. While the effects produced by the fly agaric are variable, they are distinctive and can easily be identified as part of a unique constellation of effects associated with fly agaric inebriation. I will address some of these distinctive features below, but first want to provide a handful of anecdotal accounts from individuals who have attempted to describe some of the differences between the effects of the fly agaric and those of the classic psychedelics. This will help give a general sense of the differences before we address some of the unique features of fly agaric inebriation individually.

The first account comes from an acquaintance of Terence Mckenna (1992), who characterized his fly agaric experience as:

> Unlike anything I had felt before – "psychedelic" is
> too broad a term, too all encompassing, it was not

> truly psychedelic. It was as if everything were exactly the same but totally unfamiliar – but it all looked like I knew it to be. Except that this world was about a shade (or a quantum level off) – different in an eerie, profound and unmistakable way. (p. 110)

Another individual, sharing his experience on *Erowid*, was also compelled to distinguish his experience from the "classic" psychedelics:

> As for visuals, the fly mushrooms don't have the same kind of colors and patterns as lsd or psilocybin, but objects tend to take on a sort of surreal quality, like in a dream. This quality is similar to that from lsd but not really the same. There also tends to be some visual processing errors mostly in the periphery of vision, like objects seeming to 'jump' from one place to another, although that description doesn't really accurately portray what happens. (Funk Shui, 2010)

A report made following the ingestion of 10g of dried *Amanita muscaria* var. *muscaria* was described this way:

> It's as if my mind is an apartment complex and there is a rockin party going on… [on] one or two floors, but on the rest of the floors, it's life as usual… [M]uscimol is a weird and completely different drug from all the classic entheogens. I don't know if I'd call it hallucinogenic or entheogenic. It was very interesting though. Lots of ENERGY, with some sedation. Definite ego loss, but in a way completely sober. (An encounter, 2000)

Another report on *Erowid* sought to distinguish fly agaric from *Psilocybin* mushrooms with the following explanation:

> I was having some unusual effects…not what I would call 'visuals'…I would say more like small 'visions'. Just very peculiar things like a blue pyramid floating in front of me, flashing. It had an eye on top that was very lifelike, in fact, the entire pyramid looked as if it had skin. Very weird. I had never had a vision like this. My evening was filled with these occasional

> visions. It is this that made me realize how different Amanitas are from Psilocybe mushrooms. With Psilocybe mushrooms, everything sort of waves and breathes, with colors and tracers. With Amanitas, there are occasional weird visions that come and go, accompanied by very deep thoughts. (Skandre, 2004a)

While every "psychedelic" experience is different the above accounts suggest that there are some fundamental differences between the visual effects produced by "classic" psychedelics and those produced by the fly agaric. The experiences shared here suggest that the brightly colored kaleidoscopic visions typical of psychedelics are less prominent, or even absent, but that the visual world takes on a surreal or dream-like quality that defies clear description, one that may be interrupted by visual disturbances or actual hallucinations. The differences do not stop with the visual experience, however.

The effects of fly agaric as experienced through the body are also quite different. Individuals who have ingested the mushroom may experience alternating periods of drowsiness and hyperactivity, and often report sensations and effects more akin to drunkenness than what one might typically expect from a psychedel-

Fig. 2: Untitled (ID 114970696 © Ilkin Guliyev | Dreamstime.com).

ic. One individual characterized the effects as "more like C[o]caine, alcohol and X all mixed in perfect proportions" than purely psychedelic (A most rewarding, 2000). This individual clearly recognized a psychedelic quality, as indicated by the comparison to "X" (MDMA), but suggests that feelings of stimulation (cocaine) and drunkenness (alcohol) were significant features of the experience as well. The following report, following ingestion of 11g of dried fly agaric, further corroborates the above account:

> ...within 20-30 minutes of ingestion I start to feel pill-drunk, like I've taken some codeine and drank three beers. At this point some trippy stuff happens, but I have to say, amanitas aren't psychedelic in the LSD-Mescaline-Mushrooms way. The feeling I get is as if my inner-head, or mind's eye, is no longer fixed in place inside my head. I turn, and it arrives a moment late. ... Overall, amanitas are like drinking five beers slowly with a belly full of high quality goof-balls. Not psychedelic, but real fun. (what the hell, 2003)

Finally, Tom Robbins (1981) provides a more epic characterization of the differences between the fly agaric and the classic psychedelics:

> I got gloriously, colossally drunk. I say "drunk" rather than "high" because I was illuminated by none of the sweet oceanic electricity that it has been my privilege to conduct after swallowing mescaline or LSD-25. On acid, I felt that I was an integral component of the universe. On *muscaria* I felt that I was the universe. I felt invincibly strong and fully capable of dealing with the furniture, which was breaking apart and melting into creeks of color at my feet. Although my biceps are more like lemons than grapefruit, I would have readily accepted a challenge from Muhammed Ali... (p. 120)

As demonstrated above, the body high is frequently described as more akin to drunkenness than the electrical quality of psychedelics, and is also described as imparting courage, an element also reminiscent of alcohol inebriation. These accounts should provide a brief taste of some of the general differences between the fly agaric and the classic psychedelics, but while the authors above have made a strong effort to identify and distinguish the effects of the fly agaric from other similar substances, there is much more to be said and for the reader to be aware of. In any case, if you read no further it should be clear that the fly agaric is in a class of its own.

Below, I will outline the constellation of fly agaric effects and provide some anecdotal accounts to help illustrate some of its more distinctive features. Briefly, these defining features include: Looping, Echopictures/Frame Reduction, Size Distortion, Vacillations between Vigor and Lassitude, Feelings of Strength, Visionary Dreams, Loss of Coordination, Amnesia/Blackouts, Dissociation, Muscle Twitching, Nausea and Vomiting, Imperviousness to Pain and Muscarinic symptoms. These are all classic and distinguishing features of the fly agaric experience, though they occur in varying combinations and some only at higher doses.

Looping

The first category of effects I will address is Looping, which is perhaps the most unique feature of the fly agaric experience. The term Looping is used to describe an individual who has become stuck in a looping or repetitive behavior or thought pattern. The following account of an experience with *Amanita pantherina* was collected by Andrew Weil and Winnifred Rosen (2004) and provides a good illustration of how looping is experienced by the individual as well as how it might be perceived by onlookers:

> I crawled out on a big fallen tree over a pond and, while trying to find a good position, fell off. Then I wasn't sure if it had happened or not, so I got back up on the tree and fell off again. I kept having a compulsion to repeat the fall, because I couldn't tell if it had happened or was going to happen. On about the seventh time, I hit my head on some rocks and was

> bleeding pretty badly. Some people saw me and got scared. I guess I looked bad, although I was unaware of being hurt. They drove me an hour to the nearest emergency hospital. By the time I got there I thought I had died and gone to heaven. I thought the doctors and nurses were angels and started singing hymns. They did not know what to make of me. (p. 252)

The above experience illustrates how one can find themselves stuck in repetitive behavior loops, unsure if they are playing out their destiny or repeating some event in their past. This example also demonstrates how individuals inebriated on isoxazole-containing mushrooms may become impervious or insensitive to pain and may engage or continue to engage in dangerous behavior without any awareness of having sustained significant injuries. Instances such as this provide a good reason for anyone experimenting with this mushroom to have a "sitter" to help avoid and prevent injury.

The next account comes from Paul Stamets, who shared the following experience with *Amanita pantherina* during a talk at the Mycological Society of San Francisco Fungus Fair in 2000:

> I get out of the car, and in front of all these people my camera falls -*kaboom*- on the pavement. And I go "oh no, my camera fell." I look down at my camera… it's really a nice camera… I reached down. I pick up my camera and I think to myself "did that really happen?" And I get into this memory loop. And what happens with *Amanita pantherina*… if you think about a memory you act [out] a memory. So, I dropped my camera again. The camera hits the pavement and goes *boom*. I go, "oh no, I dropped my camera." So, I picked up my camera again. Meanwhile, you know, I drop it again… three, four, five, six, seven, eight times. I notice this crowd of people are looking at us, talking back and forth [and] really, really concerned… So… I get up, I pick up my camera [to leave]. I take two steps. I drop my camera. *Boom!*… And I get into this repetitive behavior pattern that I can't break. So, every few steps I drop my camera… And the camera [is] totally destroyed. (Stamets, 2000)

While the above examples illustrate behavioral-loops, loops might also be experienced as a repetitive hallucination or as a thought loop, as reported in the following trip report:

> Now I was getting a still image smaller than what it should be, as if I was looking through binoculars backwards, and it was rapidly zooming towards me. This repeated over and over again with the same image (because I was standing still staring off into space), increasing in speed with each repetition. Then it felt like I was repeating the same moment over and over again, as if time had begun to repeat itself rather than just the image zooming at me. (Atrocitic, 2012)

Another individual, after consuming 25g dried (an extremely high dose), described their mental looping thus:

> With each repeat of the loop a new "scene" would be added at the end and I assumed the role of investigator trying to solve the mysterious loop. And as each new scene was added I felt like I became more and more aware of what I was being taught, of what the point of the whole thing was, and I began to feel that the loop was a revelation of some beginning – the beginning of time, or biologically-defined life, or the universe – the whole loop unveiled a history in reverse chronological order… The last scene came. The final card was turned over. The secret was revealed. I rejoiced in ecstasy. This was the peak of the strange journey. No more scenes were added to the loop and it began to vanish from my focus. (Chad, 2011)

Echopictures / Frame Reduction

One feature of fly agaric inebriation is the experience of Echopictures, or what might be called Frame Reduction (the concepts are similar but not entirely the same). The idea of Frame Reduction can best be understood in relation to film. Film is essentially a series of pictures that are strung together to give the appearance of motion, thus the term "motion pictures." A certain number of pictures, or frames (typically 24 frames per second), must be shown each second in order to give the impression of movement. Claymation also operates on the same principle but the number of frames per second is significantly fewer than in film (only 12 frames per second), which results in Claymation figures having characteristically jerky movements. The eye and brain work together in a similar fashion. The eye continuously captures visual information while the brain processes and registers what is seen. With frame reduction the eye continues to receive visual information but the number of frames processed and registered by the brain appears to slow down resulting in what has been termed "Echopictures," where a visual frame or snapshot remains in the mind's eye until it is updated by a new frame. The term echopicture comes from Peter G. Waser (1979), a former Professor at the University of Zurich, who reported the following effects after ingesting 15 mg of muscimol:

> After a phase of stimulation, concentration became more difficult. Vision was altered by endlessly repetitional echopictures of situations a few minutes before…I felt sometimes as if I had lost my legs, but never had hallucinations as vivid and colourful as with LSD. (p. 435)

The echopictures experienced by Waser under the influence of muscimol have also been described by others after having ingested *Amanita muscaria* mushrooms. One individual, after consuming a fresh fly agaric mushroom, described his experience as follows:

> As I sat on the toilet watching [my son] splash in the tub I had the most peculiar visual experience…As I watched him, say laying belly down, my mind captured this image so it remained in my perception, till another motion, say my son sitting upright, caught my attention, of which the new image would appear. The whole effect was sort of a still frame slide show, where an image would remain in perception till a new one burst from the center of the old one. (ChemBob, 2005)

Others have described this experience in some of the following ways: "the world around me felt like it was ramming itself into my head one still image at a time" (Booth, 2008); and "my eyesight began skipping like an old movie" (Axic, 2008). Another interesting report includes descriptions of temporary instances of what might be considered blindness. Following the ingestion of tea made from four fresh caps this individual explained:

> I noticed I was [losing] my balance, and stumbled some of the way. Also, when I'd look from one place to another, there would be no frames of vision during the movement in-between locations of sight... It was just a black blur, and I could really only see what I was statically looking at. (Existence, 2011)

Here, there may be an interaction between muscimol and muscarine, which can cause blurry vision. The combination of frame reduction with blurry vision might best explain the visual impairment described above. Here, we also see evidence of loss of coordination, another feature of fly agaric inebriation. The final example of frame reduction, below, also suggests that there may be a connection between this phenomenon and the experience of amnesia or blackouts. The following comes from an experience with 8g of dried *Amanita muscaria*:

> I could hardly walk without falling and all the while my mind was flashing and pulsing, like a strobe, between phases of total non-perception and flashes of perception. Initially these "dark" areas of non-perception were long, so I would find myself suddenly in the toilet not remembering making the decision to go there. (Pithtaker, 2015)

This individual, while experiencing frame reduction, appears to have simply experienced periods of darkness rather than echopictures, as described in some of the accounts above. These dark areas or dark periods of "non-perception" suggest that the brain is period-

Fig. 3: *Mystery Flavor* by John W. Allen.

ically failing to register information into short-term memory banks (not just visual information), leading to brief periodic blackouts.

Size Distortion

While size distortion is also reported from *Psilocybe* mushrooms, this phenomenon was first observed among Siberian users of the fly agaric and also appears to be more frequently and substantially experienced with fly agaric mushrooms, making it an important defining feature. In fact, it is believed that early reports of Siberian mushroom use are what inspired Lewis Carroll (Charles Dodgson), the author of *Alice in Wonderland*, to write about mushrooms that would cause Alice to either shrink or grow in size after ingesting them. Accounts that Carroll might have been familiar with include a report from Krasheninnikov, writing in 1755 (*see* Wasson, 1968: 236), who described how those inebriated by the fly agaric "might deem a small crack to be as wide as a door, and a tub of water as deep as the sea," or the following report from Von Langsdorf (1809, *see* Wasson 1968: 249), who observed that "[I]f one [bemushroomed] wishes to step over a small stick or straw, he steps and jumps as though the obstacles were tree trunks."

Similar descriptions of size distortions are also present in modern first-person accounts of fly agaric inebriation. One individual explained that "although nothing really grew or shrank, things (desk, window, etc.) somehow seemed as though they were bigger or smaller than they were normally" (Anonymous, 2000). Another reported that "when I looked down at my feet it seemed as though I was very small, like an elf. I would then look up and it seemed as though I was 20 feet tall!" (Skandre, 2004b). Another report, that also details other effects of the fly agaric, including fatigue and nausea, follows:

> The first visual effects I noticed was that colors became brighter and the size of objects became distorted. I looked down at my hand and it appeared to be very small and very far away from my face. I perceived what appeared to be transparent energy flowing through the air in a swirling pattern. I felt very tired and somewhat delirious. My body felt almost weightless as if I was floating. I then experienced the sensation of slowly leaving my body through the top of my head. The tea I consumed [made from 30g] gave me an intense urge to vomit. I managed to resist

vomiting although it was incredibly difficult to do so. (Space Elf, 2012)

Dissociation

The fly agaric, in high doses, can cause episodes of dissociation or delirium. This has led to some disagreements about whether it should be classified as a dissociative drug like ketamine or DXM, or should be classified as a deliriant along with tropane-containing plants like *Datura* and Belladonna. Interestingly, ibotenic acid is a potent agonist of the brain's NMDA (N-methyl-D-aspartate) receptors, the same group of receptors that dissociative drugs act upon. However, the dissociatives are NMDA antagonists, meaning they inhibit NMDA activity while ibotenic acid activates these receptors. Of course, the role of ibotenic acid in fly agaric inebriation is not entirely understood as large portions of it metabolize into muscimol through the digestive process. In any case, it is a noteworthy connection even if ibotenic acid acts upon the brain's NMDA receptors in the opposite manner of dissociative drugs.

With the fly agaric the individual does not necessarily dissociate from themselves, though this is possible through the process of ego loss, but they typically dissociate from the world around them. The state they enter is dream-like, a world with its own reality, one separate from the "waking" world. The difference between the dreamer and the individual under the influence of the fly agaric is that dreamers typically stay in their beds, a fairly safe place for one who is not entirely responsive to external stimuli. The following is a rather epic account of fly agaric inebriation (following ingestion of four dried caps), one that helps illustrate the ways in which dissociation might manifest under the influence of this mushroom:

> The next thing I knew I was running, where to I have no idea. I thought I was a deer, I literally thought and felt like I was a deer running through the forest. Things were very dream-like and somewhat clouded but I disti[n]ctly remembering running through the forest past trees and rocks, not knowing or caring where I was going. I can't remember very much else, except that when I finally came back to [awareness], I was sitting under a tree rubbing my feet. I had no idea where I was or how I got there…
>
> I looked at my feet and my shoes were miss-

Fig. 4: *Mezmerized by Mukhomors* by Timothy White.

> ing and I had cuts and scratches all over them. There was mud and leaves in my hair, my clothes were all torn up, and I was really dirty. I also had this huge scratch along the backside of my leg, it looked like some animal-claw scratch, like a racoon or something. And the weird thing is, I have no memory of getting cut or feeling pain or anything like that... after walking a few miles, I found a state highway that I recognized... I followed the highway (as distant as I could from the road) for 5 more miles and finally reached the cabin...
>
> The next day I got up and decided to try and find my shoes. So, I tried to retrace somewhat of a path and I counted 5 barb-wire fences that I must have jumped over to get where I found myself, 6 miles away from that cabin! I have no idea how I travelled those 6 miles, where my shoes ever were, how I got those scratches, or how I got over those fences in the condition I was in. (Somewhat Hazy, 2000)

In addition to dissociation the above account also illustrates the imperviousness to pain that is sometimes experienced, as well as heightened levels of energy and "apparent" increases in strength. Running six miles through the woods while hurdling barbed-wire fences under the influence of the fly agaric is an incident that recalls observations of some of the early Western explorers of Siberia. Speaking to this issue, Georg H. Langsdorf (1809, *see* Wasson, 1968: 249) explained:

> In this intense and stimulated state of the nervous system, these persons exert muscular efforts of which they would be completely incapable at other times; for example, they have carried heavy burdens with the greatest of ease, and eye-witnesses have confirmed to me the fact that a person in a state of fly-agaric ecstasy carried a 120-pound sack of flour a distance of 10 miles, although at any other time he would scarcely have been able to lift such a load easily.

While the opportunity to experience life as a deer, however briefly, would certainly be unforgettable, the dangers posed by dissociation should be clear. One who is unaware of their physical environment, including the presence of traffic, steep cliffs, or dangerous weather, could quickly find themselves in life-threatening situations. This is another facet of fly agaric inebriation that strongly suggests that anyone experimenting with high doses of these mushrooms should only do so under the supervision of someone competent and sober.

Lassitude & Vigor

One of the defining features of fly agaric inebriation is the periodic experience of drowsiness and sleep followed by periods of vigor and stimulation. It is believed that ibotenic acid contributes to feelings of stimulation while muscimol has more of a sedative effect. Either way, individuals frequently succumb to sleep at some point during the experience while at other times they may feel strong and energized. One individual has explained that "The effects aren't exactly constant. They vary; one moment a lapse of sleepiness will rise, and as I dip into it the nice smooth slipping feeling arouses my senses, and I'm back up" (what the hell, 2003). Another pair of experimenters reported "it goes up and down, we feel mentally sedated but ph[y]sically speedy, then mentally speedy but ph[y]sically sedated" (Gubi, 2007). The next individual reports a period of relaxation followed by a stimulant effect after ingesting two dried caps, an experience that also illustrates this strange and contradictory quality of the fly agaric:

> I noticed the first psychedelic effects, as I lay there with my eyes closed, when colorful scenes displaced each other in a manner that reminded me of rapidly changing TV channels. After about two hours of this, I opened my eyes, and this was when the experience took on a speedy and somewhat psychedelic quality. There were no real open-eye visuals to speak of. Instead, it occurred to me that this substance would be ideal at a rave as a sort of substitute for MDMA. I was highly energized. (Norm de Plume, 2003)

Dream States

The dream states that arise during periods of slumber are thought, by some, to be the most magnificent part of the fly agaric experience, and potentially, the most hallucinatory. Some will wake up with no memory of their dream experiences, but others recall extremely vivid and colorful visions and fantasies. Joseph Kopéc (1863, *see* Wasson, 1968: 244), a Polish brigadier, recounted the ensuing dream following ingestion of fly agaric while traveling in Siberia in the late 1700s:

> Dreams came one after the other. I found myself as though magnetized by the most attractive gardens where only pleasure and beauty seemed to rule. Flowers of different colours and shapes and odours appeared before my eyes; a group of most beautiful women dressed in white going to and fro seemed to be occupied with the hospitality of this earthly paradise. As if pleased with my coming, they offered me different fruits, berries, and flowers. This delight lasted during my whole sleep, which was a couple of hours longer than my usual rest.

Muscle Twitching

Muscle twitches are also frequently experienced under the influence of the fly agaric. Sometimes these twitches are misinterpreted as convulsions, which appears to be a mischaracterization, though they can become exaggerated at higher doses. One experimenter described how his "hands were twitching somewhat as if I had tremors and I would occasionally jerk a bit but it didn't ever get to the point where it was annoying" (Zenergy, 2007). Another reported the following experience after ingesting two fresh caps of fly agaric:

> A few of my nerves were tweaking especially in my arms, whereas my friends had this a lot more, especially in his legs. A person on the course observed how I exaggerated head movements and gestures when I spoke... I also felt incredibly spaced out - a feeling unique to this drug (for me). I didn't feel dizzy in the slightest, yet I was wobbling all over the place, obviously my balance/inner ear... was being affected. (Sam, 2004)

Here we see muscle twitching combined with coordination loss, a combination that might be somewhat discombobulating for the unprepared, and perhaps even the experienced.

Muscarinic Symptoms

While not psychoactive, muscarine can play a significant role in the experience one has following ingestion of the fly agaric and related mushrooms. Typically, the effects of muscarine are not particularly pleasant – though it adds a purging element which some with interests in shamanic or alternative healing practices might view as beneficial. The most frequent muscarinic symptoms experienced with the fly agaric are excessive perspiration and salivation, increased urination, blurry vision, and gastrointestinal distress, which may include nausea, vomiting and/or diarrhea. While there is disagreement about whether these symptoms are caused by muscarine or some other compound with muscarine-like activity, the fact remains that these are symptoms that occur with some regularity. One study has shown that muscarine-like symptoms occur approximately 20% of the time following ingestion of *Amanitas* from the Muscarioid group (*see* Feeney & Stijve, Ch. 23), though symptoms appear to be less frequent within the Pantheroid group.

Ethnobotanist and author Clark Heinrich (2002) shared the following experience with muscarinic symptoms after consuming a single fly agaric cap:

> After fifteen or twenty minutes I began to sweat and salivate excessively. In a very short time I was soaking wet; water was pouring out of me at the same time as my saliva flow was nearly drowning me. I had to keep swallowing to keep from choking. I was beginning to wonder if perhaps I'd made a miscalculation when the chills started and I became certain of my error. It was ninety degrees and I was shivering with hypothermia… (p. 204)

Heinrich's experience highlights the effects of excessive sweating and salivation which tend to be the hallmark muscarinic symptoms experienced on fly agaric mushrooms. Nausea and vomiting are not uncommon, but these symptoms can also be caused by ibotenic acid or muscimol, so are not just muscarinic symptoms. The other prevalent muscarinic symptom is blurry vision.

One individual, following ingestion of 28 dried grams (an extremely large dose), reported "[m]y vision was extremely blurred, to the point I could not see with any accuracy literally beyond 5 ft. I was salivating with increased intensity. No extreme [loss] of balance and no nausea though" (C., 2016). Blurred vision frequently accompanies excessive perspiration and salivation, as noted above and illustrated in the following experience with three dried caps:

> At this point I'm twitching and sweating, my vision goes in and out of blurriness. I'm too sick to be scared… I can't stop drooling. I feel drunken but spinny and colors are flashing everywhere… Then it happened! I SHRUNK! I shrunk and entered a different dimension with native women pressing their faces on me and technology and colors swirling all over the place! This went on forever! (Nightmare, 2011)

In the above account we see several muscarinic symptoms combined, including blurry vision, nausea, sweating and excessive salivation, but several other significant effects of the fly agaric are also described, including feelings of drunkenness and size distortion. While the muscarinic effects are typically reported as being unpleasant, there is no universal agreement on this point as some report that it is positive, or at least not disagreeable. One user, after consuming an infusion of 10 dried grams, reported the following pleasant experience:

> The effects started approximately one hour after ingestion. The best description I can come up with is that it is like having [the] flu, but without feeling sick. Profuse sweating, starting at the top of the head and gradually moving down the body was the most noticeable effect. This was soon joined by extreme cold much like the chills one gets when feverish. As strange as it sounds these effects were not unpleasant. It felt much like being drunk but without any loss of mental faculties. Other than being cold there was a general analgesia something like a narcotic but, again, without the loss of mental faculties. Throughout the evening I was relaxed and in a very good mood. After changing clothes three times because they were drenched with sweat I finally took a warm shower which chased away the chills and seemed to reduce the amount of subsequent sweating. All effects had worn off after about five hours. (Zardoz, 2003)

While muscarinic symptoms are typically only reported in about 20% of cases, they do occur. Sometimes these symptoms are mild and other times they can be experienced intensely. Anyone experimenting with these mushrooms should be prepared for the potential of experiencing some of these symptoms. You will probably want extra clothes and towels nearby, as well as access to a shower.

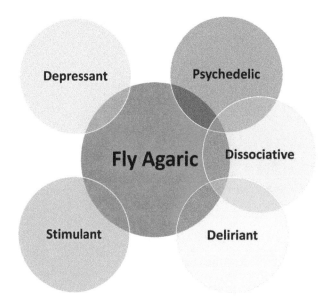

Fig. 5: Venn Diagram demonstrating the different type of drug effects exhibited by fly agaric inebriation. While these different effects are all produced by the fly agaric this combination of effects is not consistently produced following ingestion of psychoactive doses. There is a lot of variability in the effects produced by this mushroom, and some effects are dose dependent.

Further Comments

By now it should be clear that the fly agaric is unlike other psychedelics and hallucinogens, with a unique constellation of effects all its own, including looping, dissociation, and loss of coordination. While the fly agaric does not fit particularly well into any existing drug classifications, it exhibits effects that, when taken alone, could be identified as characteristic of a variety of different drug categories, including: deliriants, depressants, dissociatives, psychedelics, and stimulants (Figure 5). Because the fly agaric defies these traditionally recognized drug categories we are left with the very general term "psychoactive" to describe its effects and properties. This is somewhat unsatisfying as it can hardly convey an understanding of the numerous effects that have been discussed here, but hopefully what has been laid out in the preceding pages can help advance discussions that could lead to an appropriate classification. Until then the fly agaric remains a *psychoactive* mushroom, in contrast to the psychedelic *Psilocybes*, though both varieties clearly deserve the moniker of "magic mushrooms."

Chapter 29

The Formula?

While the fly agaric and its hallucinatory properties are well known the bright red mushroom, with its white-flecked regal appearance, has never achieved the esteem or popularity that its *Psilocybe* cousins have achieved. One reason is that the experience produced by the fly agaric is quite different from *Psilocybes* and other classic psychedelics, but the other reasons (perhaps the primary ones) have to do with confusion regarding dose and preparation. The fly agaric is notorious for its unpredictability, with potency varying from mushroom to mushroom, but many experimenters have also assumed, to their occasional detriment, that no preparation is required. While some have faired well after consuming fresh fly agarics, the most difficult and nauseating experiences tend to follow consumption of fresh specimens, so much so that some Siberians consider them deadly. While not actually deadly, consuming fresh fly agarics is foolhardy at best.

As the astute reader will have observed during the reading of this volume, preparation plays a key role in determining the effects of the fly agaric. While no perfect "formula" has yet been discovered that eliminates gastrointestinal distress and muscarinic symptoms, or that guarantees an *ecstatic* or otherworldly experience, there are some key steps that can be taken to help shape the type of experience one has. In the pages below I will outline some of the basic critical steps in preparation and will draw on some of the practices and recipes found around the world to help provide the reader with a better sense of how this mushroom has been traditionally prepared, as well as providing some pharmacological information to help explain the purpose of the various techniques presented here.

Dehydration

The first, and most critical step is to dry the mushrooms thoroughly. The drying process causes ibotenic acid to decarboxylate to muscimol, which is more highly psychoactive and is less likely to cause gastrointestinal distress. Drying is the simplest preparation method and fly agaric was once commonly used in this form among various Siberian cultural groups. While the traditional method is to sun-dry the mushrooms, a dehydrator works just as well and produces the same result. Some prefer to oven-dry, however, many ovens are not designed to heat to the optimal temperatures for drying fly agaric (between 85 and 105°F). Using a higher temperature may also lead to a liquidy mess. Typically, people will dry the mushrooms until they become brittle, or "cracker-dry." Once they are dried, they are likely to take some moisture back on (ie: lose their brittleness), but this is fine as long as they remain in a dry state. Mushrooms should be stored in a cool, dark place in an air-tight container. Mushrooms should be used within a year or two, though specimens as old as five-years have been tested and come back positive for psychoactive isoxazoles (Ott, 1993), though it is likely that this represented only a fraction of the mushroom's original content.

Homogenization

Variability in potency is one of the trickiest issues to deal with when working with psychoactive *Amanitas*. The best way to get around this is to gather and dry a large batch of mushrooms, break them up into smallish pieces, and mix them all together. This helps to standardize the potency of the batch and decreases the likelihood of dosage mishaps. For example, if one were to take two dried mushrooms on one occasion without any effect, that individual might choose to double that amount next time. However, the difference between two mildly active mushrooms and four highly potent mushrooms can be astronomical. It is best to avoid this type of mistake. Standardization also increases an individual's ability to gauge the potency of their next dose, and to repeat doses they previously found to be optimal.

It should be noted here that the psychoactive compounds are not found equally throughout the mushroom. The cap is the most potent part, followed by the basal bulb and lastly the stem (Tsunoda et al., 1993b). To ensure true homogenization it is best to keep these materials separate (they can then be recombined in equal parts if one wishes). A mixture of dried caps, for example, will have less variability than a mixture that consists of each part of the mushroom. It will, of course, also be more potent. Some people only use caps, setting aside the stem for medicinal or culinary preparations, but each part of the mushroom does contain psychoactive isoxazoles. It is up to you whether you choose to use the whole mushroom or just the cap.

Fasting

Fasting before ingesting psychoactive doses of *Amanita* mushrooms offers a couple of benefits. First, fasting for at least five-to-six hours ahead of time can help reduce the risk of severe and unpleasant nausea and vomiting. Second, it also adds a level of intent and ritual, elements that can be very grounding and which can add focus to the experience. Fasting is frequently employed by shamans of various traditions for one to three days before ingesting a powerful shamanic inebriant. While skipping a meal is probably sufficient, this type of dedicated fasting can be used to help shape your goals, focus your intent, and plan your experience. It all depends on how you prefer to approach these mushrooms and what types of outcomes you are searching for.

Start Small

The psychoactive muscarioids are typically less potent than their *Psilocybe* counterparts, but sometimes surprising things happen, and psychoactive *Amanitas* can be overwhelming in ways that *Psilocybes* are not. It is best to start small. While 5g is typically considered a low dose for muscarioids, starting with 2g of a homogenized mixture and working one's way up would be wise. There is no hurry, and thus no need to rush into these things. Someone looking to jump right into a full-blown mind-bending experience would be best advised to look elsewhere.

Below is a chart that broadly outlines the potency and respective doses that have been observed with psychoactive *Amanitas* (Table 1). This is intended as a very general reference as potency can vary dramatically within a species. For example, anecdotal reports from the state of Washington indicate that some *Amanita* "*pantherina*" specimens may produce a powerful experience with as few as two to three grams of dried cap material. This appears to be more the exception than the rule but should demonstrate the potential danger of starting with 10g of *A*. "*pantherina*" from an untested batch of mushrooms.

In any event, the combination of drying the mushrooms, homogenizing

Psychoactive Level	Dose		
	Low	Moderate	High
Mild	10 – 15 g	15 – 20 g	20 – 25 g
Moderate	5 – 10 g	10 – 15 g	15 – 20 g
High	1 – 5 g	5 – 10 g	10 – 15 g

Table 1: Potency ranges of psychoactive *Amanita*. *See* Chapter 3 for potency ranges of individual species. Weights provided refer to dried mushrooms.

them, starting with small amounts and fasting beforehand, are all simple steps that can be taken to reduce the likelihood of severe nausea and vomiting, as well as to avoid an overpowering experience that one may not be prepared for. There are no guarantees with these mushrooms, but these are the best steps to follow for the experimentally inclined. Without further ado, here are some recipes inspired by various cultural, historical, and hypothetical (*Soma*) uses of the fly agaric.

Beverages

Following simple consumption of dried caps, consuming fly agaric in some sort of beverage appears to be the second most common or popular method for ingesting this mushroom. This is also a popular method of consumption among Siberians, and if Wasson's Soma theory is correct, was also the preferred method of the Indo-Aryans. My own investigations (Feeney, 2010) suggest that making a tea of the dried fly agaric is one of the most effective ways to minimize the potential for gastrointestinal distress. Before addressing different cultural and potential "historical" preparations, the simplest way to prepare your psychoactive *Amanitas* as a beverage follows:

- 10 – 20 g dried fly agaric caps
- 2 cups of water

Boil the mushrooms for 20 – 30 minutes then strain. Keep an eye on the mushrooms as they boil as you may need to add water periodically to prevent water levels from getting too low. Once you have strained the mushrooms out you may want to reduce the total volume by further boiling. Drink no more than half of your beverage, unless you are already familiar with the potency of your material and are comfortable with consuming more. If effects are weak or mild after two hours you can choose to augment your dose with what remains of your beverage. Some people choose to add a little lemon juice before the boil. This might help with flavor, but also may assist in the further conversion of ibotenic acid to muscimol, a process that will be discussed in more depth momentarily.

Below several recipes for making beverages of the fly agaric are outlined and a brief discussion provided on each recipe, its origins, and any potential pharmacological implications that may arise from the described recipe.

Siberian Blueberry Drink

Among the Koryak of Siberia, a drink is made by boiling dried fly agaric mushrooms and blueberries together. This was once regularly prepared for an annual harvest celebration and was used to provide energy and induce a celebratory mood (Salzman et al., 1996; *see also* Ch. 6). To make this drink you will need:

- 5 quarts of water
- 10 – 15 dried fly agaric mushrooms
- Blueberries
- Sugar

Mix all ingredients together and boil for 30 minutes. One-half to one cup is the prescribed dose. Of course, this is a very general recipe since most of the measurements are inexact. One would have to experiment with these ingredients to determine the optimal ratio.

The above is one of the only traditional recipes that we have, and it is important to consider the ingredients used and whether they serve a particular purpose other than flavor. Blueberries and sugar will clearly impart a sweet flavor that may help to mask or make palatable the flavor of dried fly agarics. However, the fly agaric, when boiled, imparts a savory flavor so the choice to sweeten the beverage is interesting.

There are several studies that suggest that boiling ibotenic acid in an acidic solution prompts its decarboxylation into the more potent muscimol. Studies by Trent Austin (2014) and Elsebet Nielsen et al. (1985) have demonstrated that ibotenic acid readily decarboxylates to muscimol when boiled in solutions with a pH of 2.6 or 2.7. These studies involved boilings of 3 and 2.5 hours respectively, significantly longer than the 30 minutes recommended in the above recipe. It is also unlikely that the acidity of the blueberries is capable of dropping the pH to 2.7, nevertheless it seems likely that boiling the fly agaric mushrooms with blueberries contributes to the further decarboxylation of ibotenic acid, thus increasing the potency of the final drink.

With these considerations in mind one might adapt the above recipe as follows:

- 10 – 20 g dried fly agaric mushrooms
- 3 cups water
- 1 cup blueberry juice
- Citric acid
- Sugar

Combine the first three ingredients in a small to medium pot and bring to a boil. Check the pH using a digital pH meter (these can be found for $10 – 20 at a wine supply store or on Amazon). Add citric acid a little at a time until the pH falls below 3.0. Once you have achieved the correct pH boil for at least 30 minutes. If time allows you might boil up to 3 hours in order to maximize the conversion of ibotenic acid to muscimol, however, you will need to periodically add either water or additional blueberry juice to avoid burning the pot. Stainless steel is preferred as the low pH may damage other types of pots.

After the boil strain out the mushrooms and measure the remaining volume. If more than two cups remain, continue to boil until you have reduced your decoction to two cups. If under two cups, add blueberry juice to bring the volume back to two cups. Add sugar as you like. Take one-half to one cup.

Fly Agaric Sake

The following recipe is based on a report from Alan Phipps (2000) who encountered a man in Sanada, Japan, who described steeping fly agaric mushrooms in a bottle of *sake* and drinking the resulting beverage for its relaxing effects. While the interview was conducted in a town that was familiar with the psychoactive properties of this mushroom and where making pickles of the mushroom was once popular, it appears the *sake* recipe was specific to this one individual. Here's what you will need:

- 1 cup *sake*
- 10 – 15 g dried fly agaric caps

Soak the dried mushrooms in a sealed bottle of *sake* for 1-2 weeks then strain. Sip 1 oz for a relaxing nightcap.

"Soma"

If Wasson is correct and the *Soma* of the ancient Aryans is a beverage made from the fly agaric, then we can apply the preparation techniques outlined in the Rig Veda to come up with a beverage that approximates the Soma beverage as it was originally prepared and consumed. To prepare your Soma beverage you will need:

- 20 – 30 g dried fly agaric mushrooms
- 3 cups water
- 2 cups Acidophilus milk

Bring water to a boil and then add mushrooms. Boil for 30 minutes then strain. Add decoction back to the stove and boil until the volume drops below two cups. Add two cups of acidophilus milk and simmer on low for 30 minutes to an hour. Set aside to cool. Drink one-half to one cup.

Here, acidophilus milk is used because it contains *Lactobacillus acidophilus* which produces glutamate decarboxylase (GAD). As demonstrated in Chapter 4, GAD helps to catalyze the conversion of ibotenic acid to muscimol, and its presence in acidophilus milk should lead to a more potent final beverage, perhaps one befitting the ancient Vedic priests of the Indus Valley.

Urine-Recycling

One of the unforgettable features of fly agaric use in Siberia is the practice of urine-recycling. Because the primary inebriating compounds in *Amanita muscaria* are excreted in the urine it is possible to consume the urine of someone who is inebriated and then oneself become inebriated. So, how does it work?

For those willing to entertain this option of ingestion you should know that both ibotenic acid and muscimol begin to show up in the urine approximately one hour following ingestion (Michelot & Melendez-Howell, 2003; Ott, Wheaton &

Chilton, 1975) and both can be detected in the urine up to eight hours after ingestion (Stříbrný et al., 2012). It is not clear, however, if there is a particular window where the urine is most concentrated or if there is a typical time when isoxazole levels in the urine begin to drop off. It seems the best bet is probably to collect urine between one and three hours after ingestion. It has also been reported that fly agaric-infused urine is sometimes boiled by the Saami people of Northern Europe (Letcher, 2011). It is not clear why this is done, but one obvious benefit is that boiling reduces the volume of urine one would have to consume. Boiling the urine, which is mildly acidic, might also contribute to the further decarboxylation of ibotenic acid to muscimol, resulting in a more potent beverage. So, with this information, how does one go about urine-recycling?

To prepare for your urine-recycling experience you should plan to eliminate meats from your diet for the day as well as any other foods that might impart an unpleasant odor to your urine (ie: asparagus) and drink plenty of water. You might consider fasting for several hours before consuming your mushrooms and should plan to urinate just before you consume the mushrooms to ensure that you are starting with a relatively empty bladder. One hour following ingestion you can begin to gather your urine. You may gather it as many times as you wish since your urine will contain the psychoactive principles up to 8 hours after ingestion. Once you have finished collecting your urine you can boil it down to whatever volume you believe you can handle. Let it cool a little before drinking, then bottoms up!

Smoking Blend

Another traditional method of fly agaric use is to smoke the dried cap material, which is typically first mixed with tobacco. Reports of this method of use are few and appear to be limited to several Mayan groups in Mexico and Guatemala (Diaz, 1979; Knab, 1976-78; Rätsch, 1987; 2005). The practice is further limited to shamans or *curanderos* who use this mixture to diagnose disease, purify the air, to prophesy and to find lost objects. Some of these reports indicate that the whole dried cap is used but the Tzeltal Maya of Chiapas, Mexico, reportedly use only the red cuticle of the cap (Rätsch, 2005). Either way these practices suggest intimate knowledge of the fly agaric and the distribution of its psychoactive compounds, which are most heavily concentrated in the mushroom cap (Tsunoda et al., 1993b), though the cuticle itself tends to be only weakly active (Gore & Jordan, 1982).

Jonathan Ott (1993: 338) has demonstrated that muscimol is present in dried fly agaric smoke and has also confirmed that inhalation of smoke from the cap produces short but distinct inebriatory effects. Anecdotal reports collected from across the internet, however, are fairly divided on the effectiveness of smoking fly agaric caps. There are several possible explanations for the mixed results. First, is that the traditional method requires mixing fly agaric with tobacco, and it is possible that there is a particular synergy produced by this combination, thus failure to combine fly agaric with tobacco may account for some of the inconsistencies in results. Second, is that the nature of the psychoactive effects is fundamentally different. The effects come on rapidly and are not typically hallucinatory but rather mellow with mild changes in awareness and diminish quickly. It is possible that some are reporting no effects because they are anticipating the more intense effects typically achieved through oral ingestion. Finally, the most likely reason for the disparity in reported results is the actual disparity in potency from mushroom to mushroom. The fly agaric is not the most potent of the psychoactive *Amanitas* and a low potency specimen is likely to produce little to no effect when smoked.

So, how should the fly agaric be prepared for purposes of smoking? To begin you will need the following items:

- Fresh psychoactive *Amanita* mushrooms
- Tobacco, loose

Because the muscarioid *Amanitas* are only mildly to moderately potent it is preferable to use a pantheroid species like *A. velatipes* or *A.* "*pantherina*," since these are more likely to produce a psychoactive smoking blend. Starting with the fresh caps you can proceed in one of two ways. First, you may simply choose to peel the cuticle from the cap, which can be done by pinching the skin at the margin of the cap and pulling it back towards the middle. You might not be able to peel the skin from the central disk, but the cuticle should otherwise be easily removed. The second, and preferred method, would be to slice the caps lengthwise so you get a profile of the cap. Here, you will use a knife to separate not just the cuticle but the colored flesh beneath it from the white flesh of the remainder of the cap. This tiny layer of colored flesh is the most potent part of the mushroom and should make the smoking mixture stronger than just using the cuticle, which is the weakest part of the cap (Gore & Jordan, 1982). Once you have removed the desired part of the mushroom you will want to set this aside for dehydration.

The traditional formulation of this smoking blend would have used *Nicotiana rustica*, sometimes called Aztec tobacco, which is used in shamanic practice and in other ritual contexts throughout the Americas. *N. rustica* plants have been documented with nicotine levels up to 9%, far exceeding the levels found in commercial tobacco (*N. tabacum*), which range between 1 and 3 percent (Buchanan, 1994). *N. rustica* also contains the psychoactive harmala alkaloids harmane and norharmane (Janiger & de Rios, 1976), which almost certainly contribute to the overall effect of the smoking blend. However, unless you grow your own Aztec tobacco you will probably have to settle for using commercial rolling tobacco.

Once your mushroom material is dried you will mix equal parts of tobacco with your mushrooms. Crumble your material lightly to encourage integration and homogenization, then smoke in a pipe or as a cigarette.

Final Thoughts on the "Formula"

The "Formula" for preparing psychoactive *Amanitas* in a manner that successfully reduces or eliminates gastrointestinal distress while ensuring a successful psychoactive experience, if there is one, is elusive. While a "perfect" formula is unknown, I have strived to provide the reader with some simple steps that can be taken to reduce the likelihood of experiencing severe gastrointestinal distress and improve the likelihood of experiencing some of the unique psychoactive effects produced by the fly agaric and its psychoactive relatives. In any case, any "successful" formula that may be discovered will likely include many of the elements discussed here: dehydrating mushrooms, brewing in a low pH solution, and possibly treatment with glutamate decarboxylase (GAD). There is still much to learn about these finnicky mushrooms, but even without a perfect "formula", mindful and proper preparation can make the difference between a non-experience, a horrible one, and an ecstatic or otherwise productive experience. Preparation matters!

Some Notes of Caution

One might presume that a topic with a whole book dedicated to it, like this one on the fly agaric, might indicate that the subject matter is safe and without any significant dangers. Such a presumption, of course, would be incorrect. One can find any number of books on topics or activities that humans enjoy, such as mountain biking or hang-gliding, where issues of safety and potential dangers are naturally implicated. There is nothing inherently deadly about a mountain bike, though it can be used in ways that raise the risk of sustaining life-threatening injuries. While most bike injuries are simple scrapes and bruises, they can also include broken limbs and head injuries and at the very worst, may result in death. These are known risks and people weigh these risks against their enjoyment of the activity when deciding if, and at what level, they wish to engage in this particular sport. Of course, knowing that there are dangers involved, safety precautions have been created to reduce the risk of injury to bikers. Bikes are typically equipped with brakes and specially designed helmets are also available to help prevent serious head injuries.

Similarly, any use of the fly agaric mushroom poses potential risks. The best way to avoid these risks, of course, is to not use these mushrooms at all. Just as with mountain biking, hang-gliding, or driving a car, however, there are ways to minimize and mitigate the potential dangers one might encounter with this mushroom; but first, it is important to understand what these dangers might be. While the fly agaric is not known to be deadly poisonous the effects of the mushroom can lead one into potentially life-threatening circumstances. A 2006 review of over thirty-years of mushroom poisonings, including approximately 2,000 incidents, revealed a single death attributed, in part, to *Amanita muscaria* ingestion (Beug et al., 2006). In this case, a man became unconscious after consuming a large quantity of *Amanita muscaria* while camping in Michigan and "froze to death" (p. 48). While the details are sparse, presumably the man was prepared for the cold. However, large doses of *Amanita muscaria* can lead to unconsciousness or simple lack of awareness of one's surroundings. It can also cause an imperviousness to cold and pain, warning signs that would normally cause one to add extra layers or dive deep into their sleeping bag for warmth. Here, the individual did not die of poisoning but died because he did not have the proper mental faculties to protect himself from the cold. In the case of death, of course, this distinction provides little consolation.

The primary risk with consuming the fly agaric, or its relatives, comes at higher doses that may lead to prolonged states of unconsciousness or sleep, or that lead to delirium and dissociation. Unlike *Psilocybe* trips, where one typically retains a solid foot in reality, one under the influence of the fly agaric may become completely *ungrounded*. Under such circumstances, proximity to traffic, bodies of water, cliffs or other rugged terrain could pose life-threatening risks. Exposure to extreme weather conditions in such a state could also be dangerous. Any individual planning to take a moderate to high dose of any of the psychoactive *Amanita* species would be wise to arrange for a "sitter," someone sober who can monitor the inebriated individual and help ensure their safety. This is something that may seem burdensome to some people and I have heard of individuals simply arranging to have someone they can call at any time of the night should they encounter difficulties, such as illness or a "bad trip." This is not a bad idea but will be of limited value to those inebriated to the extent that they no longer "recognize" numbers or otherwise lack the physical coordination to use a phone. It is best to have someone on-site. Someone who can prevent an inebriated individual from walking in the freezing cold without shoes or a jacket, who can help an individual who is ill or sweating profusely, and who can prevent potential mishaps with knives, the stove, or the bathtub. This person does not have to "hover" but should generally be alert during the period of inebriation, particularly during the *peak* of inebriation. In any case, the "sitter" is to the fly agaric-eater as the helmet is to the mountain biker or the airbag to the driver. No one can make someone take this precaution, but it is perhaps the best precaution one can take when planning to experiment with these mushrooms.

Another consideration requires being mindful of where mushrooms are picked. Many mushrooms, and *Amanita muscaria* in particular, can hyper-accumulate heavy metals and toxins from the surrounding environment (Housecroft, 2019; Murati et al., 2015; Falandysz et al., 2018). While it can be hard to resist picking that perfect scarlet specimen one finds in the parking lot, along a highway, on a golf course, or in publicly landscaped spaces, any specimen picked in an area where it may be exposed to pesticides, fertilizers, or excessive amounts of car exhaust, among other pollutants, should not be consumed. Same goes for any mushroom one finds in industrial areas or places where toxins are known to have been stored and or dumped.

A final, but very important consideration, concerns children and pets. Children will not be harmed by touching or handling psychoactive *Amanitas*, but these mushrooms should never be left anywhere where a child might access them unsupervised. While the effects of these mushrooms are not known to be deadly poisonous for adults, some of the more troubling effects of these mushrooms are more pronounced and severe in children. For example, seizures are rarely reported in recorded poisonings of adults but are not infrequent in children who ingest these mushrooms (Benjamin, 1992). Children typically recover after symptoms subside but ingestion of these mushrooms by children should be considered potentially life-threatening and prompt medical attention should be sought for any child suspected of ingesting any of the psychoactive *Amanitas*.

Dogs and cats are also susceptible to poisoning with psychoactive *Amanitas* as both appear to be attracted to the smell and or flavor of these mushrooms. While fatalities of both dogs and cats have been reported, fatal outcomes appear to be rare (Beug et al., 2006; Rossmeisl et al. 2006). In some cases, veterinarians have chosen to euthanize dogs due to the severity of symptoms, however, data indicating that most dogs recover suggests that euthanized dogs were also likely to fully recover once symptoms had subsided (Beug & Shaw, 2009). Thus, euthanization numbers create a skewed picture of the degree to which fatal outcomes are likely in pets. If you suspect your pet has consumed psychoactive *Amanitas* medical assistance should be sought, but you will need to be clear with veterinary staff that euthanization is not an option. In any case, it is best to simply prevent consumption of these mushrooms by children or pets. Please keep your family safe, take proper precautions and store psychoactive mushrooms away from children and animals.

References

A Most Rewarding Experience (2000). Trips: Amanita muscaria. Lycaeum.org (defunct). Retrieved Jan 25, 2008 from http://leda.lycaeum.org/?Table=Trips&Ref_ID=47

Academy Publishers (1872). *Kungliga vitterhets historie och antiqvitets akademiens månadsblad* [The History of the Royal Vitterhets and the Academy of Antiquities] Vol. 1. Stockholm: Academy Publishers [Akademiens Förlag].

Adalsteinsson, J. H. (1978). *Under the cloak: The acceptance of Christianity in Iceland with particular reference to the religious attitudes prevailing at the time.* Stockholm: Almqvist & Wiksell International.

Afanas'ev, A. (1945). *Russian fairy tales.* New York: Pantheon.

Allegro, J.M. (1970). *The sacred mushroom and the cross.* New York: Bantam.

Allegro, J. M. (1973). *The sacred mushroom and the cross. A study of the nature and origins of Christianity within the fertility cults of the Ancient Near East.* Revised Ed. Doubleday.

Allen, T. F. (1874). *Encyclopedia of pure materia medica index*, Vol. I. New York: Boericke & Tafel.

An Encounter with Musicmol Weirdness (2000). Trips: Amanita muscaria. Lycaeum.org (defunct). Retrieved Jan 25, 2008 from http://leda.lycaeum.org/?Table=Trips&Ref_ID=47

Anati, E. (1989). *Origini dell'arte e della concettualità.* Milano: Jaca Book.

Anonymous (1993). Amanita muscaria notes. *The Entheogen Review* 2(4): 8-10.

Anonymous (2000, June 11). Conversational weirdness: An experience with Amanita muscaria (exp365). Erowid.org. https://www.erowid.org/experiences/exp.php?ID=365

Anujan Achchan, P. (1952-53). *The excavation of an umbrella stone monument.* Travancore Cochin: Ad. Report of Department of Archaeology.

Arizona Daily Star (1960). Mushrooming 'mushrooms'. *Arizona Daily Star*.

Associated Press (2006, May 4). Three hospitalized in Olympia after eating poisonous mushrooms. *Seattle Times*. https://archive.seattletimes.com/archive/?date=20060504&slug=webmushrooms04

Atrocitic (2011, Mar. 2). A trip through Hell: An experience with Amanita muscaria (exp73990). Erowid.org. https://www.erowid.org/experiences/exp.php?ID=73990

Atwood, M. A. (1960). *Hermetic philosophy and alchemy.* Jazzybee Verlag.

Austin, T. (2014). Method for producing muscimol and/or reducing ibotenic acid from Amanita tissue. *U.S. Patent No. 8,784,835.* Washington, DC: U.S. Patent and Trademark Office.

Axic (2008, July 20). The void stabbed my solar plexus: An experience with psychoactive Amanita spp. (exp41879). Erowid.org. https://www.erowid.org/experiences/exp.php?ID=41879

Babington, J. (1823). Description of the Pandoo Coolies in Malabar. *Transactions of the Literary Society of Bombay* 3: 324-330.

Bäckman, L. & Hultkrantz, A. (1978). *Studies in Lapp shamanism*. Stockholm: Almqvist & Wiksell International.

Barceloux, D. G. (2008). Isoxazole-containing mushrooms and pantherina syndrome (Amanita muscaria, Amanita pantherina). In D. G. Barceloux (Ed.), *Medical Toxicology of Natural Substances: Foods, Fungi, Medicinal Herbs, Plants, and Venomous Animals* (pp. 298–302). Hoboken, NJ: John Wiley & Sons, Inc.

Barman, B., Warjri, S., Lynrah, K. G., Phukan, P., & Mitchell, S. T. (2018). Amanita nephrotoxic syndrome: Presumptive first case report on the Indian subcontinent. *Indian journal of nephrology* 28(2): 170.

BD, IL (1994). Amanita muscaria answer. *The Entheogen Review* 3(2): 17.

Benedict, R. G. (1966). Chemotaxonomic significance of isoxazole derivatives in Amanita species. *Lloydia* 29: 333-342.

Benjamín, D. R. (1992). Mushroom poisoning in infants and children: The Amanita pantherina/muscaria group. *Journal of Toxicology: Clinical Toxicology* 30(1): 13-22.

Benjamin, D. R. (1995). *Mushrooms: Poisons and panaceas*. W.H. Freeman and Company.

Berezkin, Y. E. (2006). Folklore-mythological parallels among peoples of Western Siberia, Northeastern Asia, and the Lower Amur—Primorye region. *Archaeology, Ethnology and Anthropology of Eurasia* 27: 112-122.

Berlant, S. R. (2005). The entheomycological origin of Egyptian crowns and the esoteric underpinnings of Egyptian religion. *Journal of Ethnopharmacology* 102(2): 275–288.

Berry, R. E., Armstrong, E. M., Beddoes, R. L., Collison, D., Ertok, S. N., Helliwell, M., & Garner, C. D. (1999). The structural characterization of amavadin. *Angewandte Chemie International Edition* 38(6): 795–797.

Beug, M. W. (2006). The mushroom poisonings 2001-2004. *McIlvainea* 16(1): 56-69.

Beug, M. W. (2007). NAMA toxicology committee report for 2006: Recent mushroom poisonings in North America. *McIlvainea* 17(1): 63-72.

Beug, M. W., & Shaw, M. (2009). Animal poisoning by Amanita pantherina and Amanita muscaria: A commentary. *McIlvainea* 18(1): 37-39.

Beug, W., Shaw, M., & Cochran, K. W. (2006). Thirty plus years of mushroom poisoning: summary of the approximately 2,000 reports in the NAMA case registry. *McIlvainea* 16(2): 47-68.

BF, NC (1994). Amanita and milk thistle. *The Entheogen Review* 3(4): 18.

Bhattacharjee, A. K., Nagashima, T., Kondoh, T., & Tamaki, N. (2001). The effects of the Na+/Ca++ exchange blocker on osmotic blood-brain barrier disruption. *Brain Research* 900(2): 157-162.

Bicknell, C. (1972[1913]). *Guida alle incisioni rupestre preistoriche nella Alpi Marittime italiane*. Bordighera: Istituto Internazionale Studi Liguri.

Blöndal, S. (1924). *Islandsk-dansk ordbog*. Copenhagen: H. Aschenhoug & Co.

Bogoraz, V. G. (1904-09). *The Chukchee*. New York: Memoir of the American Museum of Natural History.

Booth (2008, Feb 27). Universal conclusions: An experience with Amanita muscaria (exp49060). Erowid.org. https://www.erowid.org/experiences/exp.php?ID=49060

Borhegyi de, S. F. (1961). Miniature mushroom stones from Guatemala. *American Antiquity* 26(4): 498-504.

Borhegyi de, S. F. (1963). Pre-Columbian pottery mushrooms from Mesoamerica. *American Antiquity* 28(3): 328-338.

Borhegyi de, S. F. (1965). Archaeological synthesis of the Guatemalan Highlands. In: G. R. Willey (Ed.), *Handbook of Middle American Indians*, Vol. 2 (pp. 3-38). Austin, TX: University of Texas Press.

Borhegyi de, S. F. (1980). The Pre-Columbian ballgame: A pan-Mesoamerican tradition. Milwaukee, WI: Milwaukee Public Museum.

Borhegyi de, S. F. & Borhegyi de, S. (1963). The rubber ball game of ancient America. Milwaukee, WI: Milwaukee Public Museum.

Briem, O. & Briem, J. (1968). *Eddu Kvædi*. Reykjavík: Skálholt.

Brown, R. C., Egleton, R. D., & Davis, T. P. (2004). Mannitol opening of the blood-brain barrier: Regional variation in the permeability of sucrose, but not 86Rb+ or albumin. *Brain Research* 1014(1–2): 221–227.

Buchanan, R. (1994). Tobacco: The most provocative herb. *Mother Earth Living* October/November: 34-38.

Buchholz, P. (1984). Odin: Celtic and Siberian affinities of a German deity. *Mankind Quarterly* 24(4): 427-437.

Buckskin, F., & Benson, A. (2005). The contemporary use of psychoactive mushrooms in Northern California. *Journal of California and Great Basin Anthropology* 25: 87-92.

Bunyard, B. A. (2018). Deadly Amanita mushrooms as food. *Fungi* 10(4): 40-48.

Byrne, A. R., Šlejkovec, Z., Stijve, T., Fay, L., Gössler, W., Gailer, J., & Lrgolic, K. J. (1995). Arsenobetaine and other arsenic species in mushrooms. *Applied Organometallic Chemistry* 9(4): 305–313.

C. (2016, May 27) Mildly altered to euphoria: An experience with Amanita muscaria (exp108578). Erowid.org. https://www.erowid.org/experiences/exp.php?ID=108578

Caesar, J. (50 BCE). *Gallic Wars*.

Calder, G. (1917). *Auraicept na n-eces: The scholars primer being the Ogham Tract from The Book Of Ballymote*. John Grant; First Scottish Edition.

Cao, R., Peng, W., Wang, Z., & Xu, A. (2007). Beta-carboline alkaloids: Biochemical and pharmacological functions. *Current Medicinal Chemistry* 14(4): 479–500.

Caputo, F., & Bernardi, M. (2010). Medications acting on the GABA system in the treatment of alcoholic patients. *Current pharmaceutical design* 16(19): 2118-2125.

Carapeto, L. P., Cárcamo, M. C., Duarte, J. P., de Melo, L. G., Bernardi, E., & Ribeiro, P. B. (2017). Larvicidal efficiency of the fungus Amanita muscaria (Agaricales, Amanitaceae) against Musca domestica (Diptera, Muscidae). *Biotemas* 30(3): 79-83.

Cárcamo, M. C., Carapeto, L. P., Duarte, J. P., Bernardi, E., & Ribeiro, P. B. (2016). Larvicidal efficiency of the mushroom Amanita muscaria (Agaricales, Amanitaceae) against the mosquito Culex-quinquefasciatus (Diptera, Culicidae). *Revista da Sociedade Brasileira de Medicina Tropical* 49(1): 95-98.

Catalfomo, R., & Eugster, C. H. (1970). Amanita muscaria: Present understanding of its chemistry.

Bulletin on Narcotics 22(4): 33–41.

Catalfomo, P., & Eugster, C. H. (1970). L'Amanita muscaria: connaissance actuelle de ses principes actifs. *Bulletin des Stupefiants* 22: 35-43.

Chad (2011, Aug 7). The expanding mobius strip: An experience with Amanita muscaria (exp68309). Erowid.org. https://www.erowid.org/experiences/exp.php?ID=68309

Chadwick, N. K. (1935). Imbas forosnai. *Scottish Gaelic Studies* 4(2): 97-135.

Chadwick, N. K. (1970). *The Celts*. Harmondsworth: Penguin.

ChemBob (2005, Dec 28). Interesting, not eager to repeat it (exp30701). Erowid.org. https://www.erowid.org/experiences/exp.php?ID=30701

Chen, S., Yong, T., Zhang, Y., Su, J., Jiao, C., & Xie, Y. (2017). Anti-tumor and anti-angiogenic ergosterols from Ganoderma lucidum. *Frontiers in Chemistry* 5: 85.

Chilton, W. S. (1975). The course of an intentional poisoning. *McIlvainea* 2: 17-18.

Chilton, W. S., & Ott, J. (1976). Toxic metabolites of Amanita pantherina, A. cothurnata, A. muscaria and other Amanita species. *Lloydia* 39(2-3): 150-157.

Chinnian, P. (1983). Megalithic monuments and megalithic culture in Tamil Nadu. In S. V. Subramanian & K. D. Thirunavukkarasu (Eds.), *Historical heritage of the Tamils* (pp. 25-62). Madras: International Institute of Tamil Studies.

Christenson, A. J. (2007). *Popol Vuh: The sacred book of the Maya: The great classic of Central American spirituality*. Translated from the Original Maya text. Norman, OK: University of Oklahoma Press.

Cleasby, R., Vigfússon, G., & Craigie, W. A. (1957). *An Icelandic-English dictionary, initiated by Richard Cleasby, subsequently revised, enlarged and completed by Gudbrand Vigfusson,... with a supplement by Sir William A. Craigie...* Clarendon Press.

Cochran, K.W. (1985). Poisoning in 1984. *Mushroom: The Journal of Wild Mushrooming* (Spring): 30-33.

Cochran, K.W. (1999). 1998 annual report of the North American Mycological Association's Mushroom Poisoning Case Registry. *McIlvainea* 14(1): 93-98.

Cochran, K.W. (2000). 1999 annual report of the North American Mycological Association's Mushroom Poisoning Case Registry. *McIlvainea* 14(2): 34-40.

Colby, J. M., et al. (2013, Oct). *Semi-quantitative GC-MS/MS method for identification of muscimol and ibotenic acid in Amanita mushrooms*. Poster at Society of Forensic Toxicology's 2013 Meeting (Orlando, Florida).

Corbett, R., Fielding, S., Cornfeldt, M., & Dunn, R. W. (1991). GABAmimetic agents display anxiolytic-like effects in the social interaction and elevated plus maze procedures. *Psychopharmacology* 104(3): 312-316.

Cowan, T. (1993). *Fire in the head: Shamanism and the Celtic spirit*. San Francisco: Harper.

Cox, E. D., Hamaker, L. K., Li, J., Yu, P., Czerwinski, K. M., Deng, L., ... & Krawiec, M. (1997). Enantiospecific formation of trans 1, 3-disubstituted tetrahydro-β-carbolines by the Pictet–Spengler reaction and conversion of cis diastereomers into their trans counterparts by scission of the C-1/N-2 bond. *The Journal of Organic Chemistry* 62(1): 44-61.

Coxwell, C.F. (1925). *Siberian and other folk-tales*. London: C.W. Daniel.

Cripps, C. L., Lindgren, J. E., & Barge, E. G. (2017). Amanita alpinicola sp. nov., associated with

Pinus albicaulis, a western 5-needle pine. *Mycotaxon* 132(3): 665-676.

Cross, T. P., & Slover, C. H. (1988). *Ancient Irish tales*. Totowa: Barnes & Noble.

Culbert, T. P. (Ed.). (1974). *The lost civilization: The story of the Classic Maya*. New York: Harper & Row Publishers.

da Silva, J. A., da Silva, J. J. F., & Pombeiro, A. J. (2013). Amavadin, a vanadium natural complex: Its role and applications. *Coordination Chemistry Reviews* 257(15-16): 2388-2400.

Davidson, H. R. E. (1969). *Scandinavian mythology*. Verona, Italy: Hamlyn Publishing Group.

D.D., CA (1999). Hyperspatial maps. *The Entheogen Review* 8(4): 133.

de Félice, P. (1936). *Poisons sacrés, ivresses divines: Essai sur quelques forms inférieures de la mystique*. Paris.

Deja, S., Jawień, E., Jasicka-Misiak, I., Halama, M., Wieczorek, P., Kafarski, P., & Młynarz, P. (2014). Rapid determination of ibotenic acid and muscimol in human urine. *Magnetic Resonance in Chemistry* 52(11): 711-714.

de la Garza, M. (2012). *Sueño y éxtasis: Visión chamánica de los Nahuas y los Mayas*. Mexico City: Fondo de Cultura Económica.

Devlet, M. A. (1982). Petroglify verchniego yenisieia. *Akademya Nauk SSRR* 2: 111-120.

Devlet, M. A. (2001). Petroglyphs on the bottom of the Sayan Sea (Mount Aldy-Mozaga). Part II. *Anthropology Archaeology in Eurasia* 40(2): 7-94.

Díaz, J. L. (1979). Ethnopharmacology and taxonomy of Mexican psychodysleptic plants. *Journal of Psychedelic Drugs* 11(1-2): 71-101.

Dikov, N. N. (1971). *Naskal'nye zagadki drevney Chukotki: Petroglify Pegtymelya*. Moskva: Isdat Nauka.

Dikov, N. N. (2004). *Early culture of Northeast Asia*. Anchorage: U.S. Department of Interior.

Doniger, W. (1977). *Siva: l'asceta erotico*. Milano: Adelphi.

Doniger, W. (2005). *The Rig Veda*. Penguin Classics.

Drugs Forum (2009, Feb 3). *Amanitas* [Online Forum]. Drugs-Forum: Addiction Help & Harm Reduction. http://www.drugs-forum.com/forum/archive/index.php/f-79.html

Dugan, F. M. (2009). Dregs of our forgotten ancestors: fermentative microorganisms in the prehistory of Europe, the steppes and Indo-Iranian Asia, and their contemporary use in traditional and probiotic beverages. *Fungi* 2(4): 16-39.

Dugan, F. M. (2011). *Conspectus of world ethnomycology: Fungi in ceremonies, crafts, diets, medicines, and myths*. St. Paul: APS Press.

Dugan, F. M. (2017). Baba Yaga and the mushrooms. *Fungi* 10(2): 6-17.

Dumézil, G., & Hiltebeitel, A. (1970). *The destiny of the warrior*. Chicago: University of Chicago Press.

Dunn, E. (1973). Russian use of Amanita muscaria: A footnote to Wasson's Soma. *Current Anthropology* 14(4): 488–492.

Durán, D. (1971). *Book of the gods and rites and the ancient calendar* (F. Horcasitas & D. Heyden, Trans.). Norman, OK: University of Oklahoma Press.

Duvivier, P. (1998). *Amanita muscaria*, ancient history. *The Entheogen Review* 7(2): 34-35.

Edmonson, M. S. (1971). *The book of counsel: The Popol Vuh of the Quiche Maya of Guatemala*. Middle American Research Institute Publication 35. New Orleans: Tulane University.

Egilsson, S. (1931). *Ordbog over det Norsk-Islandsk Skjaldesprog* (2nd Edition). Copenhagen: S. L. Mollers Bogtrykkeri.

Egli, S., Peter, M., Buser, C., Stahel, W., & Ayer, F. (2006). Mushroom picking does not impair future harvests–results of a long-term study in Switzerland. *Biological conservation* 129(2): 271-276.

Ekholm, S. M. (1968). A three-sided figurine from Izapa, Chiapas, Mexico. *American Antiquity* 33(3): 376-379.

Eliade, M. (1988). *Shamanism*. London: Penguin Books.

Ellis, P. B. (2003). *The Celts: A history*. Running Press.

Eluère, C. (1993). *The Celts: Conquerors of Ancient Europe* (D. Biggs, Trans.). New York: Harry N. Abrams.

Entheogen Dot Com (2008). *The mush room* [Online Forum]. Entheogen.com (defunct). Retrieved Jan. 27, 2008 from http://www.entheogen.com/forum/forumdisplay.php?f=140

Eriksson, J. (1991). *Kosmisk extas* (including article by B. Collinder). Uppsala: Gimle förlag.

Erjavec, J., Kos, J., Ravnikar, M., Dreo, T., & Sabotič, J. (2012). Proteins of higher fungi - from forest to application. *Trends in Biotechnology* 30(5): 259–273.

Erowid (2009, Jan 25). *Amanitas Reports* [Online Forum]. Erowid.org. http://www.erowid.org/experiences/subs/exp_Amanitas.shtml

Erowid (2018). Psychoactive *Amanitas*. Erowid.com. https://erowid.org/plants/amanitas/amanitas.shtml

Eugster, C. H. (1956). Über muscarin aus Fliegenpilzen. *Helvetica Chimica Acta* 39(4): 1002.

Eugster, C. H. (1959). Brève revue d'ensemble sur la chimie de la muscarine. *Revue de Mycologie* 24(5): 369-385.

Eugster, C. H. (1967). Isolation, structure, and syntheses of central-active compounds from Amanita muscaria (L. ex Fr.) hooker. *Psychopharmacology Bulletin* 4(3): 18–19.

Eugster, C. H. (1968). Wirkstoffe aus dem Fliegenpilz. *Die Naturwissenschaften* 55(7): 305–313.

Existence (2011, Nov 11). Infinite nothingness: An experience with Amanita muscaria var. formosa (exp66049). Erowid.org. https://www.erowid.org/experiences/exp.php?ID=66049

Fabing, H. D. (1956). On going berserk: A neurochemical inquiry. *The Scientific Monthly* 83(5): 232-237.

Falandysz, J., Mędyk, M., & Treu, R. (2018). Bio-concentration potential and associations of heavy metals in Amanita muscaria (L.) Lam. from northern regions of Poland. *Environmental Science and Pollution Research* 25(25): 25190-25206.

Feeney, K. (2010). Revisiting Wasson's soma: Exploring the effects of preparation on the chemistry of Amanita muscaria. *Journal of Psychoactive Drugs* 42(4): 499-506.

Feeney, K. (2013). The significance of pharmacological and biological indicators in identifying historical uses of Amanita muscaria. In J. A. Rush (Ed.), *Entheogens and the development of culture* (pp. 279-317). Berkeley, CA: North Atlantic Books.

Feeney, K. (2014). Peyote as medicine: An examination of therapeutic factors that contribute to healing. *Curare* 37(3): 195-211.

Feeney, K. & Stijve, T. (2010). Re-examining the role of muscarine in the chemistry of Amanita muscaria. *Mushroom, The Journal* Spring-Summer: 32-36.

Feldman, G. F. (2008). *Cannibalism, headhunting and human sacrifice in North America*. Chambersburg, Pennsylvania: Alan C. Hood & Co.

Festi, F., & Bianchi, A. (1991). Amanita muscaria: Mycopharmacological outline and personal experiences. In T. Lyttle (Ed.), *Psychedelic monographs and essays* (pp. 209–233). PM&E Publishing Group.

Festi, F., & Bianchi, A. (1992). Amanita muscaria. *Integration Journal of Mind-Moving Plants and Culture* 2(3): 79-89.

Filer, C. N. (2018). Ibotenic acid: On the mechanism of its conversion to [3 H] muscimol. *Journal of Radioanalytical and Nuclear Chemistry* 318(3): 2033-2038.

Filer, C. N., Lacy, J. M., & Peng, C. T. (2005). Ibotenic acid decarboxylation to muscimol: Dramatic solvent and radiolytic rate acceleration. *Synthetic communications* 35(7): 967-970.

Forrester, S., Goscilo, H., and M. Skoro (Eds.). (2013). *Baba Yaga: The wild witch of the east in Russian fairy tales*. Forward by J. Zipes, introduction & trans. S. Forrester, captions to illustrations H. Goscilo, selection of images M. Skoro and H. Goscilo. Jackson: University Press of Mississippi.

Fouilleux, B. & Mouchet, A. (2010). Un trait culturel du style des "Têtes Rondes" de la Tassili-n-Ajjer (Algérie): les masques dans leur environnement. *Les Cahiers de l'AARS* 14: 131-142.

Fox, J. W. (1987). *Maya postclassic state formation: Segmentary lineage migration in advancing frontiers*. New Studies in Archeology. Cambridge University Press.

Funk Shui (2010, Mar 5). Nice red mushrooms: An experience with psychoactive Amanita spp. (exp49265). Erowid.org. https://www.erowid.org/experiences/exp.php?ID=49265

Furst, P. T. (1972). *Flesh of the gods: The ritual use of hallucinogens*. New York: Praeger.

Furst, P. T. (1974). Hallucinogens in precolumbian art. In M.E. King & I.R. Traylor (Eds.), *Art and environment in native America* (pp. 55-101). Lubbock: Texas Tech.

Gantz, J. (1981). *Early Irish myths and sagas*. London: Penguin Books.

Gelling, P., & Davidson, H. R. E. (1969). *The chariot of the sun: and other rites and symbols of the northern bronze age*. Aldine Paperbacks.

Geml, J., Laursen, G. A., O'Neill, K., Nusbaum, H. C., & Taylor, D. L. (2006). Beringian origins and cryptic speciation events in the fly agaric (Amanita muscaria). *Molecular Ecology* 15(1): 225-239.

Geml, J., Tulloss, R. E., Laursen, G. A., Sazanova, N. A., & Taylor, D. L. (2008). Evidence for strong inter-and intracontinental phylogeographic structure in Amanita muscaria, a wind-dispersed ectomycorrhizal basidiomycete. *Molecular Phylogenetics and Evolution* 48(2): 694-701.

Girod, P.-A., & Zryd, J.-P. (1991). Biogenesis of betalains: Purification and partial characterization of dopa 4,5-dioxygenase from Amanita muscaria. *Phytochemistry* 30(1): 169-174.

Global Village Video (1993). The soma seminar: with Emanuel Salzman, Andrew Weil, Mark Niemoller, and Walter Johnson. Telluride Mushroom Festival, 1992 [VHS].

Gore, M. G., & Jordan, P. M. (1982). Microbore single-column analysis of pharmacologically active alkaloids from the fly agaric mushroom Amanita muscaria. *Journal of Chromatography A* 243(2): 323-328.

Goscilo, H. (2013). Caption to 'storehouse.' In: S. Forrester, H. Goscilo, M. Skoro (Eds.), *Baba Yaga: The wild witch of the east in Russian fairy tales* (p. xxvii). Jackson: University Press of

Mississippi.

Gottlieb, A. (1973). *Legal highs: A concise encyclopedia of legal herbs and chemicals with psychoactive properties*. Manhattan Beach: 20th Century Alchemist.

Gowda, U., Mutowo, M. P., Smith, B. J., Wluka, A. E., & Renzaho, A. M. (2015). Vitamin D supplementation to reduce depression in adults: meta-analysis of randomized controlled trials. *Nutrition* 31(3): 421-429.

Grant, W. B., & Holick, M. F. (2005). Benefits and requirements of vitamin D for optimal health: a review. *Alternative Medicine Review* 10(2): 94-111.

Graves, R. (1966). *The white goddess: A historical grammar of poetic myth*. Toronto: McGraw-Hill Ryerson.

Green, M. J. (1989). *Symbol and image in Celtic religious art*. New York: Routledge.

Green, M. J. (1992). *Dictionary of Celtic myth and legend*. London: Thames & Hudson.

Grieshaber, A. F., Moore, K. A., & Levine, B. (2001). The detection of psilocin in human urine. *Journal of Forensic Science*: 46(3): 627-630.

Griffith, R. T. H. (1891). *The hymns of the Rigveda, Vol. III*. E.J. Lazarus & Co.

Griffith, R. T. H. (1896). *The hymns of the Rig Veda*, 2nd Edition. Kotagiri Nilgiri.

Gubi (2007, Jul 8). The moderate path: An experience with Amanita muscaria & various (exp41814). Erowid.org. https://www.erowid.org/experiences/exp.php?ID=41814

Guzmán, G. (1997). *Los nombres de los hongos y lo relacionado con ellos en America Latina*. Xalapa: Instituto de Ecologia.

Guzman, G. (2001). Hallucinogenic, medicinal, and edible mushrooms in Mexico and Guatemala: Traditions, myths, and knowledge. *International Journal of Medicinal Mushrooms* 3(4): 10.

Guzmán, G. (2012). New taxonomical and ethnomycological observations on Psilocybe SS (fungi, Basidiomycota, agaricomycetidae, Agaricales, Strophariaceae) from Mexico, Africa and Spain. *Acta Botánica Mexicana* 100: 79-106.

Guzmán, G. (2013). Sacred mushrooms and man: Diversity and traditions in the world, with special reference to Psilocybe. In J. A. Rush (Ed.), *Entheogens and the development of culture* (pp. 485-518). Berkeley, CA: North Atlantic Books.

Hahnemann, S. (1833). *The homœopathic medical doctrine: Or "Organon of the healing art"*. Dublin: W. F. Wakeman.

Hall, A. H., & Hall, P. K. (1994). Ibotenic acid/muscimol-containing mushrooms. In D. G. Spoerke & B. H. Rumack (Eds.), *Handbook of mushroom poisoning—Diagnosis and treatment* (pp. 265-278). Boca Raton: CRC Press.

Haney, J.V. (2013). *Long, long tales from the Russian North*. Jackson: University Press of Mississippi.

Härkönen, M. (1998). Uses of mushrooms by Finns and Karelians. *International Journal of Circumpolar Health* 57(1): 40-55.

Harner, M. J. (1973). *Hallucinogens and shamanism*. New York: Oxford University Press.

Hatanaka, S. I. (1992). Amino acids from mushrooms. *Fortschritte der Chemie organischer Naturstoffe/Progress in the Chemistry of Organic Natural Products* 59: 1-140.

Hatto, A. (2017). *The world of the Khanty epic hero-princes: An exploration of a Siberian oral *tradition*. Cambridge: Cambridge University Press.

Hedrick, B. C. (1971). Quetzalcoatl: European or indigene? In C. L. Riley, J. C. Kelley, C. W. Pennington, R. L. Rands (Eds.), *Man across the sea: Problems of Pre-Columbian contacts* (pp. 255-265). University of Texas Press.

Heinrich, C. (1995). *Strange fruit: Alchemy, religion and magical foods*. London: Bloomsbury.

Heinrich, C. (2002). *Magic mushrooms in religion and alchemy*. Rochester, VT: Park Street Press.

Hering, C. (1879). *Guiding symptoms of our materia medica*. Philadelphia, PA: Press of Globe Printing House.

Hoffman, U. & Hoffman, A. (2001). Erinnerungen an den Fliegenpilz (Memories of the Fly-Mushroom). *Entheos: The Journal of Psychedelic Spirituality* 1(1): 9-12.

hÓgáin, D.Ó. (1991). *Myth, legend and romance: An encyclopedia of the Irish folk tradition*. New York: Prentice Hall.

Hoppál, M., & Von Sadovszky, O. (Eds.). (1989). *Shamanism: Past and present*. Budapest: Ethnographic Institute Hungarian Academy of Sciences.

Hosseini, M., Karami, Z., Janzadenh, A., Jameie, S. B., Mashhadi, Z. H., Yousefifard, M., & Nasirinezhad, F. (2014). The effect of intrathecal administration of muscimol on modulation of neuropathic pain symptoms resulting from spinal cord injury; an experimental study. *Emergency* 2(4): 151.

Housecroft, C. E. (2019). The fungus Amanita muscaria: from neurotoxins to vanadium accumulation. *Chimia* 73(1/2): 96-97.

Hughes, R. & Dake, J. P. (1886). *A cyclopædia of drug pathogenesy, Vol. I*. London: E. Gould & Son: Homœopathic Chemists and Publishers.

Ikeda, M., Bhattacharjee, A. K., Kondoh, T., Nagashima, T., & Tamaki, N. (2002). Synergistic effect of cold mannitol and Na+/Ca2+ exchange blocker on blood-brain barrier opening. *Biochemical and Biophysical Research Communications* 291(3): 669-674.

Ingalls, D. H. H. (1971). Remarks on Mr. Wasson's soma. *Journal of the American Oriental Society* 91 (2): 188-91.

Irimoto, T. (2004). *The eternal cycle: Ecology, worldview and ritual of reindeer herders of Northern Kamchatka*. Osaka, Japan: National Museum of Ethnology.

Irvin, J., & Rutajit, A. (2006). *Astrotheology and shamanism*. The Book Tree.

Ivanova, E. V. (2013). The problem of mysteriousness of Baba Yaga character in religious mythology. *Journal of Siberian Federal University, Humanities & Social Sciences* 12: 1857-1866.

Iyer, K. L .A. (1967). *Kerala megaliths and their builders*. Madras: University of Madras.

Janiger, O., & De Rios, M. D. (1976). Nicotiana an hallucinogen? *Economic Botany* 30(3): 295-297.

Jäpelt, R. B., & Jakobsen, J. (2013). Vitamin D in plants: a review of occurrence, analysis, and biosynthesis. *Frontiers in Plant Science* 4: 136.

Jenkins, D. T. (1986). *Amanita of North America*. Eureka, CA: Mad River Press.

John, K. J. P. (1982). New light on the Kodakkal of Malabar. In R.K. Sharma (Ed.), *Indian Archaeology. New perspectives* (pp. 148-154). New Delhi: Agam Kala.

Johns, A. (2004). *Baba Yaga: The ambiguous mother and witch of the Russian folktale*. International Folkloristics Vol. 3. New York: Peter Lang.

Jones, G. (Ed.). (1961). *Eirik the Red and other Icelandic sagas*. New York: Oxford University Press.

Jung, C. G. (1936, March). Wotan. *Neue Schweizer Rundschau*, No. 3. Zurich.

Kalberer, F., Kreis, W. & Rutschman, J. (1962). The fate of psilocin in the rat. *Biochemical Pharmacology* 11(4): 261-269.

Kaplan, R. W. (1975). The sacred mushroom in Scandinavia. *Man* 10(1): 72-79.

Karazhanova, I. (2016). Monstrous femininity in Kazakh folklore: Delineating normative and transgressive womanhood. School of Humanities and Social Sciences of Nazarbayev University. MS thesis, Astana, Kazakhstan.

Karliński, L., Ravnskov, S., Kieliszewska-Rokicka, B., & Larsen, J. (2007). Fatty acid composition of various ectomycorrhizal fungi and ectomycorrhizas of Norway spruce. *Soil Biology & Biochemistry* 39: 854–866.

Kendrick, B. (2000). *The fifth kingdom*, 3rd Edition. Newburyport, MA: Focus Publishing.

Kershaw, K. (2000). The one-eyed god and the (Indo-)Germanic "Männerbünde". *Journal of Indo-European Studies*, monograph 36. Washington: Institute for the Study of Man.

Kiho, T., Katsurawaga, M., Nagai, K., Ukai, S., & Haga, M. (1992). Structure and antitumor activity of a branched (1→3)-β-d-glucan from the alkaline extract of Amanita muscaria. *Carbohydrate Research* 224: 237–243.

Kiho, T., Yoshida, I., Katsuragawa, M., Sakushima, M., Usui, S., & Ukai, S. (1994). Polysaccharides in fungi. XXXIV. A polysaccharide from the fruiting bodies of Amanita muscaria and the antitumor activity of its carboxymethylated product. *Biological & Pharmaceutical Bulletin* 17(11): 1460–1462.

Kinsella, T. (Trans.). (1969). *The Táin: from the Irish epic Táin Bó Cuailnge*. London: Oxford University Press.

Kirchmair, M., Carrilho, P., Pfab, R., Haberl, B., Felgueiras, J., Carvalho, F., ... & Neuhauser, S. (2012). Amanita poisonings resulting in acute, reversible renal failure: new cases, new toxic Amanita mushrooms. *Nephrology Dialysis Transplantation* 27(4): 1380-1386.

Kirchhoff, P. (1943). Mesoamérica: Sus límites geográficos, composición étnica y caracteres culturales. *Acta Americana* (1): 92–107.

Kluge, W. (1989). *Etymological dictionary of the German language*, 22nd Ed. Revised. Berlin: Elmar Seebold (Edition).

Knab, T. (1976-1978). Minor Mexican pharmacogens: Context and effects. Unpublished manuscript.

Kodolány, J., Jr. (1968). Khanty (Ostyak) sheds for sacrificial objects. In: V. Diószegi (Ed.), *Popular beliefs and folklore tradition in Siberia* (pp. 103-106). Bloomington: Indiana University.

Kohlmunzer, S., & Grzybek, J. (1972). Charakterystyczne składniki chemiczne grzybów wielkoowocnikowych (Macromycetes). *Wiadomości Botaniczne* 16(1): 35–56.

Koukol, O., Novák, F., & Hrabal, R. (2008). Composition of the organic phosphorus fraction in basidiocarps of saprotrophic and mycorrhizal fungi. *Soil Biology and Biochemistry* 40(9): 2464-2467.

Kubarev, V. D. & Jacobson, E. (1996). *Répertoire des pétroglyphes d'Asie Centrale. Vol. 3. Sibérie du Sud: Kalbak-Tash I (République de l'Altai)*. Paris: Diffusion de Boccard.

Kuehnelt, D., Goessler, W., & Irgolic, K. J. (1997). Arsenic compounds in terrestrial organisms

II: Arsenocholine in the mushroom Amanita muscaria. *Applied Organometallic Chemistry* 11(6): 459–470.

Kulemzin, V. M. (1984). *Chelovek i priroda v verovaniyah khantov*. Tomsk.

Kurlander, E. (2017). *Hitler's monsters*. Yale University Press.

LaFaye, J. (1987). *Quetzalcoatl and Guadalupe: The formation of Mexican national consciousness*. Chicago, IL: University of Chicago Press.

Laurie, E. R. (1996). The cauldron of poesy. *Obsidian* 1(2): Spring 1996.

Laurie, E. R. & White, T. (1997). Speckled snake, brother of birch: Amanita muscaria motifs in Celtic legends. *Shaman's Drum* 44:53-65.

Lecouteux, C. (1999). *Hadas, brujas y hombres lobo en la Edad Media: historia del doble*. José J. de Olañeta.

Le Quellec, J.-L. (2013). Périodisation et chronologie des images rupestres du Sahara central. *Prèhistoires Méditerranéennes* 4: 2-45.

Letcher, A. (2011, Sept 17). Taking the piss: Reindeer and fly agaric. *Andy Letcher*. http://andy-letcher.blogspot.com/2011/09/taking-piss-reindeers-and-fly-agaric.html

Leto, S. (2000). Magical potions: Entheogenic themes in Scandinavian mythology. *Shaman's Drum* 54(Winter): 55-65.

Lewis, D. H., & Smith, D. C. (1967). Sugar alcohols (polyols) in fungi and green plants: I. Distribution, physiology and metabolism. *New Phytologist* 66(2): 143–184.

Lhote, H. (1968). Données récentes sur les gravures et le peintures rupestres du Sahara. In E. R. Perellò (Ed.), *Simposio de arte rupestre* (pp. 273-290). Barcelona.

Li, C., & Oberlies, N. H. (2005). The most widely recognized mushroom: Chemistry of the genus Amanita. *Life Sciences* 78(5): 532–538.

Li, X., Wu, Q., Xie, Y., Ding, Y., Du, W. W., Sdiri, M., & Yang, B. B. (2015). Ergosterol purified from medicinal mushroom Amauroderma rude inhibits cancer growth in vitro and in vivo by up-regulating multiple tumor suppressors. *Oncotarget* 6(19): 17832–17846.

Limerov, P. F. (2005). Forest myths: A brief overview of ideologies before St. Stefan. *Folklore* 30: 97-135.

Lincoff, G. (2005). "Is the Fly-Agaric (*Amanita muscaria*) an Effective Medicinal." Talk given at the 3rd International Medicinal Mushroom Conference, Port Townsend, Washington. www.nemf.org/files/various/muscaria/fly_agaric_text.html

Lindemann, B., Ogiwara, Y., & Ninomiya, Y. (2002). The discovery of umami. *Chemical Senses* 27(9): 843-844.

Lindgren, J. (2014). Trial key to the species of Amanita in the Pacific Northwest. Pacific Northwest Key Council. http://www.svims.ca/council/Amanit.htm

Linnaeus, C. (1753). *Species plantarum*. Stockholm: Impensis Laurentii Salvii.

Liu, J.-K. (2005). N-containing compounds of macromycetes. *Chemical Reviews* 105(7): 2723–2744.

Longhurst, A.H. (1979). *The story of the stupa*. New Delhi: Asian Educational Services.

Lowy, B. (1974). Amanita muscaria and the thunderbolt legend in Guatemala and Mexico. *Micología* 66: 188-191.

Lowy, B. (1980). *Ethnomycological inferences from mushroom stones, Maya codices and Tzutuhil legend*. Puerto Rico: Interamerican University of Puerto Rico.

Lumley, H., Béguin-Ducornet, J., Échassoux, J., Giusto-Magnardi, N., & Romain, O. (1990). La stèle gravée dite du "chef de tribu" dans la région du Mont Bego, Vallée des Merveilles, Tende, Alpes-Maritimes. *L'Anthropologie* 94(1): 3-62.

Lumpert, M., & Kreft, S. (2016). Catching flies with Amanita muscaria: traditional recipes from Slovenia and their efficacy in the extraction of ibotenic acid. *Journal of Ethnopharmacology* 187: 1-8.

Lundius, N. (1670). *Nicolai Lundii Lappi: Descriptio Lapponiae, from Berättelser om samerna i 1600-talets Sverige*. 1983. KB Wiklund, Umeå.

Lycaeum (2000). *Trips: Amanita muscaria* [Online Forum]. Lycaeum.org (defunct). Retrieved Jan. 25, 2008 from http://leda.lycaeum.org/?Table=Trips&Ref_ID=47

Lytkin, V. I. & Guliaev, E. S. (1999). *Kratkii etimologicheskii slovar komi iazyka* [A Concise Etymological Dictionary of the Komi Language]. Syktyvkar: Komi knizh izd.

Mac Cana, P. (1985). *Celtic mythology*. New York: Peter Bedrick Books.

MacDonald, I. (1992). *Saint Bride*. Edinburgh: Floris Books.

MacDonell, A. A. (1897). *Vedic mythology*. Strassburg: KJ Trübner.

MacDonell, A. A. (2006). *A Vedic reader for students*. Motilal Banarsidass Publishers.

Maciejczyk, E., Jasicka-Misiak, I., Młynarz, P., Lis, T., Wieczorek, P. P., & Kafarski, P. (2012). Muchomor czerwony (Amanita muscaria) jako obiecujące źródło ergosterolu [Fly agaric (Amanita muscaria) as promising source of ergosterol]. *Przemysł Chemiczny* 91(5): 853-855.

Maciejczyk, E., & Kafarski, P. (2013). Mannitol in Amanita muscaria - An osmotic blood-brain barrier disruptor enhancing its hallucinogenic action? *Medical Hypotheses* 81(5): 766-767.

Maciejczyk, Ewa, Wieczorek, D., Zwyrzykowska, A., Halama, M., Jasicka-Misiak, I., & Kafarski, P. (2015). Phosphorus profile of basidiomycetes. *Phosphorus, Sulfur, and Silicon and the Related Elements* 190(5–6): 763–768.

Mägi, K., & Toulouze, E. (2002). On Forest Nenets shaman songs. In: E. Bartha and V. Anttonen (Eds.), *Mental spaces and ritual traditions: An international festschrift to commemorate the 60th birthday of Mihály Hoppál* (pp. 417-433). Debrecen –Turku: University of Debrecen.

Mägi, M. (2005). Mortuary houses in Iron Age Estonia. *Estonian Journal of Archaeology* 9(2): 93-123.

Manilal, B. (1981). An ethnobotanic connection between mushrooms and dolmens. In S. K. Jain (Ed.), *Glimpses of Indian ethnobotany* (pp. 321-325). New Delhi: IBH.

Marley, G. (2010). *Chanterelle dreams, Amanita nightmares: the love, lore, and mystique of mushrooms*. Chelsea Green Publishing.

Marro, G. (1944-45). L'elemento magico nelle figurazioni rupestri delle Alpi Marittime. *Atti dell'Accademia delle Scienze di Torino* 81: 91-95.

Matsushima, Y., Eguchi, F., Kikukawa, T., & Matsuda, T. (2009). Historical overview of psychoactive mushrooms. *Inflammation and Regeneration* 29(1): 47–58.

Matthews, C. & Matthews, J. (1994). *Encyclopedia of Celtic wisdom: The Celtic shaman's sourcebook*. Rockport, MA: Element Books.

Matthews, J. (1991). *Taliesin: Shamanism and the bardic mysteries in Britain and Ireland*. London:

Harper Collins/Aquarian Press.

Mayer, H. K. (1977). *The mushroom stones of Mesoamerica*. Ramona: Acoma.

McGarry, G. (2005). *Brighid's healing: Ireland's Celtic medicine traditions*. Green Magic.

McIntosh, R. J. (1979). The megalith builders of South India: A historical survey. In H. Härtel (Ed.), *South Asian Archaeology 1979* (pp. 459-468). Berlin: Dietrich Reimer Verlag.

McIntosh, R. J. (1985). Dating the South Indian megaliths. In J. Schotsmans & M. Taddei (Eds.), *South Asian Archaeology 1983* (pp. 467-493). Napoli: Istituto Universitario Orientale.

McKenna, T. (1988). Hallucinogenic mushrooms and evolution. *ReVision* 10(4): 51-57.

McKenna, T. (1992). *Food of the gods: The search for the original tree of knowledge*. New York: Bantam Books.

McKenna, T. (1993). *True hallucinations*. San Francisco, CA: HarperSanFrancisco.

McKenny, M & Stuntz, D. E. (2000). *The new savory wild mushroom*. University of Washington Press.

McManus, D. (1991). *A guide to Ogam*. Maynooth, Ireland: An Sagart.

Mendaly, S. (2017). Funeral rituals and megalithic tradition: a study on some ethnic communities in South-Western part of Odisha. *Heritage: Journal of Multidisciplinary Studies in Archaeology* 5: 930-943.

Menon, A. S. (1990). *Kerala history and its makers*. Madras: Viswanathan.

Menon, A. S. (1991). *A survey of Kerala history*. Madras: Viswanatham.

Menon, M. S. (2016). The "round mound" and its structural requirements: a possible scenario for the evolution of the form of the stupa. *Heritage: Journal of Multidisciplinary Studies in Archaeology* 4: 26-46.

Mercier, N., Le Quellec, J.-L, Hachid, M., & Agsous, S. (2012). OSL dating of quaternary deposits associated with the parietal art of the Tassili-n-Ajjer plateau (Central Sahara). *Quaternary Geochronology* 10: 367-373.

Merkur, D. (2014). *Becoming half hidden: Shamanism and initiation among the Inuit*. Routledge.

Meroney, H. (1949). Early Irish letter-names. *Speculum* 24(1): 19-43.

Metzner, R. (1994). *The well of remembrance: Rediscovering the earth wisdom myths of northern Europe*. Shambhala Publications.

Michelot, D., & Melendez-Howell, L. M. (2003). Amanita muscaria: chemistry, biology, toxicology, and ethnomycology. *Mycological Research* 107(2): 131-146.

Mier, N., Canete, S., Klaebe, A., Chavant, L., & Fournier, D. (1996). Insecticidal properties of mushroom and toadstool carpophores. *Phytochemistry* 41(5): 1293-1299.

Miles, S.W. (1965). Summary of pre-conquest ethnology of the Guatemala-Chiapas Highlands and Pacific Slopes. In G. R. Willey (Ed.), *Archaeology of Southern Mesoamerica, Part 1*, (pp. 276-287). Handbook of Middle American Indians, Vol. 2. Austin: University of Texas.

Miller, H. R., & Miller, O. K. (2006). *North American mushrooms: A field guide to edible and inedible fungi*. Guilford, CT: Falcon Guide.

Millman, L. (2015). Gordon Wasson's woman of the northwest wind. *Fungi* 8(4): 26-27.

Ming Yi Tang, R., Kee-Mun Cheah, I., Shze Keong Yew, T., & Halliwell, B. (2018). Distribution and accumulation of dietary ergothioneine and its metabolites in mouse tissues. *Scientific Reports* 8(1): 1-15.

MN, IN (1996). Amanita pantherina. *The Entheogen Review* 5(2): 11.

Mochtar, S. G., & Geerken, H. (1979). Die halluzinogene muscarin und ibotensäure im mittleren Hindukush. Ein beitrag zur volksheilpraktischen mykologie in Afghanistan. *Afghanistan Journal Graz* 6(2): 62-64.

Morgan, A. (1995). *Toads & toadstools: The natural history, folklore, and cultural oddities of a strange association*. Celestial Arts.

Mori, F. (1968). The absolute chronology of Saharan prehistoric rock art. In E. R. Perelló (Ed.), *Simposio internacional de arte rupestre* (pp. 291-294). Barcelona: Diputación Provincial.

Mori, F. (1975). Contributo al pensiero magico-religioso attraverso l'esame di alcune raffigurazioni rupestri preistoriche del Sahara. *Valcamonica Symposium* 72: 344-366.

Mori, F. (1990). La fonction sacrale des abris à peintures dans les massifs centraux du Sahara: l'Acacus. *Origini: Rivista di Prehistoria e Protostoria delle Civiltà Antiche* 15: 79-101.

Moses, J. (2016). Letters to the editor (in response to Millman, 2015). *Fungi* 8(5): 3.

Muller, J., Corodimas, K. P., Fridel, Z., & LeDoux, J. E. (1997). Functional inactivation of the lateral and basal nuclei of the amygdala by muscimol infusion prevents fear conditioning to an explicit conditioned stimulus and to contextual stimuli. *Behavioral neuroscience* 111(4): 683.

Murati, E., Hristovski, S., Melovski, L., & Karadelev, M. (2015). Heavy metals content in Amanita pantherina in a vicinity of the thermo-electric power plant Oslomej, Republic of Macedonia. *Fresenius Environmental Bulletin* 24(5): 1981-1984.

Muto, T., Sugawara, R., & Mizoguchi, K. (1968). The house fly attractants in mushrooms. *Agricultural and Biological Chemistry* 32(5): 624–627.

Muto, T., & Sugawara, R. (1970). 1, 3-diolein, a house fly attractant in the mushroom, Amanita muscaria (L.) Fr. In D. Wood, R. Silverstein & M. Nakajima (Eds.), *Control of insect behavior by natural products* (pp. 189-208). Academic Press, Inc.

Muzzolini, A. (1986). *L'art rupestre préhistorique des massif centraux sahariens*. Oxford: BAR.

Muzzolini, A. (1991). Proposal for updating the rock-drawing sequence of the Acacus (Lybia). *Lybian Studies* 22: 7-30.

Mycotopia (2009, Feb 3). *Mycotopia* [Online Forum]. Mycotopia.net. http://forums.mycotopia.net/

Nagy, J. F. (1985). *The wisdom of the outlaw: The boyhood deeds of Finn in Gaelic narrative tradition*. Berkeley: University of California Press.

Narasimhaiah, B. (1995). Excavation at Cheramangad, district Trichur. *Indian Archaeology 1990–1991*: 33–34.

Natarajan, K. & Raman, N. (1983). *South Indian agaricales. A preliminary study on some dark spored species*. Veracruz: Cramer.

Navet, É. (1988). Les Ojibway et l'Amanite tue-mouche (Amanita muscaria). Pour une ethnomycologie des Indiens d'Amérique du Nord. *Journal de la Société des Américanistes* 74: 163-180.

Neuroglider (2000). I am a river: An experience with dried mushrooms. Lycaeum.org (defunct). Retrieved Jan. 25, 2008 from http://leda.lycaeum.org/?ID=6535

Nichols, B. (2000). The fly-agaric and early Scandinavian religion. *Eleusis: Journal of Psychoactive Plants & Compounds* 4: 87-119.

Nicholson, I. (1967). *Mexican and Central American mythology*. London: Paul Hamlyn.

Nielsen, E. Ø., Schousboe, A., Hansen, S. H., & Krogsgaard-Larsen, P. (1985). Excitatory amino acids: Studies on the biochemical and chemical stability of ibotenic acid and related compounds. *Journal of Neurochemistry* 45(3): 725-731.

Nightmare (2011, Jun 16). Enlightened stupid guy: An experience with Amanita muscaria (exp75249). Erowid.org. https://www.erowid.org/experiences/exp.php?ID=75249

Norm de Plume (2003, Nov 10). Delightful stimulant effect: An experience with Amanita muscaria (exp12889). Erowid.org. https://www.erowid.org/experiences/exp.php?ID=12889

Norvell, L. (1995). Loving the chanterelle to death? The ten-year Oregon chanterelle project. *McIlvainea* 12(1): 6-25.

Ó Catháin, S. (1995). *The festival of Brigit: Celtic goddess and holy woman*. Blackrock: DBA Publications.

O'Curry, E. (1878). *Lectures on the manuscript materials of ancient Irish history*. Dublin: William Hinch & Patrick Traynor.

O'Flaherty, W. D. (1981). *The Rig Veda*. London: Penguin.

O'Grady, S. H. (1892). *Silva Gaelica*, Volume 2. London & Edinburgh: Williams & Norgate.

Olmstead, G. S. (1994). *The gods of the Celts and the Indo-Europeans*. Budapest: Archeolingua.

Olsen, A. S. B., & Færgeman, N. J. (2017). Sphingolipids: membrane microdomains in brain development, function and neurological diseases. *Open Biology* 7(5): 170069.

Ooms, K. J., Bolte, S. E., Baruah, B., Choudhary, M. A., Crans, D. C., & Polenova, T. (2009). 51V solid-state NMR and density functional theory studies of eight-coordinate non-oxo vanadium complexes: Oxidized amavadin. *Dalton Transactions* 17: 3262–3269.

Ott, J. (1976a). *Hallucinogenic plants of North America*. Berkeley: Wingbow Press.

Ott, J. (1976b). Psycho-mycological studies of Amanita - From ancient sacrament to modern phobia. *Journal of Psychoactive Drugs* 8(1): 27–35.

Ott, J. 1993. *Pharmacotheon: Entheogenic drugs, their plant sources and history*. Kennewick, WA: Natural Products Co.

Ott, J. (1998). The post-Wasson history of the soma plant. *Eleusis: Journal of Psychoactive Plants & Compounds* 1: 9-37.

Ott, J., Wheaton, P. S., & Chilton, W. S. (1975). Fate of muscimol in the mouse. *Physiological Chemistry and Physics* 7(4): 381.

Parker, G. B., Brotchie, H., & Graham, R. K. (2017). Vitamin D and depression. *Journal of Affective Disorders* 208: 56-61.

Pedersen, C., & Schubert, L. (1993). Synthesis of the four stereoisomers of 4,5-dihydroxy-N,N,N-trimethylhexanaminium iodide (muscaridin) from aldonolactones. *Acta Chemica Scandinavica* 47: 885–888.

Peter, J. (2015). A study of the Umbrella Stones in Kerala. *Heritage: Journal of Multidisciplinary Studies in Archaeology* 3: 283-294.

Pfeifer, W. (1995). *Etymologisches wörterbuch des Deutschen* (Etymological dictionary of German). Munich: dtv.

Phillips, K. M., Ruggio, D. M., Horst, R. L., Minor, B., Simon, R. R., Feeney, M. J., ... & Haytowitz,

D. B. (2011). Vitamin D and sterol composition of 10 types of mushrooms from retail suppliers in the United States. *Journal of Agricultural and Food Chemistry* 59(14): 7841-7853.

Phipps, A. G. (2000). Japanese use of beni-tengu-dake (*Amanita muscaria*) and the efficacy of traditional detoxification methods. Master's Thesis. Florida International University.

Piggott, S. (1965). *Ancient Europe: From the beginnings of agriculture to Classical Antiquity*. Chicago: Aldine Publishing.

Piggott, S. (1970, October 11). Personal letter to R. Gordon Wasson. On file in the Tina and Gordon Wasson Ethnomycological Collection in the Economic Botany Library of Oakes Ames, Harvard University.

Piggott, S. (1975). *The Druids*. London: Thames and Hudson.

Piggot, S. (1987). *The Druids: Ancient peoples and places*. New York: Thames and Hudson.

Pilipenko, V., Narbute, K., Beitnere, U., Rumaks, J., Pupure, J., Jansone, B., & Klusa, V. (2018). Very low doses of muscimol and baclofen ameliorate cognitive deficits and regulate protein expression in the brain of a rat model of streptozocin-induced Alzheimer's disease. *European Journal of Pharmacology* 818: 381-399.

Pithtaker (2015, Dec 9). Heaven or hell mostly overwhelming: An experience with Amanita muscaria (exp88018). Erowid.org. https://www.erowid.org/experiences/exp.php?ID=88018

Pollock, S. (1975). The Alaskan Amanita quest. *Journal of Psychedelic Drugs* 7(4): 397-399.

Popenoe de Hatch, M. (2005). La conquista de Tak´alik Ab´aj. In J. P. Laporte, B. Arroyo & H. Mejía (Eds.), *XVIII Simposio de investigaciones arqueológicas en Guatemala, 2004* (pp. 992-999). Guatemala: Museo Nacional de Arqueología y Etnología.

Porter Weaver, M. (1981). *The Aztecs, Maya, and their predecessors: Archaeology of Mesoamerica*, 2nd Ed. Orlando, FL: Academic Press.

Power, R. C., Salazar-García, D. C., Straus, L. S., González Morales, M. R., & Henry, A. G. (2015). Microremains from el Mirón Cave human dental calculus suggest a mixed plant-animal subsistence economy during the Magdalenian in Northern Iberia. *Journal of Archaeological Science* 60: 39-46.

Prusty, R., Grisafi, P., & Fink, G. R. (2004). The plant hormone indoleacetic acid induces invasive growth in Saccharomyces cerevisiae. *Proceedings of the National Academy of Sciences* 101(12): 4153–4157.

Puharich, A. (1959). *The sacred mushroom: Key to the door of eternity*. New York: Doubleday.

Puschner, B. (2013). Mushrooms. In M. E. Peterson & P. A. Talcott (Eds.), *Small Animal Toxicology*, 3rd Edition (pp. 659-676). Elsevier Inc.

Quin, E. G. (1990). *Dictionary of the Irish language, based mainly on Old and Middle Irish materials*. Dublin: Royal Irish Academy.

Rackham, O. (1986). *The history of the countryside*. London: Dent & Sons.

Ransome, A. (1916). *Old Peter's Russian tales*. New York: Frederick A. Stokes.

Ramis, H. (1993). *Groundhog day* [Motion Picture]. United States: Columbia Pictures.

Rätsch, C. (1987). *Indianische heilkräuter: Tradition und anwendung*. Köln, Germany.

Rätsch, C. (2005). *The encyclopedia of psychoactive plants: ethnopharmacology and its applications*. Rochester, VT: Park Street Press.

Rätsch, C., & Müller-Ebeling, C. (2006). *Pagan Christmas: The plants, spirits, and rituals at the origins of Yuletide*. Rochester, VT: Inner Traditions.

Ripinski-Naxon, M. (1993). *The nature of shamanism: Substance and function of a religious metaphor.* Albany, NY: State University of New York.

Robbins, T. (1981). Superfly: The toadstool that conquered the universe. *The Best of High Times* Vol I: 67-73, 120, 138.

Rogers, R. (2012). *The fungal pharmacy: The complete guide to medicinal mushrooms and lichens of North America.* North Atlantic Books.

Rolfe, R. T., & Rolfe, F. W. (1974). *The romance of the fungus world: an account of fungus life in its numerous guises, both real and legendary.* Courier Corporation.

Rose, D. W. (2006). The poisoning of Count Achilles de Vecchj and the origins of American amateur mycology. *McIlvainea* 16(1): 37-42, 52-55.

Rose, J. (1972). *Herbs & things.* San Francisco: Last Gasp.

Ross, A. (1967). *Pagan Celtic Britain: Studies in iconography and tradition.* New York: Columbia University Press.

Ross, M. C. (1994). *Hedniska ekon.*

Rossmeisl, J. H., Higgins, M. A., Blodgett, D. J., Ellis, M., & Jones, D. E. (2006). Amanita muscaria toxicosis in two dogs. *Journal of Veterinary Emergency and Critical Care* 16(3): 208-214.

Rubel, W., & Arora, D. (2008). A study of cultural bias in field guide determinations of mushroom edibility using the iconic mushroom, Amanita muscaria, as an example. *Economic Botany* 62(3): 223-243.

Ruck, C. A. P., Hoffman, M. A. & Celdran, J.A.G. (2011). *Mushrooms, myth, and Mithras: The drug cult that civilized Europe.* San Francisco: City Lights Publishers.

Ruck, C. A. P., Staples, B. D., Celdran, J. A. G., & Hoffman, M. A. (2007). *The hidden world: Survival of pagan shamanic themes in European fairy tales.* Durham: Carolina Academic Press.

Rutherford, W. (1987). *Celtic mythology: The nature and influence of Celtic myth from Druidism to Arthurian Legend.* London: Thorsons.

Ruthes, A. C., Carbonero, E. R., Córdova, M. M., Baggio, C. H., Sassaki, G. L., Gorin, P. A. J., ... & Iacomini, M. (2013). Fucomannogalactan and glucan from mushroom Amanita muscaria: Structure and inflammatory pain inhibition. *Carbohydrate Polymers* 98(1): 761-769.

Ruthes, A. C., Smiderle, F. R., & Iacomini, M. (2016). Mushroom heteropolysaccharides: A review on their sources, structure and biological effects. *Carbohydrate Polymers* 136: 358-375.

Saar, M. (1991a). Ethnomycological data from Siberia and North-East Asia on the effect of Amanita muscaria. *Journal of Ethnopharmacology* 31(2): 157-173.

Saar, M. (1991b). Fungi in Khanty folk medicine. *Journal of Ethnopharmacology* 31(2): 175-179.

Sachse, F. (2001). The Martial Dynasties: the Postclassic in the Maya Highlands. In N. Grube (Ed.), *Maya: Divine kings of the rain forest* (pp. 356-371). Konemann Verlagsgesellschaft, mbH.

Sadoul, J. (1972). *Alchemists and gold.* Putnam.

Sahagún, B. de (1950). *Florentine Codex* (1540-1585), 12 vols. (A. J. O. Anderson & C. E. Dibble, Trans.). Salt Lake City: University of Utah Press.

Salzman, E., Salzman, J., Salzman, J., & Lincoff, G. (1996). In search of Mukhomer, the mushroom of immortality. *Shaman's Drum* 41 (Spring): 36-47.

Sam (2004, Oct 25). Sedated, euphoric, spaced out: An experience with Amanita muscaria (exp37671). Erowid.org. https://www.erowid.org/experiences/exp.php?ID=37671

Samorini, G. (1989). Etnomicologia nell'arte rupestre sahariana (Periodo delle "Teste Rotonde"). *Bollettino Camuno Notizie* 6(2): 18-22.

Samorini, G. (1992). The oldest representations of hallucinogenic mushrooms in the world (Sahara Desert, 9000-7000 BP). *Integration* 2/3: 69-78.

Samorini, G. (1995a). Kuda-Kallu. Umbrella-stones or mushroom-stones? (Kerala, Southern India). *Integration* 6: 33-40.

Samorini, G. (1995b). Sequenze lineari di punti nell'arte rupestre. Un approccio semiotico mediante psicogrammi e ideogrammi. *Bollettino Camuno Studi Preistorici* 28: 97-101.

Samorini, G. (1998). Further considerations on the mushroom effigy of Mount Bego. *The Entheogen Review* 7(2): 35-36.

Samorini, G. (2001a). New data from the ethnomycology of psychoactive mushrooms. *International Journal of Medicinal Mushrooms* 3: 257-278.

Samorini, G. (2001b). *Funghi allucinogeni: Studi etnomicologici*. Dozza, BO: Telesterion.

Samorini, G. (2012). Mushroom effigies in world archaeology: From rock art to mushroom-stones. In Proceedings of the Conference *The stone mushrooms of Thrace* (pp. 16-42), 28-30 October 2011, Alexandroupolis: Greek Open University.

Samorini, G. (2012-13). Le ninfee degli antichi Egizi: Un contributo etnobotanico. *Archeologia Africana* 18-19: 71-78.

Sansoni, U. (1994). *Le più antiche pitture del Sahara*. Milano: Jaca.

Sathyamurthy, T. (1992). *The Iron Age in Kerala. A report on Mnagadu excavation*. Thiruvananthapuram: Department of Archaeology.

Scarborough, V. L. & Wilcox, D. R. (1991). *The Mesoamerican ballgame*. Tucson, AZ: University of Arizona Press.

Schele, L., & Freidel, D. (1990). A forest of kings: The untold story of the Ancient Maya. New York: William Morrow and Company.

Schiffer, C. (2003). Liquid excretions: An experience with Amanita muscaria (exp40794). Erowid.org. http://www.erowid.org/experiences/exp.php?ID=40794

Schmiedeberg, O., & Koppe, R. (1869). *Das muscarin das giftige alkaloid des Flugenpilzes (Agaricus muscarius L.)*. Leipzig: Vogel Leipzig.

Schubeler, F. C. (1786). *Viridarium Norvegicum*.

Schulberg, L. (1968). *Historic India*. New York: Time-Life Books.

Schultes, R. E. (1977). The botanical and clinical distribution of hallucinogens. *Journal of Psychoactive Drugs* 9(3): 247–263.

Schultes, R. E. (1980). *Plants of the gods: Origins of hallucingeneic use*. Hutchinson.

Schultes, R. E. & Hoffman, A. (1992). *Plants of the gods: Their sacred, healing, and hallucinogenic powers*. Rochester, VT: Healing Arts Press.

Shapiro, M. (1983). Baba-Jaga: A search for mythopoeic identities. *International Journal of Slavic Linguistics and Poetics* 28: 109-135.

Sharer, R. J. (1994). *The Ancient Maya*, 5th edition. Stanford, CA: Stanford University Press.

Sharkey, J. (1975). *Celtic mysteries: The ancient religion*. New York: Avon Books.

Shawkat, H., Westwood, M. M., & Mortimer, A. (2012). Mannitol: A review of its clinical uses. *Continuing Education in Anaesthesia, Critical Care and Pain* 12(2): 82-85.

Shroomery (2009, Feb 3). *Trip Reports* [Online Forum]. Shroomery.org. http://www.shroomery.org/forums/ubbthreads.php

Siikala, A. L. & Ulyashev, O. (2011). Hidden rituals and public performances: Traditions and belonging among the post-Soviet Khanty, Komi and Udmurts. *Studia Fennica Folkloristica* 19: 368.

Simek, R. (1954). *A dictionary of northern mythology*. Suffolk: St. Edmundsberry Press Ltd.

Skandre (2004a, Sept 21). Shroomy cookies: An experience with Amanita muscaria & cannabis (exp26820). Erowid.org. https://www.erowid.org/experiences/exp.php?ID=26820

Skandre (2004b, Sept 21). Wisdom gained through experience: An experience with Amanita muscaria (exp36099). Erowid.org. https://www.erowid.org/experiences/exp.php?ID=36099

Soleilhavoup, F. (1978). *Les oeuvres rupestres sahariennes sont-elles ménacées?* Alger: Publication de l'Office du Parc Natiional du Tassili.

Somewhat Hazy (2000). Trips: Amanita muscaria. Lycaeum.org (defunct). Retrieved Jan 25, 2008 from http://leda.lycaeum.org/?Table=Trips&Ref_ID=47

Space Elf (2012). The gold realm and white room of crucifixion: An experience with Amanita muscaria (exp87567). Erowid.org. https://www.erowid.org/experiences/exp.php?ID=87567

Stamets, P. (1996). *Psilocybin mushrooms of the world*. Berkeley, CA: Ten Speed Press.

Stamets, P. (2000). *Psilocybin and Amanita mushrooms: An innocent discovers the infinite*. Recorded live at the 2000 MSSF Fungus Fair. Mill Valley, CA: Sound Photosynthesis.

Stellman, J. M., & Stellman, S. D. (2018). Agent orange during the Vietnam War: The lingering issue of its civilian and military health impact. *American Journal of Public Health* 108(6): 726–728.

Stépanoff, C. (2009). Devouring perspectives: on cannibal shamans in Siberia. *Inner Asia* 11(2): 283-307.

Stevenson, R. (1964). *Mary Poppins* [Motion Picture]. United States: Walt Disney Studios.

Stijve, T. (1981). High performance thin-layer chromatographic determination of the toxic principles of some poisonous mushrooms. *Mitt. Gebiete Lebensm. Hyg.* 72: 44 – 54.

Stijve, T. (1982). Het voorkomen van muscarine en muscimol in verschillende paddestoelen. *COOLIA* 25(4): 94 – 100.

Stintzing, F., & Schliemann, W. (2007). Pigments of fly agaric (Amanita muscaria). *Zeitschrift für Naturforschung* 62(11-12): 779-785.

Stříbrný, J., Sokol, M., Merová, B., & Ondra, P. (2012). GC/MS determination of ibotenic acid and muscimol in the urine of patients intoxicated with Amanita pantherina. *International Journal of Legal Medicine* 126(4): 519-524.

Strömbäck, D. (1935). *Sejd*. Copenhagen: Hugo Gebers Förlag.

Sturluson, S. (1911). *The Heimskringla*. Copenhagen: G.E.C. Gads.

Sturluson, S. (1987a). *Edda* (Edited by A. Holtsmark and J. Helgason). Copenhagen: Ejnar Munksgaard.

Sturluson, S. (1987b). *Eddan* (Edited and translated by A. Faulkes). London: The Guernsey Press Co.

Sturluson, S. (1990). *Poetic Eddan* (Translated by L. M. Hollander). Austin, TX: University of Texas Press.

Subramanian, K. S. (1995). "Koda kallu" megalithic monument in laterite, Kerala. *Journal of Geological Society of India* 46: 679-680.

Sudyka, J. (2010). The "Megalithic" Iron Age culture in South India. Some general remarks. *Analecta Archaeologica Ressoviensia* 5: 359-401.

Sumstine, D. R. (1905). Another fly agaric. *The Journal of Mycology* 11(6), 267-268.

Takemoto, T. (1964). Studies on the constituents of indigenous fungi. II. Isolation of the flycidal constituent from Amanita strobiliformis. *Yakugaku Zasshi: Journal of the Pharmaceutical Society of Japan* 84: 1186-1188.

Takemoto, T., & Nakajima, T. (1964). Studies on the constituents of indigenous fungi. I. Isolation of the flycidal constituent from Tricholoma muscarium. *Yakugaku Zasshi: Journal of the Pharmaceutical Society of Japan* 84: 1183-1186.

Takemoto, T. (1969). Seasoning compositions containing tricholomic acid and ibotenic acid as flavor enhancers. *U.S. Patent No. 3,466,175*. Washington, DC: U.S. Patent and Trademark Office.

Tatsuta, M., Iishi, H., Baba, M., Uehara, H., Nakaizumi, A., & Taniguchi, H. (1992). Protection by muscimol against gastric carcinogenesis induced by N-methyl-N'-nitro-N-nitrosoguanidine in spontaneously hypertensive rats. *International Journal of Cancer* 52(6): 924-927.

Taylor, T. (1992). The Gundestrup cauldron. *Scientific American* 266(3): 84-89.

Tedlock, D. (1985). *Popol Vuh: The definitive edition of the Mayan book of the dawn of life and the glories of gods and kings*. New York: Simon and Schuster.

Tengu (1998). Entheogenic Amanitas. *The Entheogen Review* 7(2): 33.

Theobald, W., Buch, O., Kunz, H. A., Krupp, P., Stenger, E. G., & Heimann, H. (1968). Pharmakologische und experimentalpsychologische Untersuchungen mit 2 Inhaltsstoffen des Fliegenpilzes (Amanita muscaria). *Arzneim. Forsch* 18(3): 311.

Thomas, H. (1993). *Conquest: Montezuma, Cortés, and the fall of Mexico*. New York: Simon and Schuster.

Thompson, J. E. S. (1948). An archaeological reconnaissance in the Cotzumalhuapa Region, Escuintla, Guatemala. *Contributions to American Anthropology and History*, No. 44. Washington, DC: Carnegie Institution.

Thompson, J. E. S. (1970). *Maya history and religion*. The Civilization of the American Indian Series, Volume 99. Norman, OK: University of Oklahoma Press.

Tibbett, M., Sanders, F. E., & Cairney, J. W. G. (2002). Low-temperature-induced changes in trehalose, mannitol and arabitol associated with enhanced tolerance to freezing in ectomycorrhizal basidiomycetes (Hebeloma spp.). *Mycorrhiza* 12(5): 249–255.

Tolento, M. (2008). Amanita muscaria. EmptyLife.com. Retrieved Jan. 27, 2008 from: http://www.emptylife.com/amanita2.html

Trestrail, J.H. (1995). Mushroom Poisoning Case Registry: NAMA Report 1994. *McIlvainea* 12(1): 68-73.

Trestrail, J.H. (1996). Mushroom Poisoning Case Registry: NAMA Report 1995. *McIlvainea* 12(2):

98-105.

Trestrail, J.H. (1997). Mushroom Poisoning Case Registry: NAMA Report 1996. *McIlvainea* 13(1): 68-67.

Trestrail, J.H. (1998). 1997 annual report of the North American Mycological Association's Mushroom Poisoning Case Registry. *McIlvainea* 13(2): 86-92.

Tsujikawa, K., Kuwayama, K., Miyaguchi, H., Kanamori, T., Iwata, Y., Inoue, H., ... & Kishi, T. (2007). Determination of muscimol and ibotenic acid in Amanita mushrooms by high-performance liquid chromatography and liquid chromatography-tandem mass spectrometry. *Journal of Chromatography B* 852(1-2): 430-435.

Tsujikawa, K., Mohri, H., Kuwayama, K., Miyaguchi, H., Iwata, Y., Gohda, A., ... & Kishi, T. (2006). Analysis of hallucinogenic constituents in Amanita mushrooms circulated in Japan. *Forensic Science International* 164(2): 172-178.

Tsuonda, K., Inoue, N., Aoyagi, Y., & Sugahara, T. (1993b). Changes in concentration of ibotenic acid and muscimol in the fruit body of Amanita muscaria during the reproduction stage. *Food Hygiene and Safety Science (Shokuhin Eiseigaku Zasshi)* 34(1): 18-24.

Tsunoda, K., Inoue, N., Aoyagi, Y., & T. Sugahara. (1993c). Change in ibotenic acid and muscimol contents in *Amanita muscaria* during drying, storing or cooking. *Food Hygiene and Safety Science (Shokuhin Eiseigaku Zasshi)* 34(2): 153-160.

Tuchkova, N.A, Kuznetsova, A.I., Kazakevich, O.A., Kim-Maloni, A.A., Glushkov, S.V., & Baĭdak, A. V. (2007). *Selkup mythology: Encyclopedia of Uralic mythothologies, vol. 4*. Helsinki: Suomalaisen Kirjallisuuden Seura [Finish Literature Society].

Tulloss, R. E., Caycedo, C. R., Hughes, K. W., Geml, J., Kudzma, L. V., Wolfe, B. E., & Arora, D. (2015). Nomenclatural changes in Amanita. II. *Amanitaceae* 1(2): 1-6.

Tulloss, R. E., & Lindgren, J. E. (2005). Amanita aprica: A new toxic species from western North America. *Mycotaxon* 91: 193-206.

Tulp, M., & Bohlin, L. (2005). Rediscovery of known natural compounds: Nuisance or goldmine? *Bioorganic & Medicinal Chemistry* 13(17): 5274–5282.

Tyler, J. V. (1958). Pilzatropine, the ambiguous alkaloid. *American Journal of Pharmacy and the Sciences Supporting Public Health* 130(8): 264-269.

USDA [United States Dept. of Agriculture] (2019). FoodData Central. U.S. Dept of Agriculture: Agricultural Research Service. Retrieved from: https://fdc.nal.usda.gov/

VanKirk, J. & Bassett-VanKirk, P. (1996). *Remarkable remains of the ancient peoples of Guatemala*. Norman, OK: University of Oklahoma Press.

Vetter, J. (2005). Mineral composition of basidiomes of Amanita species. *Mycological Research*. 109(6): 746-750.

Viess, D. (2012). Further reflections on *Amanita muscaria* as an edible species. *Mushroom, the Journal* 110: 42-49, 65-68.

Von Kotzebue, O. (1830). *A new voyage round the world, in the years 1823, 24, 25, and 26. Vol. II.* London: Henry Colburn and Richard Bentley; reprinted, Bremen, by GmbH & Co. KG.

Walker, K. A., Gottesman, R. F., Wu, A., Knopman, D. S., Gross, A. L., Mosley, T. H., ... & Windham, B. G. (2019). Systemic inflammation during midlife and cognitive change over 20 years: The ARIC Study. *Neurology* 92(11): e1256-e1267.

Waser, P. G. (1967). The pharmacology of Amanita muscaria. In D. Efron, B. Holmstedt & N. S.

Kline (Eds.), *Ethnopharmacologic search for psychoactive drugs* (p. 419-439). Washington D.C.: U. S. Department of Health, Education, and Welfare.

Waser, P. G. (1979). The pharmacology of Amanita muscaria. In D. Efron, B. Holmstedt & N. S. Kline (Eds.), *Ethnopharmacologic search for psychoactive drugs* (p. 419-439). New York: Raven Press.

Wasson, R. G. (1967). Fly agaric and man. In D. Efron, B. Holmstedt & N. S. Kline (Eds.), *Ethnopharmacologic search for psychoactive drugs* (pp. 405-414). Washington D.C.: U.S. Department of Health, Education, and Welfare.

Wasson, R.G. (1968). *Soma: Divine mushroom of immortality.* New York: Harcourt Brace Jovanovich, Inc.

Wasson, R. G. (1971a). *Soma: Divine mushroom of immortality.* New York: Harcourt Brace Jovanovich, Inc.

Wasson, R. G. (1971b). The soma of the Rig Veda: what was it? *Journal of the American Oriental Society* 91(2): 169-187.

Wasson, R. G. (1979). Traditional use in North America of Amanita muscaria for divinatory purposes. *Journal of Psychedelic Drugs* 11(1-2): 25-28.

Wasson, R. G., Kramrisch, S., Ruck, C. A., & Ott, J. (1986). *Persephone's quest: Entheogens and the origins of religion.* New Haven, CT: Yale University Press.

Wasson, R. G. & Pau, S. (1962). The hallucinogenic mushrooms of Mexico and Psilocybin: A bibliography. *Botanical Museum Leaflets of Harvard University* 20(2): 25-73.

Wasson, V. P. & Wasson, R. G. (1957). *Mushrooms, Russia, and history.* New York: Pantheon.

Weber, C. (2015). *Brigid: History, mystery, and magick of the Celtic goddess.* Weiser Books.

Weil, A. (1978). Reflections on psychedelic mycophagy. In J. Ott & J. Bigwood (Eds.), *Teonanacatl: Hallucinogenic mushrooms of North America.* Seattle: Madrona Publishers.

Weil, A. & Rosen, W. (2004). *From chocolate to morphine: everything you need to know about mind-altering drugs.* New York: Houghton Mifflin Company.

Weiss, B. & Stiller, R. L. (1972). Sphingolipids of mushrooms. *Biochemistry* 11(24): 4552–4557.

West, P. L., Lindgren, J., & Horowitz, B. Z. (2009). Amanita smithiana mushroom ingestion: A case of delayed renal failure and literature review. *Journal of Medical Toxicology* 5(1): 32-38.

Whalley, J. I. (1982). *Pliny the Elder, historia naturalis.* Sidgwick & Jackson.

What the hell (2003, Jul 10). Like a well-oiled hinge: An experience with Amanita muscaria (exp25115). Erowid.org. https://www.erowid.org/experiences/exp.php?ID=25115

White, K. & Mattingly, D. J. (2006). Ancient lakes of the Sahara. *American Scientist* 94: 58-65.

Whittington, E. M. (Ed.). (2001). *The sport of life and death.* London: Thames & Hudson.

Wiget, A. & Balalaeva, O. (2001). Khanty communal reindeer sacrifice: Belief, subsistence and cultural persistence in contemporary Siberia. *Arctic Anthropology* 38: 82-99.

Wilson, P. L. (1995). Irish soma. *Psychedelic Illuminations* VIII: 42-48.

Wilson, P.L. (1999). *Ploughing the clouds: The search for Irish soma.* San Francisco: City Lights Books.

Wingler, A., Guttenberger, M., & Hampp, R. (1993). Determination of mannitol in ectomycorrhizal fungi and ectomycorrhizas by enzymatic micro-assays. *Mycorrhiza* 3(2): 69–73.

Wirth, D. E. (2012). *Why 'three' is important in Mesoamerica and in the Book of Mormon*. Book of Mormon Archaeological Forum. www.bmaf.org/articles/why_three_important_wirth

Wright, B. (2009). *Brigid: Goddess, druidess and saint*. The History Press.

Yoshino, K., Kondou, Y., Ishiyama, K., Ikekawa, T., Matsuzawa, T., & Sano, M. (2008). Preventive effects of 80% ethanol extracts of the edible mushroom Hypsizigus marmoreus on mouse type IV allergy. *Journal of Health Science* 54(1): 76-80.

Zaidman, B. Z., Majed, Y., Mahajna, J., & Wasser, S. P. (2005). Medicinal mushroom modulators of molecular targets as cancer therapeutics. *Applied Microbiology Biotechnology* 67(4): 453–468.

Zandvoort, R. v. (2013). *The complete repertory: Mind to generalities*. Leidschendam, The Netherlands: Institute for Research in Homeopathic Information and Symptomatology (IRHIS) Publishers.

Zardoz (2003, Oct 30). A pleasant experience: An experience with Amanita muscaria (exp27960). Erowid.org. https://www.erowid.org/experiences/exp.php?ID=27960

Zenergy (2007, Jan 10). Forgot how to sleep: An experience with Amanita muscaria (exp58672). Erowid.org. https://www.erowid.org/experiences/exp.php?ID=58672

Zhang, Y., Mills, G. L., & Nair, M. G. (2002). Cyclooxygenase inhibitory and antioxidant compounds from the mycelia of the edible mushroom Grifola frondosa. *Journal of Agricultural and Food Chemistry* 50(26): 7581–7585.

Zipes, J. (2013). Unfathomable Baba Yagas. In S. H. Forrester, S. Goscilo & M. Skoro (Eds.), *Baba Yaga: The wild witch of the east in Russian fairy tales* (pp. vii-xii). Jackson: University Press of Mississippi.

Index

A

Agaricus muscarius 19, 132, 364, 377-378, 383-388, 391-394, 400
alcohol 75,78, 81, 99, 137-38, 142, 172, 226, 357, 361-63, 390, 395, 400, 407-8, 411, 422, 427, 430
amatoxins 9, 333
analgesic 238, 397-98, 413, 417
annulus 8-9, 19-20, 22, 26, 30, 34, 36, 38, 40, 42, 44, 46, 48
anti-inflammatory 361-62, 364, 397-98, 413, 417
anti-tumor 362, 397, 402, 417
arthritis 77, 84, 412
Aryan 72, 177, 189-191, 294, 296

B

battle-fury 128-130, 177, 178, 245-46
beheading 200, 218 (see also *decapitation*)
Berserker 128-142, 245-246, 385
birch 12, 19, 22, 72, 78, 85, 103, 122, 124-25, 144, 153, 156-58, 171-72, 178, 190, 209, 215, 308, 351, 398
Brahman 190, 202, 373
Brahmin (see *Brahman*)
Brigid 159-161, 164, 195-219
Bronze Age 100-01, 103, 273, 284
Brothers Grimm 97-98, 247

C

Cannabis 109, 111, 172, 408
cauldron 106, 116, 119-123, 139, 143, 155-56, 161-63, 167, 170, 175, 180, 208, 210, 373
Celtic 142-47, 149-156, 158-59, 161-174, 177-78, 182, 186, 188-191, 195-97, 199-200, 202-04, 207-08, 212-215, 217-19
chitin 342

Christian 66, 96, 105, 144-46, 149, 161, 171, 173, 187, 197, 200, 203-04, 214, 219
Christianity 96, 105, 130, 205, 223, 240
Christmas 63, 65, 68, 81, 102, 146, 187, 248, 266
convulsions 78, 80, 388-89, 394, 441

D

Datura 111, 132, 134, 198, 425, 427, 437
Death Cap 7, 9, 169, 332-33
decarboxylation 52-53, 59, 138, 174, 353-54, 357, 422, 449, 450, 452
decoction 59, 333, 344-45, 420-21, 450-451
deliriant 437, 444
dissociation 41, 437, 439, 444
dissociative 437, 444
DMSO 422
dopamine 425
Druid 143-45, 147, 149-150, 153, 156, 159, 162-63, 165, 167, 171, 173, 178, 186-87, 189-191, 197-204, 206-10, 212, 214, 216-19

E

echopictures 431, 433-35
ergocalciferol 342, 362
ergolines 425
ergosterol 342, 361-63

F

fasting 110, 113-15, 192, 301-02, 447-48, 452
Finno-Ugric 93-96, 99, 107
flycidal 360, 423
flying 64, 68, 122, 181, 382-83
frame reduction 431, 433-35
fucomannogalactan 362, 401-02, 412

G

GABA 52-53, 59, 353, 401, 408-09, 411, 427
GABAergic 11, 52, 427
gemmatoid 36
genus 7-11, 17-18, 271, 280, 292, 358, 369
glutamate 52-53, 59, 427
glutamate decarboxylase 51, 59, 60, 62, 451, 454
Guzmán, Gaston 272, 298

H

Haoma 138
Heinrich, Clark 113-14, 191, 196, 217, 442
Hindu/Hinduism 71, 91, 177, 223, 239-240, 271
honey 108-09, 118-19, 121-22, 136-138, 255, 423
human sacrifice 297, 302, 317, 327

I

Imbolc 160, 206-07, 210-11, 213, 214-18
immortality 109, 139, 143, 152, 157, 174, 299, 308, 315
India 72, 131, 142, 145, 157, 177-178, 189, 191, 202, 219, 269, 278, 281, 288-89, 291-96
indigenous 12, 64, 75, 176, 191, 219, 231, 299, 302
Indo-Aryan 145, 448
Indo-European 100, 111-12, 137, 142, 145-157, 156-57, 161, 175, 177-78, 189, 191-92, 202, 219, 296
Indra 54, 57-59, 108-09, 119, 239, 373-74
Indus Valley 54, 72, 451
infusion 51, 59, 62, 132, 138, 277, 400, 420-21, 443
insecticide 241, 360, 422-23
Ireland 143-44, 146-47, 151, 157, 159-61, 164, 171-72, 177-78, 187-191, 195-96, 204-05, 209, 212

K

Kamchatka Peninsula 71, 74, 92, 132, 148
Khanty 73, 81, 95, 97, 174, 176, 398-399

Koryak 71-73, 75-83, 86-88, 90-91, 117, 121, 123, 152, 174, 176, 183, 185, 370-71, 398-99, 449
kuda-kallu 278, 288-296
Kvasir 120-23, 137-39

L

Lactobacillus 51, 58, 59, 61, 62, 451
Lapland 65, 68
lightning 211-12, 273-74, 323, 325-26
looping 431, 433, 444
lycanthropy 137, 141
Lyme disease 397, 413

M

Mansi 81, 95, 97
McKenna, Terence 110-12, 114-15, 427
mead 106, 117, 120-22, 124, 135, 137-141, 143, 152, 172
Mediterranean 145, 157
Mesoamerica 297-99, 303-04, 308-09, 311, 314-15, 317-18, 320, 325, 327-28
Mimir's Well 119, 139, 176
Minerva 202-04
mistletoe 190-91, 198-201, 206-07, 218
Mount Bego 161, 271, 273-75
MSG 87, 340
mukhomor 64, 71-94, 176, 277, 438
muscarioid 14-15, 19, 26, 30, 36, 38, 235, 420, 442, 447, 453
mushroom stones 270, 278, 293, 297-99, 304-306, 311, 320-21, 323, 325-26
mycelium 5, 125, 207, 210, 361
mycorrhizal 12, 72, 90, 144, 190, 199, 217, 376, 398

N

Nagano Prefecture 346-47
nahualism 136-37
nausea 55, 64, 72, 90, 112-14, 174, 200, 234, 345, 367, 371-72, 392, 398, 401, 414, 431, 436, 442-43, 445, 447-48
Noaidi 66-67 (see also *Noide*)
Noide 105-07 (see also *Noaidi*)
noradrenaline 425

O

oak 22, 28, 35, 45, 142, 191, 198-199, 201, 207, 209, 212, 217, 335

occult 104, 147, 157, 159, 178, 196-197, 208, 215, 219, 239, 245-46
Odin 101, 104-05, 107, 116-17, 119, 121-22, 132-33, 135-39, 141, 199, 245
Olmec 299, 304-05, 313-15, 317, 319-320
"one eye" 139, 146, 155, 169-170, 172-173, 176, 199, 208-09 (see also "*single eye*")
otherworld 143-44, 147, 149, 153-54, 158, 160, 162-63, 166, 168, 171, 173, 176, 201-02, 208-09, 213, 242, 244, 399, 445

P

pagan 96, 101, 105, 130, 145-47, 151, 159-160, 173, 187, 197, 214, 239
pantheroid 14-15, 42, 46, 234-35, 442, 453
partial veil 8, 13
Pegtymel River 65, 276-78
perspiration 185, 367, 371, 374, 401, 414, 442
petroglyph 65, 101, 161
phenethlyamine 425
Pilzdivision 245
Popol Vuh 117, 303, 306, 308, 310-12, 321-26
pre-Christian 96, 101-02, 104, 106, 126-27, 150, 171
Psilocybe 6, 11-12, 57, 64, 99, 102-03, 109-112, 114-17, 120, 123-27, 190, 221, 223, 226, 271, 279, 288, 295, 405, 415, 425, 427, 429, 436, 444-45, 447, 455
Psilocybin 6, 11, 223, 271, 278, 295, 299, 306-7, 353, 425, 428

Q

Quetzalcóatl 297, 299-302, 304-05, 307-308, 310-320, 322-24

R

raven 83, 200, 208, 215
reincarnation 200, 208
reindeer 63, 66-69, 77-82, 85-86, 90, 132, 335, 398-99
Rg Veda (see *Rig Veda*)
Rig Veda 51, 53-58, 61, 71-72, 79, 88, 91, 108-111, 116, 118-19, 122, 124, 126, 146, 153, 163, 172, 177, 186, 189, 191, 193, 240, 373, 451
Roman 144, 149, 173, 196-98, 200, 203-04
"Round Heads" 281-83, 285-88
Rudra 191-92
Russia 64, 71-75, 77, 82, 86, 88-89, 91-99, 120, 132, 134, 136, 138, 148, 245, 276, 332, 335, 360, 393, 395, 397-400, 404, 412, 417

S

saliva 66, 79, 114, 120-22, 138, 371, 442
salivation 88, 116, 121, 352, 367, 371, 374-75, 401, 414, 442-43
Salvia divinorum 230, 425
Saami (see *Sami*)
Sami 66-68, 105-07, 126-27, 452
Samoyed 98, 132
Sanskrit 71, 108, 118, 177, 190, 293
Santa Claus 63, 65, 81, 85, 242
Scandinavia 101-08, 111-12, 116-19, 122-127, 135, 137, 139, 142, 155, 198, 458
serotonergic 11, 425
"set and setting" 91, 115
Siberia 12-13, 51-52, 56, 61, 63, 65, 67, 72-73, 80-81, 92, 94, 97-98, 104-07, 109-112, 114, 117-18, 121-24, 126-27, 139, 141, 149-153, 157, 162-63, 172, 174, 177, 183-85, 190, 218, 231-33, 271-72, 274-78, 308, 332, 335, 370, 375, 395, 397-98, 402, 410-12, 417, 436, 439, 441, 445-46, 448-49, 451
sidh (see *sidhe*)
sidhe 143, 147, 156, 165, 171, 173, 182
"single eye" 108, 119, 172, 176, 196, 219 (see also "*one eye*")
sleep 66-67, 78-80, 104, 112-17, 132, 137, 143, 148, 150, 152-53, 167, 170-71, 181-85, 208, 221, 257, 301, 364, 387-88, 392, 394, 398, 411, 440-41, 455
Soma 51-59, 61-62, 71-72, 86, 88, 91, 92, 102, 108-111, 116, 118-124, 126, 138, 140, 142, 145-57, 153, 163, 172-74, 177-78, 184, 186-87, 189-192, 232-33, 239-240, 276, 295-96, 373, 448, 451
Stamets, Paul 115, 432

stimulant 86, 354, 394, 397-98, 410, 427, 440, 444
sweating 72, 87-90, 136, 214, 352, 373, 382, 442-43, 456

T

Telluride Mushroom Festival 223, 231
Thor 101, 119, 198, 212
tincture 364, 387-88, 390, 400, 406-407, 412-15, 417, 419-422
tobacco 78, 88, 131, 226, 452-54
Toltec 297, 300-304, 310-11, 314, 316-17, 322
trance 65, 82, 91, 105-06, 116, 121, 126, 152, 170, 190, 214
tryptamines 230, 357, 425
twitching 72, 364, 379, 383, 385, 392, 394, 431, 441, 443

U

umami 339-340, 347
universal veil 8
urine 51, 56-57, 60, 65, 79-80, 88-89, 98, 109-111, 113-14, 122, 132, 140, 185-86, 357, 387, 394-95, 451-52
urine recycling 51, 56, 58, 61, 232, 451-52

V

Vedic 56-58, 61, 102, 108-111, 114, 116, 120, 138, 145-46, 172, 174, 184, 189-192, 202, 217, 232-33, 239, 276, 295, 373, 451

Venus 297, 302-03, 308-09, 315-17, 323-24
Viking 101, 103, 106, 126, 135, 137, 140, 185
Viking Age 103, 106, 122, 125, 139
Vitamin D 340-42, 362, 401-02, 409-410, 413
volva 8
volval material 8, 20, 26, 34, 36, 38, 40
vomiting 55, 72, 89, 113-14, 367, 371-374, 389, 392, 398, 401, 414, 431, 437, 442, 447-48

W

Wasson, R. Gordon 51-58, 61-62, 71-74, 77-80, 86, 91, 96, 98, 102-03, 106, 108-112, 118, 121, 123, 137, 140-42, 144-46, 148, 151, 155-57, 172-75, 184-86, 189, 222, 232-33, 240, 293-96, 298, 302, 305-06, 310, 320, 323, 373, 448, 451
Weil, Andrew 192, 231, 431
witch 93-99, 100, 145, 242
World Tree 118, 139, 156, 161-63, 170, 176, 308, 328, 376
Wotan 138, 245-46, 370, 376 (see also *Odin*)

Y

Yggdrasil 104, 116, 118-120, 122-23, 135, 139, 176, 376
Yule 146, 187

Lightning Source UK Ltd.
Milton Keynes UK
UKHW051955051120
372859UK00004B/46